PLANT
DIVERSITY
IN MALESIA III

PLANT DIVERSITY IN MALESIA III

PROCEEDINGS OF THE
THIRD INTERNATIONAL
fLORA MALESIANA
SYMPOSIUM 1995

Edited by J. Dransfield, M.J.E. Coode, D.A. Simpson

Published by the Royal Botanic Gardens, Kew

First published 1997

ISBN 1 900347 42 3

Production Editor: S. Dickerson

Cover design and page make-up by Media Resources,
Information Services Department, Royal Botanic Gardens, Kew

Printed in The European Union
by
Continental Printing, Belgium

Contents

Introduction

The third International Flora Malesiana Symposium held at the Royal Botanic Gardens, Kew in July 1995 was attended by over 150 researchers from all over the world, sharing a common interest in Flora Malesiana and its goals. The range of subjects of the papers presented was wide, including systematics, floristics, ecology, conservation, ethnobotany, information technology and the basic politics of the grand Flora project, Flora Malesiana. The papers stimulated lively discussion, both in the lecture theatre and in specialist workshops. Most of the papers presented are included in this volume. It has taken longer than we had hoped to publish these proceedings, the editing of the volume having had to compete with other commitments. We hope that what we publish here will be found useful as a record of the proceedings of a very successful and enjoyable symposium, of permanent interest to the evolution of the Flora Malesiana project.

The keynote address was provided by the Earl of Cranbrook, the Director of English Nature, and a biologist with a life time's experience as a naturalist in the Malesian region; this address, concerning nature conservation in the United Kingdom, might at first sight be thought not to be relevant to the Malesian region. However, the Earl of Cranbrook provides a strategy for nature conservation, developed for the United Kingdom, that could be applied widely elsewhere, particularly in Malesia.

As appropriate to a symposium on Flora Malesiana, a progress report is provided by Roos. Progress reports are also provided on two floristic projects within the Flora Malesiana region — by Madulid on the Flora of the Philippines and Soepadmo on the Tree Flora of Sabah and Sarawak — and of an area outside Malesia but of relevance to floristic projects in general by Father Mathew on the Flora of Tamilnadu. Rifai provides insights into the value of Flora Malesiana, seen by a user within the region, and highlights some shortcomings that will need to be addressed.

Stevens, in his paper on general systematic methodology, discusses aspects of how the practising Flora writer should cope with the consequences of phylogenetic studies.

Many papers that we have included deal with the systematics, morphology or natural history of particular plant groups in the Malesian region. These include papers highlighting needs for further study such as that of Widjaja (on bamboos), on details of morphology and anatomy and their systematic implications such as that of Boyce (on *Pothos*), Puff (on *Rubiaceae*) and Conran (on *Cordyline*), broad surveys of families or genera in the Malesian region such as that of de Wilde (on the *Cucurbitaceae* of Malesia) and Camus (on *Selaginella*), and detailed systematic studies aimed at understanding the evolution of a group both within Malesia and elsewhere, such as the papers by Pannell (on *Aglaia*), Ridder-Numan (on *Spatholobus* and its allies), Schot (on *Aporosa*) and Parris (on *Grammitidaceae*) and the paper on the systematics of *Gentianales* by Struwe that has major implications for the delimitation of families and genera in the Flora Malesiana region and elsewhere.

The subject that stimulated perhaps most discussion was information technology and its application in the study of the Malesian flora. Papers related to this include the local floristic studies on Mt Kinabalu by John Beaman and his associates in Malaysia and elsewhere. The detailed recent collecting and the development of the database for the flora of Mt Kinabalu has not only highlighted many new records for the mountain, yet again emphasising the extraordinary richness of this part of Malesia, but has also provided data for analysing the local distributions of plants on the mountain and has provided a basis for trying to understand the evolution of the vegetation of the mountain. We include papers by John Beaman, Reed Beaman and Barkman on Kinabalu. Jarvie presented his electronic Flora of another mountain national park in Borneo, Bukit Raya-Bukit Baka in Kalimantan, making pleas that Flora Malesiana should be written to be relevant and usable by as wide a readership as possible. Allkin discusses the principles for data management and exchange in large projects involving several institutes.

Two papers are devoted to aspects of the ecology of the forest canopy, that of Gardette on the ecology of epiphytic ferns in Sumatra and Nagamasu on the canopy biology programme at Bukit Lambir in Sarawak, while in two general ecological papers Pipoly discusses aspects of the forest ecology of Mt Kinasalapi in the Philippines and Schmitt the continuing story of the recolonisation of the Krakatau Islands in Indonesia.

While many of the systematic papers provide some discussion of distribution and biogeography, two papers are specifically aimed at understanding aspects of biogeography. Whitmore discusses the distribution patterns of members of the *Euphorbiaceae* in Malesia and van Welzen investigates the possible relationships of tectonic activity and increased speciation in New Guinea.

Two papers deal with the ethnobotany of fruit trees in Malesia. Van Valkenburg deals with the enormous range of fruits sold in markets in Kalimantan while Wirawan examines the role played by seafaring peoples in the dissemination of mango varieties in Sulawesi.

During the symposium we dedicated separate sessions to the memory of five particularly important botanists who died since the last Flora Malesiana Symposium: Rob Geesink of Leiden, a major contributor to new ways of thinking in systematics, to our knowledge of the legumes of Malesia and a friend to many; Ben Stone, polymath, charismatic teacher and major contributor to our understanding of *Pandanaceae, Myrsinaceae, Araliaceae, Rutaceae* and other families in the Malesian region; Ru Hoogland, expert in *Dilleniaceae, Cunoniaceae* and *Saurauiaceae*, whose knowledge of the Flora of New Guinea and the bibliography of phytography was encyclopaedic and freely shared; Karl Kramer, expert in Malesian and other Old World fern floras; and the egregious Professor Kostermans, who died in 1994, exactly a year before the symposium. Of these Kostermans is celebrated in Rifai's candid paper. Dok, as Kostermans was affectionately known, wielded an enormous influence on the development of floristic studies in Indonesia, not only through his own collections and publications, but perhaps even more importantly through his legacy of a cadre of trained committed botanists, many of whom attended the symposium.

We are pleased to acknowledge financial support from the Flora Malesiana Foundation and the Royal Botanic Gardens, Kew. British Airways provided several free flights that enabled some participants from Indonesia to attend. Pieter Baas was unstinting in his helpful advice at all stages of the organising of the symposium. We should also like to thank all the other members of the Symposium International Advisory Committee. The Director of the Royal Botanic Gardens, Kew, Professor Sir Ghillean T. Prance, Professor Charles Stirton and Professor Gren Lucas gave us their full support throughout the organising of the Symposium. Many of the administrative chores were looked after voluntarily by Mr Tim Nodder of the Friends of Kew. Together, he and a team of volunteers from the Friends of Kew ran the process of registration and day-to-day manning of the information desk with admirable friendliness and efficiency. We are most grateful to them. Barrie Blewett organised the reception with great aplomb. Anne Morley-Smith handled a substantial component of the securing of accommodation locally in Kew, enabling particpants to take advantage of budget accommodation, a service for which we are most grateful. Peter Boyce coordinated the production of the Abstracts volume and much of the day to day running of the symposium. Aaron Davis and Shirley Lucas both contributed in several ways to the smooth running of the symposium. We are grateful to all the speakers, session chairpersons and workshop leaders and the entire Southeast Asian Team at Kew. Tim Whitmore gave us entertainment in his after-dinner speech at the conference banquet. We thank the staff of Media Resources at Kew for their expertise in publication design and production. Finally, we are most grateful to Suzy Dickerson for the efficient role that she has played in the editing of this volume.

As this volume goes to press, our colleagues in Malesia are busy preparing the programme for the Fourth International Flora Malesiana Symposium that will be held in Kuala Lumpur in 1998. Already there promises to be a packed programme, testifying to the renewed enthusiasm and commitment for Flora Malesiana.

JOHN DRANSFIELD
MARK COODE
DAVID SIMPSON

Data management within collaborative projects

R. ALLKIN[1]

Summary. Botanists increasingly work in collaborative projects which span continents, plant groups and research disciplines. The desire to share knowledge with colleagues has implications for the strategies and techniques which we use to manage our data.

Meaningful data exchange requires that common data standards be used and that mechanisms be designed to enable data from different databases to be merged whilst maintaining the logical integrity of the resulting data set. Contrary to commonly held belief, it is not necessary or sufficient for all participants to use the same type of computer or software.

Ownership of data and protection of authors' rights are sensitive issues and emotion can cloud the debate — to the detriment of those most vulnerable. A pre-requisite for data security is that database software be chosen which does not bind authors into entering their data using pre-defined formats or data standards. Having achieved control of their own data, botanists building databases are free to enter into mutually beneficial agreements with colleagues. Such agreements can define what type of access is to be provided to the data and by whom, how information is to be published and distributed and how the data sets are to be maintained.

Rarely will a single computer program meet all of an individual scientist's data management needs. Collaborators from different institutes or projects will have still more diverse needs. A 'tool-kit' approach, in which different software are selected to fulfil particular needs, is a far better alternative, provided that the programs selected are able to communicate with one another.

These issues are discussed in the light of experiences in a multi-disciplinary Anglo-Brazilian programme: 'Plantas do Nordeste' (PNE). PNE aims, in the long term, to improve the quality of life of local people in Northeast Brazil by facilitating and fomenting better use, management and conservation of the native plant resources. This is being achieved by coordinating all available information about these plants, improving the quality of that information and ensuring that it be used through an extensive dissemination programme.

INTRODUCTION

Botanical research has always involved the exchange of knowledge and observation amongst scientists from different countries and disciplines. Easier travel and improved communications have, more recently, increased the possibilities for such collaboration. The pressures upon individual institutions to have a breadth of expertise among their staff increases the likelihood that a scientist will collaborate and exchange data with colleagues in other institutes or laboratories rather than those in adjacent offices. This desire to share knowledge amongst projects, institutes, countries and disciplines has implications for the ways in which we should gather, store and manage our data and poses intellectual, practical and diplomatic challenges to biologists, information scientists and managers.

---

[1] 'Plantas do Nordeste', Centre for Economic Botany, Royal Botanic Gardens, Kew, Richmond, Surrey, TW9 3AE.
email: r.allkin@rbgkew.org.uk

As botanists we generally make poor use of the technology which is available for managing our information and often behave irresponsibly with our data and research results. In particular, we still tend to be obsessed with the technology itself, to be too ready to grasp at new products as panaceas rather than being prepared to devote the necessary time and effort to curate our data responsibly or to plan data management projects with sufficient care. What is it that we want to achieve? Who are our target audience(s)? What questions will we need to answer and what reports will we need to produce? How long will it take to gather the data? What resources will be required? How will the data and/or software be maintained?

Botanists generally show great concern for the scientific quality of their research results and yet are often cavalier about how they store and curate these same data. The simplest of consistency checks or data security measures are felt unnecessary or somehow never quite important or urgent enough to be undertaken. Commitment of the resources necessary to enable a professional approach to data curation, particularly in collaborative projects, is rarely given a sufficiently high priority. The sad truth is, consequently, that while many of the data gathered by botanists today are stored electronically, the information is less accessible to colleagues, employers and future generations than are Bentham's hand scribbled notes stored in the basement at Kew. A few general guidelines for project leaders and funding agencies initiating botanical database projects are given in Allkin & Maldonado (1991).

A lack of planning or of any control of data quality are particularly debilitating in the context of collaborative research programmes. This paper explores some of the difficulties inherent in sharing data and describes how we are overcoming them in a collaborative project: 'Plantas do Nordeste', based in Northeast Brazil. The geographical location is not directly relevant to readers of this volume but the problems that we face and the information management techniques that we have adopted are relevant to collaborative projects, wherever they may be in the world. One unfortunate tendency, as common in developing countries as in first world nations, is for the fortunate few to forget the difficulties that others may have in gaining access to a particular technology.

PLANTAS DO NORDESTE (PNE)

Context & background

Plantas do Nordeste (PNE) is an Anglo-Brazilian collaborative programme that has been running since 1992 in the semi-arid Northeastern region of Brazil. It is an interdisciplinary programme that, through studies of biodiversity and applied research programmes, is generating new information about the native plant resources of the region and their use. Eleven projects are currently being run by one or more Brazilian institutions. Many, but not all of these, involve some participation from staff at the Royal Botanic Gardens, Kew, UK (RBG Kew). The programme has the full support and collaboration of the Brazilian National Scientific Research Council ('Conselho Nacional de Desenvolvimento Científico e Tecnológico', CNPq) and more than twenty Brazilian universities, government research institutes and non-governmental organisations are participating.

Northeast Brazil is characterised by social and environmental problems resulting, in part, from decades of unsustainable exploitation of natural resources and inappropriate land use. The region is over 1.5 million square km. (slightly more than six times the size of the UK) and is home to about 30% of the Brazilian population. It is semi-arid and the rural economy is severely affected by periodic droughts. This situation has led to large-scale migrations of people from Northeast Brazil to major cities in the south of Brazil and to the Amazon causing further social and environmental pressures in those places.

Northeast Brazil has the greatest variety of ecosystems of any region in Brazil and is one of the world's centres of biodiversity, with an estimated 20,000 plant species. Before ecosystems are further degraded and valuable genetic resources lost, it is essential that studies of biodiversity and vegetation dynamics be undertaken, so that sustainable management techniques can be developed. PNE is concentrating on the semi-arid areas which occupy more than 50% of the Northeast region. Currently the focus is on three vegetation types: 'caatinga' (deciduous thorn forest), 'brejos' (relict high altitude moist forest) which are the lifeline of the rural economy in times of drought, and the Chapada Diamantina (montane rocky grasslands of especially high biodiversity).

Activities within PNE include biodiversity studies, applied research into native plant use and ecosystem management, economic botany, the synthesis and dissemination of plant information and training.

PNE's wider goal is to promote the rational and sustainable use of the plants of the region for the benefit of local communities. Our guiding theme is 'local plants for local people'. To achieve this goal we seek to make available information of practical use about the native plants of the region and their use and management, to all those responsible for development and planning of the region, as well as to those working directly to improve the lives of local communities, particularly the rural poor (Allkin 1997). Sustainable development requires a basic knowledge of the resources available, yet agencies currently working within the region have limited access to information in the highly dispersed literature. Neither have the results of research undertaken within the region necessarily been available to foresters and agronomists working there. In a major project funded by the World Bank, for example, considerable sums were invested planting a tree known throughout the region by a single common name. Unfortunately the foresters involved were unaware of systematic research showing that this 'tree' is in fact three congeneric species. These do not, of course, have identical properties and yet the foresters were unaware of which of the three species was being grown or whether it was a mixture of all three.

To deliver information about the plant resources of the region effectively requires that we first coordinate all available information. A Plant Information Centre is being established in Recife, Pernambuco for this purpose. Existing data taken from the international literature and from the collections and databases stored at RBG Kew will be combined with much of the new information being generated by the projects working within PNE. Bringing together reliable information and co-ordinating it within a common framework will:

- enhance access to information
- ensure greater flexibility when searching the data: it is possible, for example, to combine chemical, physiological and ecological questions in a single database query, and
- provide greater flexibility in the presentation of information by combining data from different sources, using different media and presentation formats as appropriate for different audiences.
- enable a wide ranging information and extension service to be offered through academic, governmental and non-governmental extension agencies.

Information flow

PNE illustrates how critical an effective flow of information about biodiversity can be to scientists in achieving their scientific and technological objectives and in making their information more widely available. Within PNE we need to pass information between:

- databases built by different people in the same project (on one or more computers)
- different projects at different sites
- different computer programs
- different disciplines (e.g. systematics and agronomy)
- government and non-government agencies
- scientists, technicians and local farmers (or their representatives)
- Brazil, the UK and other international agencies

Technology, including satellites and the internet, make it ever easier to pass data electronically. Unfortunately, technology is not the primary obstacle to effective and meaningful communication — humans are! Data can only be meaningfully exchanged within a context where there is common agreement about precisely defined goals, data formats and exchange protocols.

ACHIEVING COMPATIBILITY

Classes of compatibility

The participants in collaborative database projects need to attain compatibility between their databases — but what precisely does compatibility mean? More hogwash is spoken about compatibility than any other aspect of data management — and that is saying something! Databases have many independent attributes which may be compared (see Fig. 1), such as "Do they use the same sort of computer?" or "Were they built using the same database management systems?" To which of these many attributes should those working in collaborative projects pay particular attention in order to maximise compatibility? What should their priorities be?

Achieving compatibility between the two databases for each of these attributes will have quite different significance to the degree and facility with which data can be successfully shared in meaningful ways. Certain features are far more important than others. Fig. 1 presents these independent attributes in a sequence that shows their relative contribution to data compatibility. Two databases on the same computer using the same program can be (and often are) 100% incompatible. Conversely two databases built on quite different machines and using different software can be fully compatible. How can this be?

Plantas do Nordeste:

Compatibility & Meaningful Exchange

6) hardware	superficial
5) software	
4) database engine	
3) database design	
2) data structures/rules	
1) terminological controls	fundamental

FIG. 1 Attributes for comparing databases indicating relative importance to meaningful data exchange.

A simple example will suffice. Imagine a database 'A' containing details of the habit of all plants recorded using four alternative terms 'tree', 'shrub', 'sub-shrub' and 'herb' while database 'B' records habit using the terms 'tree', 'shrub', 'liana' and 'herb'. If these two databases were built on the same computer using the same program (i.e. it were physically possible to transfer data between them), such a data transfer would be largely meaningless. Author B does not recognise the existence of sub-shrubs and must therefore use a wider definition of the term 'shrub' than does author A. Author B separates lianas which author A must have described as either shrubs or trees. Of course there exist better classifications of habit than either of these two schemes but this is not relevant. The point is that wherever alternative schemes are adopted by different authors, there will rarely be a direct mapping from one to the other (a possible exception would be the case where such terms were simply translated from English to French whilst maintaining their same botanical definitions). If such incongruences exist between the data sets then it will not be possible to automate data transfer since a botanist must refer back to the plants themselves to interpret and transform the data before they can be merged. This would be equally true of different terminological schemes describing plant use, leaf shape, vegetation type, chemical components, toxicity or whatever.

Even terms that are apparently fixed in their interpretation can still be stored in incompatible ways. The latitude and longitude of a specimen locality may be recorded to different levels of precision, for example, thus complicating the process of merging specimen data records. Species distribution records for states, countries and continents are also regularly stored in incompatible ways since different authors may use the same term to refer to different areas. Species X (actually found in Southern Mexico) might be recorded as 'present in Central America' in one database which could not be transferred meaningfully to a database which defines Central America to exclude Mexico. Databases containing data from older specimen labels may refer to places whose names and political boundaries have changed.

Making data exchange meaningful

If two databases do use a common terminology then exchange can be further improved by establishing compatibility at a higher level (see Fig. 1). Compatibility at any one level can only be achieved, if compatibility also exists at all lower levels.

Space does not permit an adequate exploration of all factors but in the following exploration of some of these 'levels of compatibility', examples are given to illustrate the sorts of decision that project managers must resolve to maximise compatibility and to assess the 'costs' should this not be feasible. The levels are described in decreasing order of significance: from the most important to the least important.

Level 1) Common data terminology

What types of information are to be stored and by whom ? How is each type of information to be represented and by how many different data variables? (e.g. should 'habit' be recorded as a single descriptor or several separate descriptors each covering a different aspect of habit?). Which alternative terms are to be adopted to describe each particular variable or descriptor? (e.g. standard colour charts or lists; for geographical areas see Brummitt & Hollis 1992).

Level 2) Common logical and data integrity rules

A myriad of common sense and logical rules underlie data. Scientists normally manage these intuitively when working with paper, but to make the best use of database technology this 'intelligence' must be built into the database (Watson 1971, Allkin *et al.* 1992). Unless these rules are written down, understood and agreed by all parties in a collaborative project there will be unexpected and unwelcome consequences once data sets built upon different premises are merged.

Rules are necessary for all types of data. They may be simple — so simple that the tendency will be to assume that your colleagues must think in exactly the same way as yourself. However, defining the rules may be more demanding. It will come as no surprise to taxonomists that the rules governing nomenclatural and taxonomic relationships within databases can become particularly complex. A useful example set of standard rules for plant names is found in Bisby (1994).

Example rules for building a database might include:

* A plant recorded from any geographical or political subdivision must be recorded for the country in which that subdivision occurs
* Characteristics of bark and trunk will be used only for plants that have been recorded as being trees or shrubs
* A species can have many scientific names but only one preferred name
* A scientific name should normally occur only once in the database — unless it has been misapplied in the literature, in which case it can occur once in its proper sense and any number of times as misapplications.

Level 3) Common data structures

Databases containing exactly the same information may store it in different ways. How the data are stored will alter how one is able to use those data and may make it impossible, or at least more difficult, to exchange with others.

A common example is that of scientific names which are often represented as single text strings (e.g. '*Calamus balanseanus* Becc. var. *castaneolepis* (Wei) Pei & Chen'). In other databases this may be split into separate fields ('*Calamus*', '*balanseanus*', 'Becc.', 'var.', '*castaneolepis*', '(Wei) Pei & Chen'). The latter, incidentally, gives much greater flexibility in use and permits 'look-up' lists of genera or authors to be referred to thus avoiding unnecessary typing and improving data consistency. Data that are stored as separate fields can be automatically joined together when needed. Automatic separation of scientific names stored as long text strings is not straightforward because of the variable number of possible data elements. As a result, transferring names from a database where they are stored as single strings to one using multiple fields may prove to be impractical.

As well as the number of fields used to represent a data item, the arrangement of these fields amongst different data tables may vary. There is a direct link between the logical and integrity rules that can be enforced within a database and how the fields are arranged into tables. The tabular (logical) structure of a database is a matter of great significance to the quality of the data and the flexibility of use, and is also relevant to data exchange.

Level 4) Common database platform

Database programs will be written in different languages, generally those available with commercial database systems (Access, xBase, Advanced Revelation, Oracle etc.). Programs may be written in the same language, even though based on different commercial products. Programs written in xBase, for example, can be used with Clipper, Foxpro, dBase and Quicksilver compilers, and may store data in files that share the same physical structure. Data from an Oracle datafile, nevertheless, can easily be passed to an Access or dBase file. Data in standard delimited ASCII files can be passed to database or word processing programs. The technology does not limit us. Compatibility of programming language has some relevance to those wishing to develop their own software systems since they may seek to share programming libraries or more direct interaction between their systems but otherwise is quite irrelevant to data compatibility. Databases that share a common platform may find data transfer slightly easier but any benefits will be trivial in comparison with the benefits of achieving compatibility of terminology, logical consistency and data structure.

Level 5) Common software

Specialist database software packages designed for biological purposes have been built around carefully designed data structures which permit these systems to enforce a range of data integrity and logical checks. Botanists in collaborative projects that adopt the same software will benefit, therefore, since their databases will be built according to the same logical and structural rules even if they collate different subsets of information. Again, however, such benefits are worthless if terminological compatibility is not enforced.

Those sharing common software can further benefit from shared experience in the use of a system, joint training exercises and, possibly, through reduced purchase and maintenance costs.

Level 6) Standardising on common hardware

Of least importance is the hardware used. Some software packages are available for a range of different machines and even those projects that use different software will be able to pass data to one another provided that compatibility has been achieved at all lower — more fundamental — levels.

'Compatibility' between databases thus depends more upon features that are the responsibility of the botanist than it does on features under control of the programmer or analyst. Common software (level 5) and common hardware (level 6) are in fact properties of the database management system chosen to build the database and have least relevance to meaningful data exchange. Despite prejudices to the contrary, it is actually very easy to transfer data between one type of computer and another or between different types of software package. Data can be passed among database packages (e.g. dBase, Access etc.) or from them to spreadsheets (e.g. Lotus, Excel) or word processors without information loss — provided that the records are structured appropriately and that the botanists have agreed to use the same terms and definitions. Incompatibility exists between botanical databases mostly as a result of decisions taken by botanists.

Terminological controls within PNE

Within PNE we have established small groups of specialists that, through often lengthy debate, have created a number of terminological standards for our needs. These standards are not used proscriptively (we do not prevent projects storing other data that interest them) but will resolve incompatibilities in key areas that have been identified, by these committees, as being of common interest.

For example, descriptive standards have been designed for:
* the vegetation types of NE Brazil
* habit descriptions
* use properties for forage plants
* use properties for medicinal plants

These standards cannot be rigidly fixed. Some have been in use for three years and experience has shown where changes would be beneficial. For example, we are now developing two parallel data standards for medicinal plants. The first is to meet the needs of pharmacologists and chemists concerned with scientifically proven results and descriptions of diseases, disorders and symptoms using technical terminology and concepts. The second standard aims to meet the needs of ethnobotanists and those working in community projects where popular language and concepts are used to describe symptoms and remedies. These two standards will, of course, share some common ground. Our task is to achieve the best balance of flexibility and compatibility.

Within the project's databases, the evolution from using one version of a particular data standard to the next will require careful planning.

Mechanisms for data transfer

The physical medium

In PNE each project will own and manage its own data on a local computer. Many participants in our programme do not yet have access to communication networks. The Brazilian national scientific communications network is expanding all the time and PNE will use this as far as possible but we do not wish to depend upon it. Currently we use it for communication (email) between some projects and to transfer data from some sites to others but data transfer can be carried out throughout the region by sending computer diskettes through the post. It is envisaged that in the future greater use will be made communication networks and that we will offer some of our information services via the World-wide Web.

Though the physical transfer of data from one computer to another is not problematic, the meaningful exchange of data can be. In addition to addressing issues of compatibility, as seen above, we must also find mechanisms to merge data from different sources which potentially express different opinions.

Transfer formats

Writing programs to transfer data is an expensive operation and the expense will be directly proportional to the complexity of the data and the differences in data structure between the two systems. Writing a program to undertake a single, one-off data transfer between two databases, is rarely cost-effective and it is cheaper to re-enter the data.

One approach to making regular automatic data transfer between databases or between different software feasible is adoption of common transfer formats. Developers of system A write an export program to transfer data from their own internal structures to some pre-defined external transfer format 'X'. The physical representation of this is irrelevant — it could be in ASCII, dBase, Access or whatever since these are interchangeable but the logical representation is critical and will generally be much less complex than the internal structure of a database. They also need to develop a program to import data from that same external format back into their own system. To enable data exchange with System A, therefore, those responsible for database B must write programs that pass data between their own system and transfer format 'X'. They need not know the complexity of the internal workings of system A, nor worry about changes to it. They do need to be familiar with the structure of the external format, which should be agreed and documented. This has two important advantages:

- Many people can participate using different databases or systems. Provided that they have access to and from the transfer format they will be able to communicate with all other participants. They do not need to write different transfer programs for Systems B, C, D and so forth which they would never achieve.
- The developers of Systems A and B and others are free to develop their own system as they need, making changes to their own internal structures without this having expensive consequences for their collaborators using other systems.

Maintaining data consistency and integrity when merging data

Imagine two data sets created using the same software and terminological controls and each of which shows complete internal data integrity and logical consistency.. The authors of these two data sets however, may have conflicting opinions in which case the databases will contain inconsistent records. For example the two authors may use different scientific names for the same plant or have divergent opinions about the distribution of a plant or its merit for medicinal purposes. In such cases it is not possible simply to join together all data from both sets and expect to create a third data set that is itself internally consistent.

To detect and resolve manually all the individual inconsistencies between two data sets would be painfully slow. This is inconceivable for PNE where data exchange is envisaged on a regular basis. We aim, therefore, to automate data merges whilst recognising that in many cases the resolution of a conflict which has been detected automatically, will require human (expert) judgement before a decision can be reached.

Together with the International Legume Database and Information Service (Zarucchi *et al.* 1993), another multi-site user of the Alice system, PNE has worked with the developers of the system to design and implement a program for the automatic detection of inconsistencies between data sets and their intelligent handling during data merges. Data categories have been classed as additive (where two or more alternative observations can be recorded for a species without causing problems for integrity) or non-additive (where more than a single observation for a given species will cause conflicts). The most complex problems are those associated with potential conflicts between lists of scientific names and their synonyms.

The person merging two data sets can currently assign priority to one data set or the other (i.e. the final data set will in all cases of inconsistency follow the opinion expressed in the prioritised data set). Future versions of the merge software will enable us to assign priority to one or other data set for each data type independently. This, for example, would allow us to follow the botanical nomenclatural decisions embedded in one data set while incorporating the medicinal use records from the other.

SOFTWARE SELECTION CRITERIA

In PNE we made three assumptions when selecting software to manage our data:

- our requirements for data management software are complex and we do not have the resources required to develop new systems
- biological data management systems already exist
- no single software system can possibly meet all of our needs.

These assumptions, which are equally relevant to most other botanical database projects (Allkin & Maldonado 1991), led us to conclude that:

- PNE should carefully select a few complimentary systems, to cover each of our major data management tasks.

- individual projects within PNE would be free to use other software if this were necessary for political or other reasons, but that data transfer to and from those systems would be required and be the responsibility of that project, and not of PNE.
- PNE would offer technical support for project databases built using the chosen systems. This would include installation, advice, training and maintenance of software and databases. Such support will not be offered for other software systems adopted by individual projects.

It was decided at an early stage that PNE projects would build their own databases locally on PC computers — either linked in networks or as stand-alone machines. We needed, therefore, to select systems available for this sort of environment. Our selection criteria (Fig. 2), have been rigorously enforced — our primary concern being to ensure the safety and quality of each project's data.

Plantas do Nordeste:

Choice of Software

- "generic" software — data independence
- data quality/integrity
- mature products — support
- links to other software
- data security

FIG. 2. Software selection criteria.

1) Data independence ('generic software')

Our first requirement was that project leaders controlled what data were to be stored in their databases and how it was to be entered. They needed to be able to define their own species descriptors and to modify these during the life span of the project. We could not use any system that pre-defined the sorts of data that are to be entered according to the needs of one particular institutional database. To achieve this meant that we could only use software showing a high degree of data independence. The major benefits of data independence are:

a) control over your own data

The companies that produce word processing programs such as 'WordPerfect' have no control over what we write in our letters and certainly make no claim of ownership over the documents that we create using their program. Data independence means that database software can be used in the same way. The author of the database controls and owns the data contained. Institutional systems have been developed and provided to users elsewhere, in some cases free of charge. This, unfortunately, imposes that institute's own data standards on others and enables it subsequently to incorporate other people's data into its own database.

b) flexibility to meet future needs

Data independent software also offer far greater flexibility in the kinds of data which can be contained in a database and how these might be entered. Thus within the PNE umbrella the same software is used to meet the divergent needs of quite different projects. This flexibility also meant that we did not need to define all of our database needs at the outset, since new descriptors can be defined, or old ones modified, as each project develops.

c) cost

Buying data independent software is a much cheaper alternative than developing a home-grown system. Since development costs are shared amongst many users, the price paid by any one user is very small relative to the overall development costs. By selecting programs already used in many different projects elsewhere PNE was able to afford more functionality that we could otherwise have achieved. We can also anticipate further improvements being made to the systems by their developers.

2) Data quality

We needed systems to help control the consistency and quality of our data. It is easy to fill computer disks with data that are so 'dirty' as to be of no practical use. Unless rigorous validation and integrity checks are imposed at data entry, data sets become worthless since searches are only partially successful. Thus, for example, to obtain a complete list of all species in the *Palmae* requires first of all that membership of that family has been recorded for all species and secondly that *Palmae* is always used (not *Arecaceae*) and that it is always spelt the same way. Searching a database using '*Palmae*' will not retrieve species recorded for the family 'Palmeae'!

Data validation involves checking the data values entered against predefined lists of 'allowed' values (e.g. a standard list of vegetation types for NE Brazil). These 'look up' lists can also, reduce typing since the user simply selects terms from a list.

Data integrity involves more complex checks of the logical consistency within a database using the common sense and biological rules built into the data structures and software. Biological data are complex in structure (Allkin 1988, Allkin & Bisby 1989, Allkin *et al.* 1992) and biologists using simple databases can very quickly find that — despite their best efforts — their data show all manner of inconsistencies: trees recorded as annuals, Brazilian plants recorded as absent from South America, species lacking any proper scientific name and so on. The more checks against such inconsistencies that are built into a system, the cleaner the data will be, and the more useful and secure they will be.

In order to control the validation and integrity of our biological data sets within PNE better we sought database systems designed from the outset to incorporate biological intelligence.

3) Maturity and reliability

So that each project within PNE can produce results (databases, printed lists, leaflets, books or other products) within a two or three year funding period, we needed software that worked from the outset and would work reliably throughout the project. Many botanical database projects have floundered because they chose to use software that was not yet developed and which did not materialise by the promised date. Software development is notoriously unreliable and the costs frequently under-estimated.

We needed systems that were already tested and in use by others — as many as possible — to ensure that the software would work and that any errors that we found would be corrected by its authors.

4) Providing for data exchange

Since we had assumed that no single system could meet all of our needs, it was necessary to choose systems that could pass data to one another. Thus an individual project could choose different systems to do different things (e.g. build synonymised checklists, curate collections, draw maps or generate dichotomous keys) and not require data to be re-entered time and time again. Apart from saving labour this also helps ensure data consistency since if the same data are stored in two different places then there is always a risk that they will become unsynchronised as data are added or edited in one copy alone.

Increasingly, biological software are being developed that import and export data through common data exchange formats. One example is the Delta data format for morphological descriptive data used today by many different identification systems (Dallwitz 1993) another is the ITF format, used to exchange cultivated plant records among botanical gardens (Botanic Gardens Conservation Secretariat 1987). Work has also gone on to incorporate biological intelligence in such formats to provide greater flexibility and to permit more of the knowledge stored in more complex databases to be successfully transferred (White & Allkin 1993).

5) Ensuring data security

The scientists participating in PNE are investing much effort and expertise in generating and collating their data. It would be irresponsible not to be concerned with the long-term security of these data. Whether, in practice, they have a future and can be used again by the author or others depends upon two factors: the quality of the data and the ability of the software to exchange data with other systems. Both of these conditions have already been discussed above.

Data that are full of inconsistencies and errors will be far more difficult to sort, order and retrieve and the effort of finding and correcting errors may often mean that it is cheaper to start again (i.e. the data have no future). Careful planning, precise data entry and software that ensures data quality are necessary pre-conditions to data security.

No software, however new and 'state-of-the-art' it might be, will survive for more than a few years. The computing industry is in constant evolution. Systems may be replaced with new versions or they may not. Systems built 'in-house' are no exception to this rule and management must recognise that software development brings long-term commitments to maintenance costs which generally match or exceed those of initial development. Therefore, it is necessary when entrusting data to any system — whoever developed it — that one first ensures that simple user options allow the export of data into commonly used formats. The more complex the data, the richer the database structures underlying the database will need to be and the more complex will be any data exchange format that needs to be considered to preserve the data effectively.

THE SOFTWARE USED BY PNE

Projects within PNE use a range of different programs. Although some programs are used only at a single site, we supply and support centrally the use of a few standard packages which are used widely by most of our projects. Three key systems used within PNE are Alice, Brahms and Delta. These systems have different, complementary functions and are compatible in the sense that it is possible to pass data between them (Fig. 3). All three satisfy our selection criteria as outlined above.

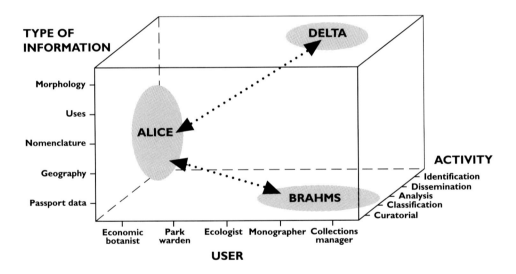

FIG. 3. The major software systems used by PNE, which exchange data with one another.

Alice is a system (actually eight different programs intended for different sub-classes of user) that enables storage, retrieval and publication of biodiversity data about species (Alice Software, 1996). Alice handles nomenclature (synonymy), distribution, morphology, ecology and — important given PNE's particular focus — economic importance. An important factor in the selection of Alice was the high priority which it gives to ensuring data quality. The system allows users to define any number of their own ecological, morphological or use-related descriptors which was also important to PNE. Alice is consequently equally as good for simple checklist projects (lists of species with synonymy and distribution together with some habit, habitat and ecological descriptors) as it is for databases containing detailed descriptions of the medicinal use of species, for example. Another factor is that Alice allows any individual observation or data item to be linked to one or many data sources — a publication, specimen or personal observation for example. This is important to any database, but particularly so in a collaborative venture. It allows us, for example, when querying a database, to know who was responsible for a particular data entry and to whom credit should be given.

Alice also provides flexible options for retrieval and publication, in formats designed by the user themselves. We may use traditional formats such as books or simple printed lists or electronic media. The latter options, of particular interest to PNE, include data distributed on disk with easily used query programs (in various languages). Thus we can send subsets of our data to those whom we wish to use it without being able to change it. Entire Alice databases, or subsets of them, can also be published on the internet (the World-wide Web) or on intranet as sets of inter-linked hypertext HTML pages (Dalcin 1997; White 1997).

Brahms is a specimen and herbarium management system that allows the storage, use and publication of data about plant specimens: collectors, localities, label data and curation details (Filer 1996). Brahms is used by those PNE projects undertaking major collecting exercises or reviewing herbarium collections. Since it is possible to pass data between the two systems, data stored for all collections of a particular species in a Brahms databases can be passed to any appropriate Alice database, where they will be combined with all sorts of other species data (descriptions of morphology or use). Specimen records are passed from Brahms into the Alice Transfer Format and the Alice import program SAM summarises this information as taxon records within the target Alice database. In return, Brahms databases within PNE will benefit from receiving updated nomenclatural and taxonomic records passed from appropriate Alice databases, enabling herbaria to update their specimen records with nomenclatural changes made by monographic specialists.

Both Alice and Brahms provide mechanisms to export data from one database to another and, more significantly, to merge two data sets into one whilst detecting and resolving data conflicts (see above). Thus both systems support PNE's need to exchange data between projects and sites.

Delta is, strictly speaking, not a database system. A number of programs generate identification aids (diagnostic descriptions, dichotomous keys and interactive identification systems) from descriptions of taxa in a particular logical format stored in ASCII (Dallwitz 1980). Currently only one PNE project is using Delta but all those projects that have Alice databases which contain morphological and habit descriptions as well as nomenclatural and geographic data, will be able to pass these descriptions automatically into Delta to generate identification guides (Dallwitz *et al.* 1993). A database of about 250 species of legumes of potential use as a forage crop in Caatinga vegetation from the state of Bahia for example will generate identification manuals for use by local farmers and agronomists.

COLLABORATIVE DATA MANAGEMENT

The ownership issue

Research scientists are often sensitive about the risks of losing 'credit' for a piece of work or a new finding. Fear of losing control of their own data may be an obstacle to exchange and collaboration. Such reticence is unfortunate from all perspectives. The individual concerned suffers from not maximising the benefits of publishing widely and from losing the benefits described below. Science itself is held back wherever information is not exchanged completely freely. Without the full participation of scientists, development projects such as PNE cannot succeed and scientists will continue to be portrayed as being uninterested in benefiting society. Failure to collaborate through fear of data loss is all the more unfortunate since it is both avoidable and counter-productive. Informed debate within the botanical community is needed to improve appreciation of the issues and to dispel the misconceptions which frequently cloud the debate.

If unpublished research results are to be gathered and managed then it is critical that all issues of the ownership of data, division of intellectual property rights and potential returns from such data be fully discussed and agreed upon at the outset. Agreements reached should be written down to avoid future misunderstanding.

Benefits to participation

What are the incentives, therefore, to scientists to contribute to programmes such as PNE that aim to offer biological information services to the community at large? It clearly would be impractical to create an infrastructure which only benefits the users of the information. Something has to be offered in exchange — some tangible benefits for the providers of the information.

There are many explicit advantages to participation in PNE. Most obviously the programme offers a vehicle by which research projects can be funded, equipment bought and studentships provided. Secondly, the programme permits exchange visits and brings those with new skills to participate directly in projects or to teach courses. But there are also advantages in participating for a project's own data management. Scientists mostly wish to take advantage of modern data management techniques and also recognise that to work effectively they need ready access to the accumulated body of existing knowledge. PNE offers training (e.g. in the use of software packages) and advice on how to design databases and make the best use of the available technology to advance each project's own research goals. Regular site visits provide technical assistance with data management. Future plans also provide for a documentation centre to provide access to literature not currently available within the region and a number of thematic regional databases coordinating data from international sources.

Distributed data ownership

Scientists should be in control of their own data and appreciate the benefits to be gained by exchanging as much of their data as possible with their colleagues.

Within PNE we use the following guidelines (Fig. 4):
- each project leader has ownership and control over all of the data collated locally
- PNE offers them support and advice on building databases containing their own research results — thus each project database is different all others
- each project is provided with the tools needed to manage its own data
- each project selects that subset of the data stored in its own databases which it wishes to share with the common pool and exports these into a commonly agreed exchange format
- in return each project receives copies of those centrally held databases to which it contributes
- PNE seeks ways and means to assist individual projects to publish and disseminate their own research results.

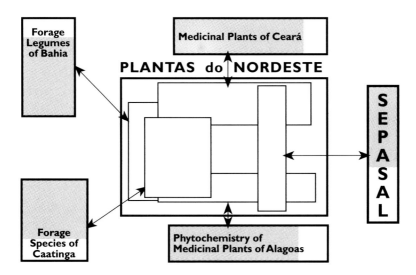

FIG. 4. The PNE data model: sharing data through distributed ownership.

Existing data from the literature and from databases available at RBG Kew are being pooled centrally in a number of central thematic databases together with data gathered by PNE projects. An important source of information for PNE is the SEPASAL database at Kew (Davis *et al.* 1995) which brings together information from the literature on useful plants of the semi-arid parts of the world.

Since published information is already in the public domain PNE does expect participating projects to contribute their published data. It is, therefore, not only in a project's own interests to publish new data quickly (to benefit the careers of the scientists involved, to help raise further research funding etc.) but also in PNE's interest.

The benefits of pooling data

Having ready access to available up-to-date knowledge is important to those undertaking research or development. The collation of data from different disciplines and sources within a common framework also makes possible analyses, data retrieval and reports which could not be produced otherwise. An important benefit for scientists contributing data, not immediately apparent to those unaccustomed to working with databases, is the ability to explore their own data within a context in which it has been integrated with data from many other sources and disciplines. This enable questions to be asked about the plants or plant properties on which the scientist works — combining data from several people's work in ways which would be quite impossible were the data still stored in separate databases. In the longer-term we feel that this will become an increasingly important factor in encouraging more people to collaborate with the programme. (Allkin 1997).

Providing access to information

It is as important to regulate and control who will have access to data held in common and how they will be able to use it as it is to be clear about data ownership.

Project leaders within PNE are responsible for the fate and accessibility of the data contained within their database. Centrally held databases however, contain data which are mostly already in the public domain and are, in that sense, already available to everybody. Nevertheless, the Board of the Association Plantas do Nordeste (a legal body set up in Brazil in 1995 to administer and oversee the work of the PNE programme) has statutory powers to decide our priorities for information provision and exactly what information services are offered, how data are to be made available and what charges might be made for publications or use of data or services provided. We are at an early stage in building and exploring different types of information service (Allkin 1997). We do not anticipate that funds generated from 'publication' (in the broadest sense) aimed at more wealthy clients will cover all of the future research costs of the programme, but it will contribute significantly towards further information gathering and help pay the costs of providing more innovative information services aimed at less fortunate sectors of society.

Conclusions

1. The Plantas do Nordeste programme is taking an innovative approach to the organisation, exchange and dissemination of scientific information which it is hoped will provide a successful working model for other international collaborative projects striving to provide local benefits through the sustainable use of plant resources.

2. When botanical database projects have failed it is mostly because of poor planning by botanists rather than through technological shortcomings. Successful data management requires that scientists turn their attention away from the technology itself and to their own responsibilities as data stewards. Time and effort are required to decide who, precisely, are the intended users of any information system and how they will use it and to obtain precise specifications of the needs of all involved.

3. Collaborative data management requires particularly careful planning. All contributors need to agree upon principles and plan their own contribution.

4. Effective data exchange requires agreement to adopt common data standards and exchange protocols. Discussions about which existing data standards to use or about the design of new standards will help focus debate upon the project's goals and purpose. Common software or hardware are not precursors for such data exchange.

5. It cannot be assumed that all partners within a collaborative agreement will want, or be able, to work in identical ways or use the same software. Flexibility is necessary.

6. Within PNE, 11 projects of the most varied kinds have successfully built databases using existing generic biological database software systems. Projects with fixed time scales and funding periods have been able to collate their data and publish their results. They have not failed as many botanical projects do through reliance on homegrown or yet-to-be developed software.

7. The long term safety, accessibility and usability of the data gathered within these projects are secure. Careful selection of software has ensured the consistency and internal integrity of the data sets stored and that export to future generations of software and databases is possible.

8. Collaboration and data exchange adds significant overhead costs to information management but bringing different data sets from different sources together within the same structure provides enormous long term benefits both to the producers and users of the information.

REFERENCES

Alice Software (1996). Alice Software home page: *http://dspace.dial.pipex.com/alice.*
Allkin, R. (1988). Taxonomically intelligent database programs, In: D. L. Hawksworth (ed.), Prospects in Systematics, Systematics Association Special Volume No. 36: 315 – 331. Oxford University Press.
—— (1998). Effective management and delivery of biodiversity information. In P. Bridge, P. Jeffries, D. R. Morse and P. R. Scott (eds), Information Technology, Plant Pathology and Biodiversity, pp. 87 – 102. CAB International, Wallingford, U.K.
—— & Bisby, F. A. (1989) The structure of monographic databases. Taxon 37(3): 756 – 763.
—— & Maldonado, S. (1991) Recommendations for the application of botanical databases within Latin America. Taxon 40(3): 527 – 529.
——, White, R. J. & Winfield (1992). Handling the structure of biological data. In: Mathematical and Computer Modelling 16(6): 1 – 9.
—— & Winfield,P. J. (1993). Cataloguing biodiversity: new approaches to old problems. Biologist 40(4): 179 – 183.
—— & —— (1993). Software development strategies for Global Plant Information Systems. In: F. A. Bisby, R. J. Pankhurst & G. R. Russell (eds), Designs for a Global Plant Species Information system. Oxford University Press, Oxford.

Bisby, F. A. (1994). 'Plant Names in Botanical databases'. Plant Taxonomic Database Standards No. 3 Version 1.00. Published for the International Working Group on Taxonomic Databases for Plant Sciences (TDWG) by the Hunt Institute for Botanical Documentation, Carnegie Mellon University, Pittsburgh.

Botanic Gardens Conservation Secretariat (1987). The International Transfer Format for Botanic Garden Plant Records. Plant Taxonomic Database Standards No. 1 Version 1.00. Published for the International Working Group on Taxonomic Databases for Plant Sciences (TDWG) by the Hunt Institute for Botanical Documentation, Carnegie Mellon University, Pittsburgh.

Brummit, R. K. & Hollis, S. (1992). World Geographic Scheme for Recording Plant Distributions. Plant Taxonomic Database Standards No. 2 Version 1.00. Published for the International Working Group on Taxonomic Databases for Plant Sciences (TDWG) by the Hunt Institute for Botanical Documentation, Carnegie Mellon University, Pittsburgh.

Cook, F. E. (1995). Economic Botany Data Collection Standard. Royal Botanic Gardens Kew.

Dalcin, E. (1997). "Surfing" your data with Alice Web: a new program for Alice users. In: Alice Software News 5 (ed. R. Allkin & P. J.Winfield) Alice Software, London.

Dallwitz, M. J. (1980). A general system for coding taxonomic descriptions. Taxon 29 ( ): 41 – 46

——, Paine, T. A. and Zurcher, E. J. (1993). Users guide to the DELTA System: a general system for processing taxonomic descriptions. 4th Edition. CSIRO Division of Entomology, Canberra.

Davis, S. D., Sinclair, N. J. & Cook, F. E. M. (1995) The work of Kew's Centre for Economic Botany and the Survey of Economic Plants for Arid and Semi-Arid Lands (SEPASAL). In: N. E. West (ed.), Rangelands in a Sustainable Biosphere: Proceedings of Fifth International Range Congress. Society for Range Management, pp. 111 – 112. Colorado, USA.

Filer, D. (1996). BRAHMS — Botanical Research and Herbarium Management System Version 4.0, Dept. of Plant Sciences, Oxford.

Watson, L. (1971). Basic taxonomic data the need for organisation over presentation and accumulation. Taxon 20, 131 – 136.

White, R. J. (1997). The ILDIS Legume Web service. In: Alice Software News 5 (ed. R. Allkin and P. J.Winfield) Alice Software, London.

—— & Allkin, R. (1993). A strategy for the evolution of database designs. In: F. A. Bisby, R. J. Pankhurst & G. R. Russell (eds), Designs for a Global Plant Species Information system. Oxford University Press, Oxford.

Zarucchi J. L., Winfield, P. J., Polhill, R. M., Hollis, S., Bisby, F. A. & Allkin, R. (1993). The ILDIS project on the world's legume species diversity. In: F. A. Bisby, R. J. Pankhurst & G. R. Russell (eds), Designs for a Global Plant Species Information system. Oxford University Press, Oxford.

A biogeographic analysis of orchid distributions on Mt Kinabalu

Todd J. Barkman[1], Rimi Repin[2], Reed S. Beaman[3] & John H. Beaman[4]

Summary. Evidence presented in this paper suggests that inferred divergence of the present-day orchid flora at various delimited sites on Mt Kinabalu may be explained, in large part, through vicariant events resulting from historical geologic processes proposed by Collenette (1965). These processes include the initial uplift of the granitic pluton which produced the first high-elevation habitats. Final uplifting of the pluton in the mid- to late Pleistocene allowed recent colonisation of the highest elevation habitats. If the timing of Collenette's geologic events are accurate for the formation of Mt Kinabalu, then much of the flora is quite young, particularly at the highest elevations. Results of this analysis also suggest that ultramafic sites such as Marai Parai and Pig Hill are habitat patches acting as islands with respect to plant colonisation. Indeed, the occurrence of relatively high numbers of endemic species on these islands may be due to evolutionary processes analogous to those resulting in the formation of endemics on true oceanic islands. Because of the endemism as identified by our analyses, continued conservation efforts should focus on the high-elevation areas and ultramafic sites surrounding the mountain. Difficulty in distinguishing between ecological and historical factors is discussed with respect to the biogeographic patterns observed in this study.

Introduction

Identifying areas of high endemism is an important outcome of biogeographic studies (Morrone 1994, Morrone & Crisci 1995). Well supported biogeographic analysis may also allow inference of potential factors responsible for the formation of endemic floras and faunas. These two outcomes of biogeographic research can be used subsequently to prioritise areas for conservation which include regions of high endemicity and putative source areas which may have contributed propagules to the endemic biota.

The results of our biogeographic analysis of Mt Kinabalu, Sabah, Malaysia presented below allow both an identification of areas of endemism and their interrelationships as well as inference of processes leading to the formation of the flora there.

Mt Kinabalu (4,101m) is a granitic batholith putatively uplifted in the Pleistocene (Jacobson 1970) to its present position, nearly 1,200 m higher than any other mountain in Borneo. Kinabalu is unique in its geology and diverse vegetation types, as described by Jacobson (1970) and Kitayama (1991) respectively. The major substrates found on the mountain include ultramafics, granodiorites, sandstones,

---

[1] Department of Botany, University of Texas, Austin, TX.
[2] Park Botanist, Sabah Parks, Sabah, Malaysia.
[3] Department of Botany, University of Florida, Gainesville, FL.
[4] Department Botany and Plant Pathology, Michigan State University, East Lansing, MI. Present address: The Herbarium, Royal Botanic Gardens, Kew, Richmond, Surrey TW9 3AB, U.K

and Pinosuk gravels, as described by Jacobson (1970). Beaman & Beaman (1990) and Kitayama (1991) have suggested that substrate is an important factor affecting species distributions and vegetation types found on Kinabalu. More than 15 plant community types were identified by Kitayama (1991), including lowland, lower and upper montane, and sub-alpine forests, with half of these communities restricted to ultramafics, one of the major substrate types identified by Jacobson (1970). From the standpoint of both species composition and physiognomy, some of the most interesting plant communities are found growing on ultramafic substrates. Furthermore, these areas have higher percentages of endemic taxa than are found on non-ultramafic substrates (Wood *et al.* 1993).

The ultramafic substrates of Kinabalu are rich in chromium, cobalt, iron, magnesium, and nickel and poor in calcium, nitrogen, phosphorous, and potassium (Jacobson 1970 quoting from Reinhard and Wenk 1951). This composition is comparable to that reported for other ultramafics (Krause 1958). The latter group of elements are essential for normal plant growth; a deficiency may thus affect plant growth on these sites, possibly to the extent of excluding certain species (Brooks 1987). Ultramafic outcrops, commonly encountered on Mt Kinabalu, are of relatively recent geologic origin and are patchy in distribution (Jacobson 1970). Floristic composition may be substantially different on the ultramafic sites but is not completely distinct from that found on adjacent non-ultramafic sites. Some species found on both ultramafic and adjacent non-ultramafic sites include *Phyllocladus hypophyllus, Leptospermum recurvum,* and *Schima wallichii.* Species such as these were termed "facultative" by Brooks (1987) because their distribution is unaffected by substrate type. Other species, such as *Nepenthes villosa, N. rajah, Coelogyne rupicola,* and *Platanthera kinabaluensis* are virtually restricted to ultramafic sites and can be considered "obligate" ultramafic species *(sensu* Brooks 1987). Because the ultramafic areas on Mt Kinabalu are unusual in terms of the flora, which is partially restricted to them, the relationships among these regions is of interest in the context of a biogeographic hypothesis for the entire mountain. Such biogeographical information should complement the ongoing floristic and evolutionary studies of the Kinabalu region.

The inference of biogeographic patterns and processes can be achieved using various methodological approaches as reviewed by Morrone & Crisci (1995). Current methodologies, such as Brooks Parsimony Analysis (BPA) (Wiley 1988), rely on phylogenetic hypotheses from multiple groups of unrelated organisms for estimation of a general-area cladogram. In the absence of phylogenetic relationships another approach may be employed, Parsimony Analysis of Endemism (PAE), which uses the principles of cladistics whereby areas are treated as taxa and species occurring in the areas as characters (Rosen 1988). The analysis identifies cladistic patterns among areas based upon known species distributions. Implicit in PAE are assumptions that historical events are unique, that there has not been subsequent dispersal of taxa between areas since their origin, and that the resulting nested sequences reflect a time sequence. The main difference between BPA and PAE is that ancestors, identified by a phylogenetic hypothesis, are not used in the latter method. Although phylogenetic approaches for inferring biogeography may be preferable, PAE makes use of floristic data of the kind that is often available in most regions of the world and thus may have more general applicability. Parsimony Analysis of Endemism is advantageous over distance-based methods relying on similarity/dissimilarity measures and clustering algorithms like UPGMA, because instead of estimating a single branching diagram summarising area relationships, PAE will generate multiple equally parsimonious cladograms. This is preferable because multiple biogeographic patterns often exist for a group of areas, because some members of a biota will be likely to have experienced different colonisation histories. In addition, reconstruction of synapomorphies on a PAE cladogram allows researchers to see which taxa provide support for a particular branching relationship.

Previous work (Beaman & Beaman 1993, Kitayama 1991, Parris *et al.* 1992, Wood *et al.* 1993) and field experience of the authors indicate that certain plant species are found only on widely separated ultramafic areas that may differ somewhat in elevation. Based on this observation we hypothesised that the ultramafic areas are functioning as islands between which species may be dispersing. To test this prediction in terms of a PAE-estimated area cladogram, we expect to see a single lineage comprised of the ultramafic areas only. To produce expected area cladograms that could be compared to those observed, it was necessary to make an assumption concerning the process of colonisation of ultramafic habitat patches. We assumed that many plant species growing on adjacent non-ultramafic areas, although tolerant of the similar elevation, may or may not be tolerant of the edaphic conditions. If intolerant, these species may be closed out of the site due to competitive exclusion by tolerant species. This theoretical phenomenon was described by MacArthur & Wilson (1967 p.122) in their seminal work on island biogeography, noting that "Species already present on a habitat island are faced with constant pressure of high immigration of less-well adapted species drawn from the surrounding habitats. Simultaneously, species attempting to colonise a habitat island should find it harder to do so because of the greater diversity of competitors opposing them at any given moment." Empirical support for this phenomenon has been reported by Huenneke *et al.* (1990) on California serpentine and non-serpentine grasslands. Because the ultramafic areas may be closed to nonultramafic immigrants, we would expect the immigrants to come from other places, particularly other ultramafic areas, sites at which the plants may already have developed tolerance to the conditions there. This prediction may be tested by estimating PAE area cladograms and comparing them to the expected area relationships.

METHODS

In order to estimate relationships among areas, we first delimited regions with homogeneous geology within a defined elevational range. Global Positioning System (GPS) units helped provide precise locations, while a vegetation map (Kitayama 1991), a geological map (Jacobson 1970) and GIS software used in conjunction with a satellite image of the mountain (R. Beaman & Thomas, unpublished data) facilitated determining boundaries of these areas. Several areas were delimited which included ultramafic and geographically proximate non-ultramafic regions at similar elevations. Table 1 lists the delimited areas and their approximate upper and lower elevations. A location map of Mt Kinabalu showing the circumscribed areas used in the following analyses is available upon request from the senior author. Morrone (1994) suggested the use of quadrats for delimiting areas in a PAE analysis. We avoided this practice however, because of the possibility of having sites of mixed geology. Because of the observed effect of geology on plant composition our approach seems justified.

TABLE 1. Areas queried from KINABALU database represented by the standardised locality name, elevation in metres, and substrate type found within the region.

Standard locality	Elevation (m)	Substrate
1. Hempuen Hill	500 – 1000	Ultramafic
2. Dallas	500 – 1000	Sedimentary
3. Poring	500 – 1000	Sedimentary-granitic
4. Penataran River	500 – 1000	Ultramafic
5. Peniguppan Ridge	1300 – 1500	Ultramafic
6. Marai Parai	1600 – 2000	Ultramafic
7. Pinosuk Plateau	1000 – 2000	Quaternary gravel
8. Lower Gurulau	1000 – 2000	Sedimentary
9. Summit Trail	1800 – 2400	Sedimentary
10. Pig Hill	2000 – 2300	Ultramafic
11. Upper Marai Parai	2000 – 3200	Granitic
12. Kinabalu Lipson	2600 – 3000	Ultramafic
13. Summit Area	3100 – 4101	Granitic

Figure 1 shows expected area relationships among the delimited regions of Mt Kinabalu which will be compared to observed area cladograms. These relationships are expected if substrate alone is the most important factor determining colonisation. Although this may represent an oversimplified explanation of the process of plant colonisation on Mt Kinabalu, comparison of this cladogram to those observed allows an explicit test of this assumption.

In order to compile lists of orchid species to be used for estimating area cladograms, we have performed queries of the relevant localities from the orchid subset of the KINABALU database (Beaman & Beaman unpublished data, but see Wood *et al.* 1993). In some cases, a particular delimited area had more than one name in the database. In these cases we queried using both names with the elevational restrictions listed in table 1. The areas delimited for querying are similar in part to the standardised localities of Beaman & Regalado (1989, p. 36).

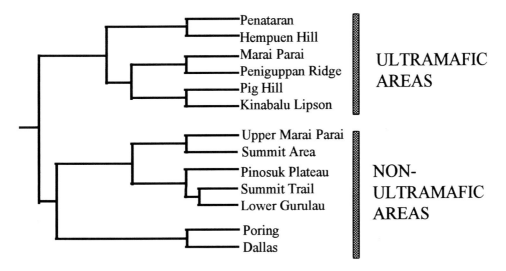

FIG. 1. Hypothetical cladogram depicting expected relationships among areas on Mt Kinabalu.

RESULTS AND DISCUSSION

From the 13 areas delimited, 466 taxa were recorded. In many cases different areas had very different numbers of species. This may have resulted from differences in size of the standardised localities recognised. However, diversity between like-sized areas also differed in some cases, and some of the localities are less well collected than others. A matrix was constructed whereby the areas were coded for the presence or absence of only those species found in two or more areas and is shown in the Appendix. Parsimony algorithms from PAUP (version 3.0, Swofford 1991) were used to estimate branching relationships among the 13 areas using 203 informative species listed in the data matrix. Outgroup rooting was attempted in this analysis to provide polarity to the tree. It is difficult to decide on an outgroup area for use in PAE analyses; however, in our case it was assumed that Sabah, excluding Mt Kinabalu, should be considered for two reasons. First, Wood & Cribb (1994) have enumerated orchid species present in Sabah. Second, the area surrounding Mt Kinabalu is the most likely source for the Kinabalu flora due to its proximity. We scored one outgroup area, Sabah-low, for the presence and absence of species found below 1,000 m elevation and a second, Sabah-mid, for species found at elevations higher than 1,200 m. An analysis performed with both of these areas as outgroups did not maintain the monophyly of the ingroup. In this analysis, Sabah-low was part of a lineage consisting of low elevation areas and Sabah-mid was part of a lineage of mid elevation areas. This result suggests that the areas of Kinabalu are "non-monophyletic" and that the low-elevation areas of Kinabalu have closer relationships with surrounding low-elevation areas in Sabah, while those areas above 1,200 m are related to surrounding montane regions of Sabah. This result is not surprising, because Mt

Kinabalu is contiguous with surrounding lowland and montane areas; however, it does not provide an unambiguous root for the cladogram.

As an alternative, the unrooted cladogram was rooted according to the relative timing of geologic events leading to the formation of Mt Kinabalu as proposed by Collenette (1965). The most prominent feature of the rooted cladogram, shown in Fig. 2, is the presence of three lineages corresponding to low-, mid- and high-elevation areas. This result suggests that historical, ecological, or both processes in combination are responsible for the relationships observed between the areas. Several studies have suggested the importance of ecological factors related to elevation in determining the distribution patterns of plant species. In a rigorous analysis by Wolf (1994) it was reported that altitude above sea level explained most of the variation in epiphytic species distribution in the northern Andes. In addition, Kitayama (1992) reported that several climatic factors on Mt Kinabalu explain species distributions along an altitudinal transect. He thus defined "critical altitudes" whereby groups of species are displaced by others along an elevational gradient. The three lineages defined by our cladogram are roughly bounded by the same critical elevations identified by Kitayama. Pendry & Proctor (1996) similarly reported that altitude is primarily responsible for zonation between lowland and lower montane rain forest in Brunei.

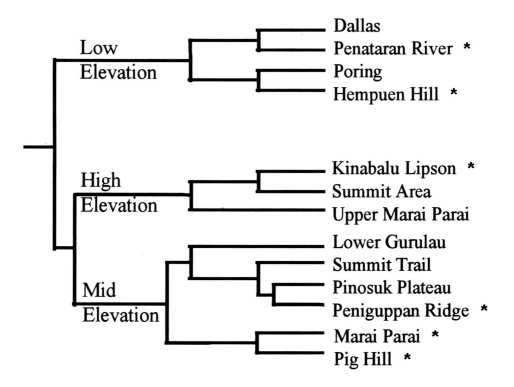

* = Ultramafic area

FIG. 2. Single cladogram obtained from PAE analyses. Rooting is based upon relative timing of inferred geologic events leading to the formation of Mt Kinabalu. Three lineages are identified based upon elevational differences.

Historical factors could also explain the pattern of area relationships observed. In this tree (Fig. 2), Dallas and Penataran River as well as Poring and Hempuen Hill are always found as sister areas and this clade is also separated into an eastern lineage (Poring and Hempuen Hill) and a western lineage (Dallas and Penataran River). These areas are expected to be divergent from areas of higher elevation due to the assumed uplift of the granitic pluton through the surface of overlying sedimentary substrates which would have produced an altitudinal disjunction in the Late Pliocene to early Pleistocene times (see Fig. 3). In addition, this geologic event would produce a barrier between the east and west low-elevation areas where there previously had been none. The branching relationships also suggest that the low-elevation ultramafic areas have a shared biogeographic history with other low elevation but non-ultramafic sites. This implies that, while the low-

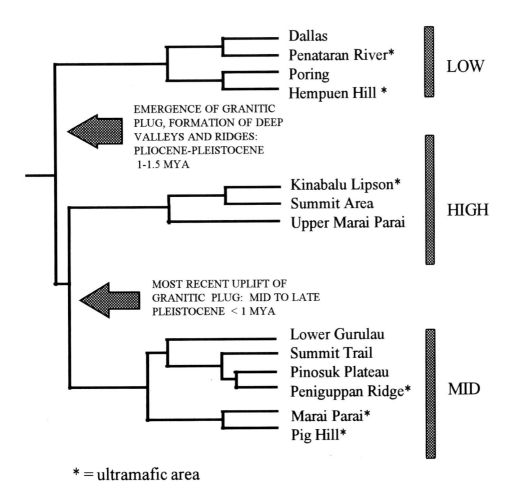

FIG. 3. Putative historical events resulting in the formation of the Kinabalu orchid flora are indicated on the single most parsimonious cladogram of Fig. 2.

elevation ultramafic areas may have a somewhat unique species composition, many taxa are shared with non-ultramafic sites, such as *Grammatophyllum speciosum* and *Plocoglottis lowii* (both of which appear to be facultative showing no preference for a particular substrate type).

The high-elevation areas (Summit Area, Kinabalu Lipson, and Upper Marai Parai) comprise a single lineage in addition to one comprised of mid-elevation areas including Pinosuk Plateau, Summit Trail, Lower Gurulau, Marai Parai, Pig Hill, and the Peniguppan Ridge. The sister-area relationship of these two lineages suggests that a historical vicariant event caused their divergence. Further uplifting of the granitic pluton would have produced altitudinal separation of these areas, resulting in their inferred divergence during mid- to late- Pleistocene times, as shown in Fig. 3. Many orchid species shared among the three high-elevation sites are restricted to elevations above 2800 m, and are endemic to Mt Kinabalu. Interestingly, species such as *Thrixspermum triangulare* and *Eria grandis* are examples of high-elevation endemics which show no preference for substrate type, because they occur both on Kinabalu Lipson (ultramafic) as well as Upper Marai Parai and the Summit Area (granitic).

An important result bearing upon our original prediction is the preservation of one lineage, Marai Parai and Pig Hill, which is comprised of ultramafic areas only. This result is significant as it suggests that there are floristic ties between these relatively distant areas, which are separated by expanses of non-ultramafic forest. Obligate ultramafic species such as *Dendrobium tridentatum* and *Dilochia cantleyii* may be found at these sites only. The presumed wind-dispersed seeds of these orchid species provide an opportunity for their exchange between these regions. It is interesting to note that the ultramafic area Peniguppan ridge, which is of comparable elevation to Marai Parai and Pig Hill, is unrelated. This is surprising because it is adjacent to Marai Parai and both are ultramafic. This unexpected result may be explained by different soil conditions present, possibly due to differential weathering and soil formation between the two. Peniguppan ridge possesses a well developed soil profile whereas Marai Parai is conspicuous because of its shallow soil layers and abundant ultramafic boulders similar to conditions on Pig Hill. Our result may not be totally unexpected as Fitch (1965) suggested the flora at a site may be controlled more by soil depth than by chemical composition. Rigorous analyses would be necessary, however, to verify whether different chemical or physical characteristics of the soils between the Marai Parai and Peniguppan sites were responsible for the observed species differences. The separation of the two extreme ultramafic areas from all others suggests that propagules which occur at the similar elevation areas such as Pinosuk Plateau and Summit Trail are not common colonisers of the ultramafic patches, presumably because they are non-adapted to the stringent edaphic conditions found there. Because these sites are closed to some proportion of colonists, we suggest that the habitat patches, Marai Parai and Pig Hill, are acting as islands and that these two areas are exchanging propagules, although it is not clear if this movement is bi-directional or biased in one direction or the other.

The identification of areas of endemism from a PAE cladogram may be achieved by recognising lineages of areas which have at least two "synapomorphic" endemic species (Morrone 1994). Two lineages of areas were identified as areas of endemism. One was the high-elevation lineage and the other was the mid-elevation ultramafic lineage. In addition, two other ultramafic areas, Hempuen Hill and Penataran River, were identified as areas of endemism although they were not recognized as sister areas. In the context of conservation, these results suggest that the ultramafic as well as the high-elevation sites of Mt Kinabalu are areas of endemism and would benefit from continued conservation efforts. Fortunately, all of these areas, except one, are currently within the boundaries of Kinabalu Park.

The observed cladogram in Fig. 2 differs markedly from that which was expected in Fig. 1. This difference is probably due to a violation of our original assumption that substrate type is the most important factor affecting plant distribution. Indeed, from an analysis of the observed cladograms it appears that elevation is a strong factor affecting plant distribution as evidenced by the presence of lineages comprised of areas of similar elevation, but differing substrate. One other factor to consider is the use of epiphytic orchids in the present analysis. One hundred and sixteen species in the matrix were solely epiphytic while 87 were either only recorded as terrestrial or both epiphytic and terrestrial. As epiphytes, these species are not in direct contact with the substrate; however, their presence may reflect a preference for the substrate through a preference for trunks and branches of the tree species growing there. In addition, seemingly obligate ultramafic epiphytes may prefer the open scrubby stature of the forests present on extreme ultramafic sites.

Distinguishing between historical and ecological factors in biogeography or their relative importance is difficult, because both could produce similar patterns as demonstrated by Endler (1982). Clearly, our data suggest that the ecological factor of substrate type is important in determining those species present in the extreme ultramafic areas of Marai Parai and Pig Hill. In the other areas, however, it is not possible to discern between the two possibilities. It seems likely that the two are inextricably tied, because the geological (historical) event of the uplift of the granitic batholith provided novel ecological conditions, and both factors therefore are likely responsible for the area relationships observed. Further studies involving phylogenetic analyses of the taxa found on Mt Kinabalu may allow discrimination of the relative importance of these processes. While the results of this study are significant, they are preliminary and several limitations are acknowledged. It is important for patterns of relationships among these areas to be corroborated by evidence obtained from distributions of other plant groups. The inclusion of these groups awaits further revision of the floristic database. Rooting PAE cladograms is a difficult exercise due to problems in identifying appropriate outgroups. Conclusions from PAE cladograms are limited unless a root can be identified. In addition, the delimitation of areas within Mt Kinabalu was a subjective exercise although based upon explicit criteria and included only those areas which are well represented in the database. While the KINABALU database is extensive, some species present in a particular area may remain uncollected or unrepresented. Although this problem is inherent as a result of our chosen methodology, this approach represents a novel use for the floristic database and as it is developed further, other uses will arise.

ACKNOWLEDGEMENTS

   We thank Datuk Lamri Ali, Director, and Francis Liew, Deputy Director of Sabah
Parks for the support they have provided throughout the duration of this study.
Kanehiro Kitayama contributed many useful discussions concerning Mt Kinabalu.
Vicki Funk and Beryl Simpson offered valuable suggestions to improve the
manuscript. Answ Gunsalam and Alim Biun were most helpful in the field, and
the Kinabalu Park research staff provided extensive logistical help. This research
was supported in part by NSF grant DEB-9400888 to Michigan State University, J. H
Beaman, Principal Investigator, and a research grant awarded to the senior author
by the Orchid Digest.

REFERENCES

Beaman, J. H. & Beaman, R. S. (1990). Diversity and distribution patterns in the
    flora of Mount Kinabalu. In: Baas, P., Kalkman, K. & Geesink, R., eds. The Plant
    Diversity of Malesia. Kluwer Academic Publishers, Dordrect/Boston/London, pp.
    147 – 160.
—— & Beaman, R. S. (1993). The gymnosperms of Mount Kinabalu. Contr. Univ.
    Michigan Herb. 19: 309 – 340.
—— & Regalado, J. C. Jr. (1989). Development and management of a
    microcomputer specimen-oriented database for the flora of Mount Kinabalu.
    Taxon 38: 27 – 42.
Beaman, R. S., Thomas, A. G. & Beaman, J. H. (1997). A geographic information
    system (GIS) for Mount Kinabalu. This volume.
Brooks, R. R. (1987). Serpentine and its Vegetation. In Ecology, Phytogeography, &
    Physiology Series, vol. 1. Ed. T. H. Dudley. Dioscorides Press, Portland Oregon.
Collenette, P. (1965). A short account of the geology and geological history of Mt
    Kinabalu. Proceedings of the Royal Society of London 161: 56 – 63.
Conran, J. G. (1995). Family distributions in the Liliiflorae and their biogeographical
    implications. J. Biogeogr. 22: 1023 – 1034.
Endler, J. A. (1982). Problems in distinguishing historical from ecological factors in
    biogeography. Amer. Zool. 22: 441 – 452.
Fitch, F. H. (1965). Discussion. Proceedings of the Royal Society of London 161: 87.
Huenneke, L. F., Hamburg, S. P., Koide, R., Mooney, H. A., & Vitousek, P. M.
    (1990). Effects of soil resources on plant invasion and community structure in
    Californian serpentine grassland. Ecology 71(2): 478 – 491.
Jacobson, G. (1970). Gunong Kinabalu Area, Sabah, Malaysia: Geological Survey
    Malaysia, Report 8. 111 pp. Government Printing Office, Kuching, Sarawak,
    Malaysia.
Kitayama, K. (1992). An altitudinal transect study of the vegetation on Mt Kinabalu,
    Borneo. Vegetatio 102: 149 – 171.
—— (1991). Vegetation of Mount Kinabalu Park, Sabah, Malaysia. Map of
    physiognomically classified vegetation. East-West Center, Honolulu, Hawaii.
Krause, W. (1958). Andere Bodenspezialisten. In: Encyclopedia of Plant Physiology,
    vol . 4. Springer, Berlin. pp. 755 – 806.

MacArthur, R. H. & Wilson, E. O. (1967). The Theory of Island Biogeography. Monographs in Population Biology. 1. Princeton Univ. Press, Princeton. 203 pp.

Morrone, J. J. (1994). On the identification of areas of endemism. Syst. Biol. 43: 438 – 441.

—— & Crisci, J. V. (1995). Historical biogeography: introduction to methods. Ann. Rev. Ecol. Syst. 26: 373 – 401.

Parris, B. S., Beaman, R. S. & Beaman J. H. (1992). The Plants of Mount Kinabalu. 1. Ferns and Fern Allies. Royal Botanic Gardens, Kew. 165 pp.

Pendry, C . A. & Proctor, J. (1996) . The causes of altitudinal zonation of rain forests on Bukit Belalong, Brunei. J. Ecol. 84: 407 – 418.

Reinhard, M. & Wenk, E. (1951). The geology of the colony of North Borneo. Brit. Borneo Geol. Survey Bull. 1.

Rosen, B. R. (1988). From fossils to earth history: Applied historical biogeography. In: Myers, A., & Giller, P. (eds). Analytical Biogeography: An Integrated Approach to the Study of Animal and Plant Distribution. Chapman and Hall, London. Pp. 437 – 481

Swofford, D. L. 1991. PAUP Phylogenetic Analysis Using Parsimony. Version 3.0s. Computer program distributed by the Illinois Natural History Survey, Champaign, Illinois.

Wiley, E. O. (1988). Parsimony analysis and vicariance biogeography. Syst. Zool. 37: 271 – 290.

Wolf, J. H. D. (1994). Factors controlling the distribution of vascular and non-vascular epiphytes in the northern Andes. Vegetatio 112: 15 – 28.

Wood, J. J., Beaman, R. S., & Beaman, J. H. (1993). The Plants of Mount Kinabalu: 2. Orchids. Royal Botanic Gardens, Kew. 411 pp.

—— & Cribb, P. J. (1994). A checklist of the orchids of Borneo. Royal Botanic Gardens, Kew. 409 pp.

APPENDIX. Data matrix used in PAE analyses. Taxa scored as 1 (present) or 0 (absent) for areas 1 – 13.

	1	2	3	4	5	6	7	8	9	10	11	12	13
Acanthephippium javanicum	1	0	1	0	0	0	1	0	0	0	0	0	0
Acriopsis indica	1	0	0	0	1	0	0	0	0	0	0	0	0
Acriopsis javanica	1	1	0	0	0	0	0	0	0	0	0	0	0
Adenoncos sp. 1	0	0	1	1	0	0	0	0	0	0	0	0	0
Aerides odorata	1	0	1	1	0	0	0	0	0	0	0	0	0
Agrostophyllum majus	1	0	0	0	0	0	1	0	0	0	0	0	0
Apostasia odorata	0	0	1	0	0	0	1	0	0	0	0	0	0
Apostasia wallichii	0	1	0	1	0	0	1	0	0	0	0	0	0
Appendicula anceps	1	1	0	0	0	0	0	0	0	0	0	0	0
Appendicula congesta	1	1	0	0	0	0	1	0	0	0	0	0	0
Appendicula cristata	1	0	0	1	0	0	1	0	0	0	0	0	0
Appendicula foliosa	0	1	0	0	0	0	1	0	0	0	0	0	0
Appendicula linearifolia	0	0	0	0	0	0	0	1	0	0	0	1	0
Appendicula longirostrata	0	0	0	0	0	0	1	0	0	0	0	0	1
Appendicula pendula	1	1	0	0	0	0	1	0	0	0	1	1	0
Appendicula torta	1	1	1	1	0	0	1	0	0	0	0	0	0
Ascidieria longifolia	0	0	0	0	1	0	1	0	0	0	0	0	0
Bromheadia scirpoidea	0	0	0	0	0	0	1	0	0	0	1	0	0
Bulbophyllum alatum	0	0	0	0	1	1	0	0	0	0	0	0	0
Bulbophyllum biflorum	0	0	1	1	0	0	0	0	0	0	0	0	0
Bulbophyllum caudatisepalum	0	0	0	0	1	0	1	0	0	0	0	0	0
Bulbophyllum coniferum	0	0	0	0	1	1	0	0	0	0	0	0	0
Bulbophyllum coriaceum	0	0	0	0	1	1	1	1	1	1	0	0	0
Bulbophyllum deltoideum	0	0	0	0	0	0	1	0	0	0	0	1	0
Bulbophyllum disjunctum	0	0	0	0	1	1	1	0	0	0	0	0	0
Bulbophyllum flavescens	0	1	0	0	1	1	1	0	0	0	0	0	1
Bulbophyllum gibbsiae	0	0	0	0	1	0	1	0	0	0	1	0	0
Bulbophyllum lobbii	1	1	0	1	1	0	0	0	0	0	0	0	0
Bulbophyllum longimucronatum	1	0	0	0	0	1	0	0	0	0	0	0	0
Bulbophyllum macranthum	0	0	0	1	1	0	0	0	0	0	0	0	0
Bulbophyllum mandibulare	1	0	1	0	0	0	0	0	0	0	0	0	0
Bulbophyllum montense	0	0	0	0	0	1	0	1	1	1	0	0	0
Bulbophyllum mutabile	0	0	0	0	1	0	1	0	0	0	0	0	0
Bulbophyllum obtusum	0	0	0	0	0	0	1	0	0	0	1	0	0
Bulbophyllum odoratum	1	0	0	0	0	0	1	0	0	0	0	0	0
Bulbophyllum pocillum	0	0	0	0	1	1	1	0	0	0	0	0	0
Bulbophyllum purpurascens	1	0	0	1	0	0	0	0	0	0	0	0	0
Bulbophyllum sopoetanense	0	0	0	0	0	1	1	0	0	0	0	0	0
Bulbophyllum trifolium	0	0	0	1	0	0	1	0	0	0	0	0	1
Calanthe pulchra	1	0	0	0	0	0	1	1	0	0	1	0	0
Calanthe triplicata	0	0	1	1	1	0	0	0	0	0	0	0	0

	1	2	3	4	5	6	7	8	9	10	11	12	13
Ceratochilus jiewhoei	0	0	0	0	1	0	1	0	0	0	0	0	0
Ceratostylis ampullacea	0	0	0	0	0	0	1	0	1	1	0	0	1
Ceratostylis crassilingua	0	0	0	0	0	0	1	0	0	0	0	0	1
Ceratostylis radiata	0	0	0	0	1	0	1	0	0	0	0	0	0
Chelonistele amplissima	0	0	0	0	0	1	1	0	0	0	1	0	0
Chelonistele kinabaluensis	0	0	0	0	0	0	0	0	1	1	0	0	1
Chelonistele lurida	0	0	0	0	0	1	0	1	0	0	0	0	1
Chelonistele sulphurea	0	0	0	1	0	1	1	0	0	0	0	0	0
Chroniochilus virescens	0	0	1	1	0	0	0	0	0	0	0	0	0
Cleisocentron merrillianum	0	0	0	0	1	0	1	0	0	0	0	1	1
Cleisostoma discolor	0	0	1	1	0	0	0	0	0	0	0	0	0
Cleisostoma striatum	1	0	1	0	0	0	0	0	0	0	0	0	0
Coelogyne clemensii	1	0	0	0	0	1	1	0	0	0	0	1	0
Coelogyne compressicaulis	0	1	0	0	0	1	0	1	0	0	0	1	1
Coelogyne craticulaelabris	1	0	0	0	0	1	0	0	0	0	1	0	0
Coelogyne dayana	1	1	0	0	0	0	0	0	0	0	0	0	0
Coelogyne hirtella	0	0	0	0	1	0	1	0	0	0	0	0	1
Coelogyne kinabaluensis	1	0	0	0	1	0	1	0	0	0	0	0	0
Coelogyne longibulbosa	0	0	1	0	0	0	1	0	0	0	0	0	0
Coelogyne moultonii	0	0	0	0	1	0	0	1	0	0	0	0	0
Coelogyne papillosa	0	0	0	0	0	0	0	1	1	1	0	0	0
Coelogyne planiscapa	0	0	0	0	0	1	1	0	0	0	0	0	0
Coelogyne plicatissima	0	0	0	0	0	0	0	1	1	1	0	0	0
Coelogyne radioferens	0	0	0	0	1	0	0	0	0	0	0	1	1
Coelogyne reflexa	0	0	0	0	0	0	0	1	0	0	0	0	1
Coelogyne rupicola	0	0	0	0	0	0	0	0	1	0	0	1	1
Coelogyne tenompokensis	0	0	1	0	1	1	0	0	0	0	0	1	0
Coelogyne venusta	1	0	0	0	0	1	1	0	0	0	0	0	0
Corybas pictus	0	0	1	1	0	1	1	0	0	0	0	0	0
Corymborkis veratrifolia	1	0	0	0	0	0	1	0	0	0	0	0	0
Cryptostylis acutata	0	0	0	0	0	1	0	0	0	0	1	0	0
Cryptostylis arachnites	0	1	0	0	0	1	0	0	0	0	0	0	0
Cymbidium elongatum	0	0	0	0	0	0	0	0	0	0	0	0	1
Cystorchis variegata	0	1	0	0	0	0	1	0	0	0	0	0	0
Dendrobium alabense	0	0	0	0	0	1	0	1	0	0	0	1	1
Dendrobium beamanii	0	0	0	0	1	0	1	0	0	0	0	1	0
Dendrobium bifarium	0	0	1	1	1	0	0	0	0	0	0	0	0
Dendrobium cymbulipes	1	0	0	0	0	1	1	0	0	0	0	1	0
Dendrobium kiauense	0	0	0	1	0	0	1	0	0	0	0	0	0
Dendrobium lamelluliferum	0	0	0	0	1	1	0	0	0	0	0	0	0
Dendrobium maraiparense	0	0	0	0	0	0	1	0	0	0	0	1	1
Dendrobium oblongum	0	0	0	0	0	0	1	0	0	0	1	0	0
Dendrobium orbiculare	0	0	0	0	0	0	1	0	0	0	1	0	0
Dendrobium pachyanthum	0	0	1	1	0	0	0	0	0	0	1	0	0

	1	2	3	4	5	6	7	8	9	10	11	12	13
Dendrobium patentilobum	0	1	0	1	0	0	1	0	0	0	0	0	0
Dendrobium tridentatum	0	0	0	0	0	0	0	0	0	0	0	1	1
Dendrobium ventripes	0	0	0	0	0	1	1	0	0	0	0	0	0
Dendrochilum acuiferum	0	0	0	0	0	0	0	0	1	1	0	0	0
Dendrochilum alatum	0	0	0	0	0	1	0	0	0	1	0	1	1
Dendrochilum conopseum	1	0	0	0	0	0	1	0	0	0	0	1	0
Dendrochilum dewindtianum	0	0	0	0	0	1	0	0	1	0	0	1	0
Dendrochilum gibbsae	1	0	0	0	1	0	1	1	0	0	0	0	1
Dendrochilum exasperatum	1	0	0	0	0	0	1	0	0	0	0	0	0
Dendrochilum graminoides	0	1	0	0	0	0	1	0	0	0	0	1	0
Dendrochilum grandiflorum	0	0	0	0	0	0	0	1	1	1	0	0	0
Dendrochilum haslamii	0	0	0	0	0	0	0	0	1	1	0	0	0
Dendrochilum imbricatum	0	0	1	0	0	0	1	0	0	0	0	0	0
Dendrochilum kamborangense	0	0	0	0	0	1	0	0	1	0	0	0	0
Dendrochilum lancilabium	0	0	0	0	0	1	0	0	0	0	0	0	1
Dendrochilum longirachis	0	0	1	0	0	0	1	0	0	0	1	0	0
Dendrochilum pterogyne	0	0	0	0	0	0	0	1	1	1	0	0	0
Dendrochilum stachyodes	0	0	0	0	0	0	0	1	1	1	0	0	0
Dilochia cantleyi	0	0	0	0	0	0	0	0	0	0	0	0	1
Dilochia rigida	0	0	0	0	1	1	0	0	0	0	1	0	0
Entomophobia kinabaluensis	0	0	0	0	0	0	1	1	0	0	0	0	0
Epigeneium kinabaluense	0	0	0	0	0	1	1	1	0	1	0	1	1
Epigeneium longirepens	0	0	0	0	1	0	1	0	0	0	1	0	0
Epigeneium tricallosum	1	0	0	0	0	0	1	0	0	0	0	0	0
Epipogium roseum	0	0	1	1	0	0	0	0	0	0	0	0	0
Eria biflora	0	0	0	0	1	0	0	0	0	1	1	0	0
Eria brookesii	0	0	0	0	1	0	0	0	0	0	0	1	0
Eria cymbidifolia	0	0	0	0	0	1	1	0	0	0	1	1	0
Eria farinosa	0	0	0	0	1	0	1	0	0	0	0	0	0
Eria floribunda	0	0	1	0	1	0	0	0	0	0	0	0	0
Eria grandis	0	0	0	0	0	0	0	1	1	1	0	0	0
Eria iridifolia	0	0	0	1	1	0	0	0	0	0	0	0	0
Eria jenseniana	0	0	1	1	0	0	0	0	0	0	0	0	0
Eria kinabaluensis	0	0	0	0	0	1	0	0	1	1	0	0	0
Eria latiuscula	0	1	0	0	1	0	1	0	0	0	0	0	0
Eria leiophylla	0	0	0	0	1	0	1	0	0	0	1	0	0
Eria magnicallosa	1	1	0	0	1	0	1	0	0	0	0	0	0
Eria major	0	1	0	0	0	1	1	1	0	0	0	1	1
Eria nutans	1	0	0	0	0	0	1	0	0	0	0	0	0
Eria ornata	1	0	1	1	0	0	0	0	0	0	0	0	0
Eria pseudocymbidiformis	0	0	0	0	1	1	1	0	0	0	0	1	1
Eria pseudoleiophylla	1	0	0	0	0	0	0	0	0	0	1	0	0
Eria robusta	1	0	0	0	0	1	1	0	0	0	0	1	1
Eria xanthocheila	0	0	0	1	1	0	0	0	0	0	0	0	0

	1	2	3	4	5	6	7	8	9	10	11	12	13
Flickingeria aff. xantholeuca	1	0	0	1	0	0	0	0	0	0	0	0	0
Flickingeria fmbriata	1	0	0	1	0	0	0	0	0	0	0	0	0
Flickingeria pseudoconvexa	1	0	0	0	1	0	1	0	0	0	0	0	0
Goodyera hylophiloides	1	0	0	0	0	1	0	0	0	0	0	0	0
Goodyera kinabaluensis	1	0	0	0	0	0	0	0	0	0	1	0	0
Goodera ustulata	1	1	1	0	0	0	0	0	0	0	0	0	0
Habenaria damaiensis	0	0	0	0	1	0	1	0	0	0	0	0	0
Habenaria setifolia	0	0	0	0	1	0	1	0	0	0	0	0	0
Hetaeria angustifolia	1	1	0	0	0	0	0	0	0	0	0	0	0
Lepidogyne longifolia	1	0	0	0	1	0	1	0	0	0	0	0	0
Liparis aurantiorbiculata	0	0	0	0	1	0	1	0	0	0	0	0	0
Liparis caespitosa	1	0	0	1	0	0	1	0	0	0	0	0	0
Liparis elegans	1	0	0	1	0	0	0	0	0	0	1	0	0
Liparis kamborangensis	0	0	0	0	0	0	0	1	1	0	0	0	0
Liparis kinabaluensis	0	0	0	1	0	0	1	0	0	0	0	0	0
Liparis lingulata	1	0	0	0	0	1	0	0	0	0	0	0	0
Liparis lobongensis	0	0	0	1	0	0	1	0	0	0	0	0	0
Liparis mucronata	0	0	0	1	1	0	0	0	0	0	0	0	0
Liparis parviflora	1	0	0	0	0	1	0	0	0	0	0	0	0
Liparis viridiflora	1	0	0	0	0	1	0	0	0	0	1	0	0
Macodes lowii	0	1	0	1	0	0	0	0	0	0	0	0	0
Macodes petola	0	1	0	0	0	0	1	0	0	0	0	0	0
Malaxis commelinifolia	1	1	1	1	0	0	0	0	0	0	0	0	0
Malaxis lowii	1	1	0	1	0	0	0	0	0	0	0	0	0
Malaxis punctata	0	1	0	0	0	0	1	0	0	0	0	0	0
Micropera callosa	0	0	1	1	0	0	0	0	0	0	0	0	0
Nabaluia angustifolia	0	0	0	0	1	1	1	0	1	0	0	0	1
Nabaluia clemensii	0	0	0	0	0	1	0	1	0	0	0	0	0
Nephelaphyllum pulchrum	0	0	0	0	1	1	0	0	0	0	0	1	0
Neuwiedia veratrifolia	1	0	1	1	0	0	0	0	0	0	0	0	0
Neuwiedia zollingeri	0	1	0	1	0	0	0	0	0	0	0	0	0
Oberonia affinis	0	1	0	0	1	0	1	0	0	0	0	0	0
Oberonia kinabaluensis	1	1	0	0	0	0	0	0	0	0	0	0	0
Oberonia patentifolia	1	0	0	1	0	1	0	0	0	0	1	0	0
Paphiopedilum dayanum	1	1	0	0	0	0	1	0	0	0	0	0	0
Paphiopedilum hookerae	0	1	0	1	0	0	0	0	0	0	0	0	1
Paphiopedilum javanicum	1	0	0	0	1	0	0	0	0	0	0	0	0
Paphiopedilum rothschildianum	0	1	0	1	0	0	0	0	0	0	0	0	0
Phaius baconii	0	0	0	0	1	0	1	0	0	0	0	0	0
Phaius reflexipetalus	0	1	0	1	0	0	0	0	0	0	0	0	0
Phalaenopsis amabilis	1	0	1	1	0	0	1	0	0	0	0	0	0
Phalaenopsis modesta	1	0	0	1	0	0	0	0	0	0	0	0	0
Pholidota carnea	0	0	0	0	0	0	1	0	0	0	1	0	0
Pholidota clemensii	0	0	0	0	1	1	1	0	0	0	0	0	1

	1	2	3	4	5	6	7	8	9	10	11	12	13
Pholidota ventricosa	1	0	0	0	0	0	1	0	0	0	1	0	0
Phreatia listrophora	0	0	0	0	0	0	1	0	0	0	0	1	0
Pilophyllum villosum	0	0	0	0	0	0	1	0	0	0	1	0	0
Platanthera kinabaluensis	0	0	0	0	0	0	0	0	1	1	0	1	1
Plocoglottis gigantea	1	0	1	0	0	0	0	0	0	0	0	0	0
Poaephyllum cf. podochiloides	0	1	0	1	0	0	0	0	0	0	0	0	0
Podochilus lucescens	0	1	0	0	0	0	1	0	0	0	0	0	0
Podochilus microphyllus	0	1	0	0	0	0	1	0	0	0	0	0	0
Podochilus sciuroides	0	0	0	1	0	0	1	0	0	0	0	0	0
Podochilus tenuis	0	0	0	1	0	0	1	0	0	0	1	0	0
Pteroceras fragrans	0	0	1	1	0	0	0	0	0	0	0	0	0
Pteroceras teres	1	0	1	0	1	0	0	0	0	0	0	0	0
Robiquetia crockerensis	0	0	0	1	1	0	1	0	0	0	0	1	1
Robiquetia pinosukensis	0	0	0	0	1	0	1	0	0	0	0	0	0
Spathoglottis gracilis	0	1	0	1	0	0	1	0	0	0	0	1	1
Spathoglottis microchilina	1	0	0	0	0	0	0	0	0	0	0	1	0
Taeniophyllum indet.	0	0	0	0	0	0	0	0	1	1	0	0	0
Thelasis micrantha	0	1	0	0	1	0	0	0	0	0	0	0	0
Thelasis variabilis	1	0	0	0	0	0	1	0	0	0	0	0	0
Thrixspermum centipeda	1	0	0	0	0	1	0	0	0	0	0	0	0
Thrixspermum triangulare	0	0	0	0	0	1	1	1	1	1	0	1	1
Trichoglottis kinabaluensis	0	0	0	0	0	0	1	0	0	0	1	0	0
Trichoglottis winkleri	0	0	1	1	0	0	0	0	0	0	0	0	0
Trichotosia aurea	0	0	0	0	1	1	1	0	0	0	0	0	0
Trichotosia brevipedunculata	0	0	0	0	1	1	1	0	0	0	0	0	0
Trichotosia ferox	0	0	0	0	1	1	0	0	1	0	0	0	0
Trichotosia mollicaulis	1	1	0	0	0	0	1	0	0	0	1	0	0
Tropidia pedunculata	1	1	0	1	0	0	0	0	0	0	0	0	0

Key to areas listed above

1. Dallas
2. Penataran River
3. Poring
4. Hempuen Hill
5. Pinosuk Plateau
6. Summit trail
7. Peniguppan Ridge
8. Upper Marai Parai
9. Kinabalu Lipson
10. Summit Area
11. Lower Gurulau
12. Marai Parai
13. Pig Hill

Botanical inventory of Mount Kinabalu:
a progress report

JOHN H. BEAMAN[1], REED S. BEAMAN[2] & CHRISTIANE ANDERSON[3]

Summary. The project to inventory the flora of Mount Kinabalu has been in progress for nine years, during which time enumerations of about two-fifths of the flora, currently known to include *c.* 4,500 species, have been published. The north side of the mountain has received scant botanical attention, and local collectors are being employed to obtain specimens from that area. A specimen database of over 20,000 records has been developed and made available on the Internet. Other relational specimen and taxon files are being prepared, from which camera-ready copy of segments of the enumeration are produced. A geographic information system (GIS) is in preparation, which includes coverages of the topography, hydrography, species distributions, satellite imagery, vegetation, geographic locations, geology, and land use. Morphological and molecular data are being used to test biogeographic and evolutionary hypotheses and examine the evolution and speciation of nine well-represented exemplar genera in eight families. An ethnobotanical project (Projek Etnobotani Kinabalu, PEK) was initiated in 1992, and employs local collectors from communities around the mountain to collect the useful plants in their areas and gather information about plant names and uses. More than 3,500 ethnobotanical collections thus far have been obtained and the data entered into a database. The research focuses on the value and sustainability of non-timber forest products, including biochemical prospecting for medicinal plants, assessment of the impact of harvesting on selected commercial species such as rattan palms, and the role of non-marketed useful plants in the health care and diet of local communities. One ultimate objective will be an analysis of Dusun botanical classification and its correspondence to scientific classification.

INTRODUCTION

The inventory of vascular plants of Mount Kinabalu has five major components or objectives as follows: (1) Intensive collecting on the north side of the mountain, from which very few specimens are currently available; (2) development of a database of the species, which will be used initially to produce a detailed checklist; (3) preparation of a geographical information system (GIS) for the mountain, including coverages of the topography, hydrography, species distributions, satellite imagery, vegetation, geographic locations, geology, and land use; (4) analysis of evolutionary processes in the Kinabalu flora, using selected exemplar taxa; and (5) making a detailed ethnobotanical study in the area.

The basic concept of the database for the Kinabalu flora was outlined by Beaman & Regalado (1989). Diversity and distribution patterns in the flora were discussed by Beaman & Beaman (1990), during the first Flora Malesiana Symposium, and

[1]Institute of Biodiversity and Environmental Conservation, Universiti Malaysia Sarawak, 94300 Kota Samarahan, Sarawak, Malaysia. Present address: The Herbarium, Royal Botanic Gardens, Kew, Richmond, Surrey TW9 3AB, U.K
[2]Department of Botany, University of Florida, Gainesville, FL 32611, U. S. A.
[3]University of Michigan Herbarium, North University Building, Ann Arbor, MI 48109, U. S. A.

hypotheses concerning the evolution of the flora were outlined in that paper. An enumeration of the pteridophytes was published by Parris, Beaman & Beaman (1992), and enumerations of the orchids (Wood, Beaman & Beaman 1993) and gymnosperms (Beaman & Beaman 1993) are also published. The purpose of this paper is to describe the current status of the project.

COLLECTING ON THE NORTH SIDE

The traditional approach to Mount Kinabalu from 1851, when Hugh Low became the first westerner to climb the mountain, to the present has been from the west and south. This route was chosen partly because Kinabalu is relatively close to the west coast of Sabah and partly because a series of ridges on the south side provides the least difficult access to the summit. In the 19th century, Kiau on the southwest side was the settlement where the final climb to the summit began. In the 20th century Bundu Tuhan and Tenompok on the south side have been important base camps, and the establishment of Kinabalu Park headquarters near Tenompok further focussed collecting in that area. These accidents of history and geography, as well as the forbidding terrain of precipitous ridges and valleys to the north, relate to the virtual lack of collections from the north side of Kinabalu, as shown in the map of pteridophyte collections (Map 1).

We are now attempting to remedy this situation by employing local collectors from several kampungs around the north side to collect in the unexplored areas. General collecting in the area has been facilitated by the fact that for about three years we have sponsored ethnobotanical collecting in kampungs around the mountain. These collectors are now being asked to make general as well as ethnobotanical collections, and additional collectors have been hired for general collecting. The strategically positioned kampungs of Nalumad, Menggis, Serinsim, Sayap, and Melangkap Tomis around the north side are now represented by collectors. Collecting activities are being coordinated by Luiza Majuakim, Sabah Parks Ethnobotanist. Todd Barkman from the University of Texas recently spent nearly four months helping train the north-side collectors.

ENUMERATION OF THE FLORA

The research methodology for enumeration of the Kinabalu flora is similar to that of a taxonomic revision, except that many more taxa are involved and keys and descriptions of species generally are not included (for the orchids and gymnosperms keys to genera were provided). Almost all specimens in the herbaria at Kew and Leiden (over 20,000) relevant to the project have received preliminary examination and have been entered into a specimen database. As family treatments are completed, additional computer files incorporating taxon, type-specimen, synonymy, and bibliographic data are prepared, and many specimens are re-examined. The three instalments already published (pteridophytes, orchids, gymnosperms), include 1,359 taxa, about 30% of the total flora of close to 4,500 species. An example of the enumeration of the orchids (Wood *et al.* 1993) is shown by a specimen page (Fig. 1).

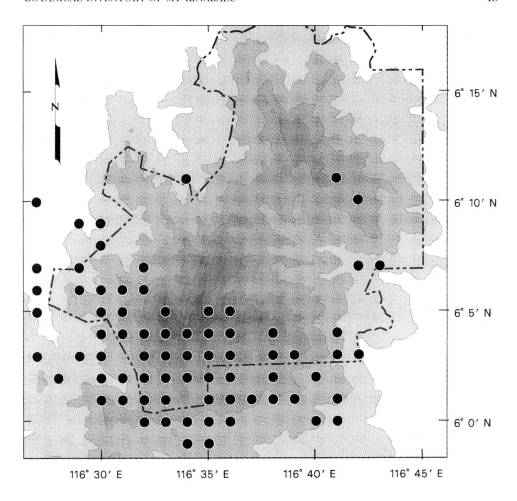

MAP 1. Locations to nearest degree and minute of latitude and longitude from which any pteridophyte collections have been made on Mount Kinabalu. The contour lines are at 500 m intervals.

Rather than accepting in an uncritical manner previous specimen identifications or copying information from the literature, we are attempting to evaluate the status of each specimen and species. In some cases an authoritative name cannot be assigned to a particular taxon. In the orchid enumeration, for example, 596 of the 711 taxa were fully named; 53 were included as 'sp. 1', 'sp. 2', etc. In 46 instances they were designated as 'aff.' some other species, and seven taxa were questionably identified with 'cf.' separating the generic name and specific epithet. Specimens sometimes need to be studied from throughout the range of a species, not just those from Kinabalu, which we attempt to do to the extent practicable.

An up-to-date version of the Kinabalu database is being developed and maintained by the first author. The second author is responsible for computer program development and has access to the database through the Internet. The third author

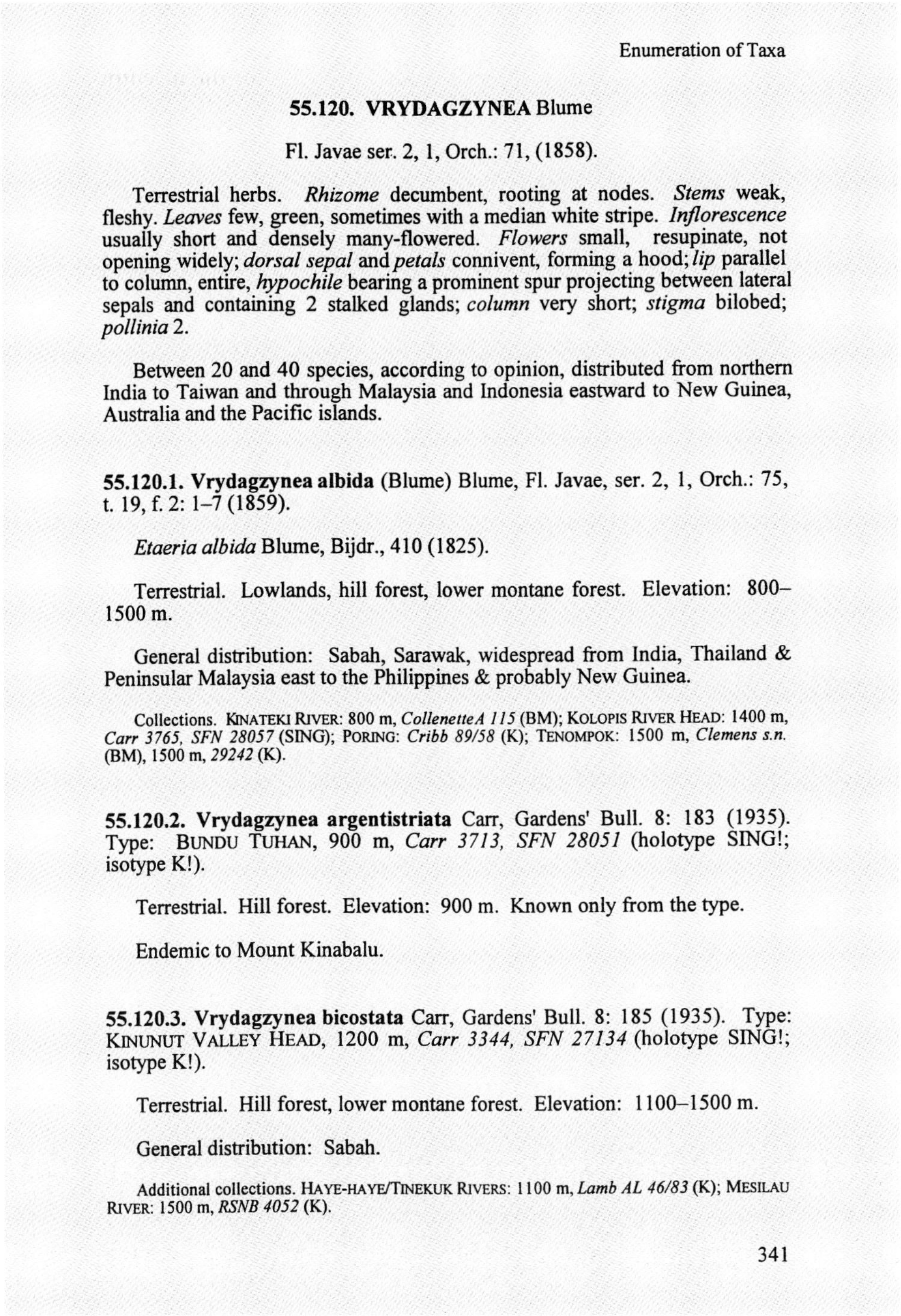

Enumeration of Taxa

55.120. VRYDAGZYNEA Blume

Fl. Javae ser. 2, 1, Orch.: 71, (1858).

Terrestrial herbs. *Rhizome* decumbent, rooting at nodes. *Stems* weak, fleshy. *Leaves* few, green, sometimes with a median white stripe. *Inflorescence* usually short and densely many-flowered. *Flowers* small, resupinate, not opening widely; *dorsal sepal* and *petals* connivent, forming a hood; *lip* parallel to column, entire, *hypochile* bearing a prominent spur projecting between lateral sepals and containing 2 stalked glands; *column* very short; *stigma* bilobed; *pollinia* 2.

Between 20 and 40 species, according to opinion, distributed from northern India to Taiwan and through Malaysia and Indonesia eastward to New Guinea, Australia and the Pacific islands.

55.120.1. Vrydagzynea albida (Blume) Blume, Fl. Javae, ser. 2, 1, Orch.: 75, t. 19, f. 2: 1–7 (1859).

Etaeria albida Blume, Bijdr., 410 (1825).

Terrestrial. Lowlands, hill forest, lower montane forest. Elevation: 800–1500 m.

General distribution: Sabah, Sarawak, widespread from India, Thailand & Peninsular Malaysia east to the Philippines & probably New Guinea.

Collections. KINATEKI RIVER: 800 m, *CollenetteA 115* (BM); KOLOPIS RIVER HEAD: 1400 m, *Carr 3765, SFN 28057* (SING); PORING: *Cribb 89/58* (K); TENOMPOK: 1500 m, *Clemens s.n.* (BM), 1500 m, *29242* (K).

55.120.2. Vrydagzynea argentistriata Carr, Gardens' Bull. 8: 183 (1935). Type: BUNDU TUHAN, 900 m, *Carr 3713, SFN 28051* (holotype SING!; isotype K!).

Terrestrial. Hill forest. Elevation: 900 m. Known only from the type.

Endemic to Mount Kinabalu.

55.120.3. Vrydagzynea bicostata Carr, Gardens' Bull. 8: 185 (1935). Type: KINUNUT VALLEY HEAD, 1200 m, *Carr 3344, SFN 27134* (holotype SING!; isotype K!).

Terrestrial. Hill forest, lower montane forest. Elevation: 1100–1500 m.

General distribution: Sabah.

Additional collections. HAYE-HAYE/TINEKUK RIVERS: 1100 m, *Lamb AL 46/83* (K); MESILAU RIVER: 1500 m, *RSNB 4052* (K).

341

FIG. 1. Specimen page from the Kinabalu orchid enumeration (Wood *et al.* 1993). Seven relational data files were involved in the printing of this page by the KINABALU programs from the orchid database. Used by permission.

is helping prepare treatments for various dicot families, over 50 of which are completed or nearly so. We have obtained the collaboration of about 40 specialists who have contributed or will contribute data and expertise for the inventory.

The general working procedure is that we give collaborators a preliminary draft inventory of the taxon or taxa in their area of expertise, produced from the Kinabalu database. They provide the taxon data including citation of the place of original publication, data on relevant types, synonymy, characterisation of habit, habitat, and special notes about species. This information is entered into the database by the first author, and a semifinal draft is produced by computer for review by the specialist.

Two software technologies have been developed (Beach *et al.*, 1993) to make the specimen database available interactively to the scientific community. Both utilize TCP/IP network communications. (1) A client/server architecture was implemented using the Ingres 4GL and C programming languages, whereby a remote Ingres site could query the holdings of the Kinabalu database over the network and receive copies of the specimen records for import back into a local Ingres server. (2) We also developed an e-mail specimen-data server based in Ingres, that responds to incoming network mail messages containing key-word search statements. Specimen-data records that meet query criteria are mailed back to the sender of the query in one of several specified formats. This was the first natural history specimen database to be made internationally available over networks. The specimen database has also been Internet accessible through the Biological Collections and Biodiversity Gopher. Recently, the orchids, ferns and gymnosperms have been placed on the World Wide Web, and additional taxa can be made available in this manner as the treatments are completed.

Geographical information system

This component of the Kinabalu project is discussed in greater detail by Beaman *et al.* (1997), so only a brief overview will be provided here. Coverages that are presently under development for the geographical information system (GIS) include topography, hydrography, species distributions, satellite imagery, vegetation, geographic locations, geology, and land use. Geodetic control of these coverages is being enhanced through linkage with a database of global positioning system (GPS) points provided by the Sabah Department of Lands and Surveys and additional points collected by Reed Beaman. The geo-referenced locality records represented in the specimen database will be mapped to associate diversity at various taxonomic levels with environmental parameters. A virtually cloud-free Landsat TM image from June 14, 1991, donated by the EOSAT Corporation, has been incorporated into the GIS.

In addition to the research questions we will address with GIS technology, the maps and map overlays, including those showing land-use polygons by class, will have a significant role in conservation and land-use planning on the mountain (cf. Morain 1993). We should be able to quantify accurately the size, distribution, and short-term historical changes in specific natural and disturbed vegetation types. Our collaboration with Malaysian scientists and students should result in the transfer of the products of our work and of the technology itself to government and planning officials and other interested parties in Malaysia. Given the ongoing conversion of natural vegetation for timber, recreation, and agriculture, it will be valuable and appropriate for this project to provide sophisticated analyses of existing biodiversity for decision makers and local university scientists. Although the GIS is currently in ARC-INFO running on a Unix platform, we intend to place all files on a CD-ROM. Digital coverages for the Kinabalu physiographic region include *c.* 2,146 km².

An early product of the GIS will be a location map of Mount Kinabalu. This map will be at a scale of 1:50,000, with 500 m contour intervals, showing over 350 localities, many of which were derived from a database of locality names in the Dusun language (Beaman *et al.*, 1996). The positioning of plant collection localities, which is unsatisfactory when based on latitude-longitude coordinates to the nearest minute, can be improved through use of more precisely registered data, some of which is being obtained through use of GPS units.

Alexis Thomas is using the Landsat image to identify areas of ultramafic geology. Biologically meaningful remote sensing analysis in tropical forest areas has proven difficult owing to high species and ecosystem diversity. The correlated GIS and plant collection data for Kinabalu should facilitate this analysis. The distinctness of vegetation on ultramafic geology will allow discrimination of plant associations by spectral signatures developed in accordance with remote sensing techniques of image classification.

GIS information can be incorporated into phylogenetic analysis in a biogeographical context. In addition, predictions of taxon distributions are possible based on modelling of known specimen-collection data. Biogeographic relationships among ultramafic areas on Mount Kinabalu and other ultramafic outcrops in Borneo will be analyzed in the context of phylogenetic relationships among ultramafic and non-ultramafic taxa. Phytogeographic patterns of related taxa will be analyzed and the taxa assigned colour signatures. Spatial correlation techniques (see Legendre 1993) will be applied and evaluated for their usefulness in biogeographic systematics.

EVOLUTION AND SPECIATION OF EXEMPLAR TAXA

Beaman & Beaman (1990) emphasized that a high percentage of the Kinabalu species have extremely localised distributions, frequently associated with ultramafic outcrops. Ideal conditions for a diverse flora and rapid evolutionary rates apparently result from several factors, including a vast range of environments from wet tropical lowlands to alpine conditions, numerous geologically recent habitats on a diversity of substrates (particularly ultramafic outcrops), regularly recurring El Niño droughts that may drive catastrophic selection, precipitous topography resulting in strong reproductive isolation over short distances, and small population size of many species, which may be susceptible to rapid natural selection or genetic drift. Numerous apparent neo-endemics in many genera and families, and numerous genera with large numbers of species suggest that frequent speciation events in the recent past may have contributed significantly to the great diversity of the flora. Short- and long-range dispersal of some plants pre-adapted to montane environments also may have contributed to the high floristic diversity.

The principles and methods of vicariance biogeography will be applied to determine levels of congruence between phylogenetic and biogeographic relationships (Brooks 1985; Page 1988; Wiley 1988). Morphological and molecular characters will be used to construct phylogenies for selected exemplar taxa, which in turn will be used to construct area cladograms.

Biogeographic relationships between ultramafic areas on Kinabalu and with other ultramafic areas in Borneo will be considered, as will phylogenetic relationships between ultramafic and non-ultramafic taxa. These analyses will be used to determine whether or not non-random biogeographic patterns of distribution are exhibited by ultramafic taxa (see Nelson & Platnick 1978). The Kinabalu specimen database provides a starting point for this analysis.

Evolution and speciation of the exemplar taxa are being studied using both morphological and DNA data. Given the Pleistocene age of the mountain and the fact that a high proportion of the species under study are endemic or nearly so to Kinabalu, many of these are believed to have originated *in situ* (Beaman & Beaman 1990). Ultramafic outcrops are common but irregular in distribution on the mountain. The occurrence of these has been likened to islands by Barkman *et al.* (1997). Isolation of the ultramafic sites results from precipitous topography and variable elevations within which the outcrops are found. Populations on these "islands" probably have mimimal genetic exchange with related populations on other ultramafic patches.

The exemplar taxa for the studies are from various families. Particular emphasis is being placed on genera containing rare or endemic taxa with ultramafic distributions. The following issues are being addressed concerning their evolution: (1) If the endemics evolved *in situ*, they must represent the most recently derived taxa within their genus. (2) If the progenitors of endemic species gave rise to other extant taxa, this should be detectable. (3) If these taxa have deducible historical lineages, then we can investigate the evolution of endemic species.

The following exemplar taxa are currently under study within the framework described above: *Dendrochilum* (Orchidaceae), *c.* 34 species, by Todd Barkman; *Elatostema* (Urticaceae), *c.* 35 species, and *Polyosma* (Escalloniaceae), *c.* 12 species, by Reed Beaman; *Sonerila* (Melastomataceae), *c.* 11 species, by Nicoletta Cellinese; *Medinilla* (Melastomataceae), *c.* 17 species, by Gudrun Clausing; *Cyathea* (Cyatheaceae), *c.* 20 species, by David Conant; *Lithocarpus* (Fagaceae), *c.* 35 species, by Charles Cannon; *Rhododendron* (Ericaceae), *c.* 26 species, by Katherine Kron; and *Carex* subg. *Indocarex*, *c.* 12 species, by Alan Yen. Whether or not studies in these disparate groups yield similar results, they will help to explain some of the evolutionary processes operating in the Kinabalu flora.

ETHNOBOTANY

The Projek Ethnobotani Kinabalu (PEK) was initiated in the summer of 1992 with an initial emphasis on palms. It has subsequently been expanded to include all useful plant groups. PEK aims to link ethnobotanical research with conservation action by strengthening the link between Kinabalu Park and local communities. Interpretive programs are being developed that reach many of the 200,000 visitors who come to the Park each year. Emphasis is being placed on helping the local Dusun people to use botanical resources in a sustainable way, and the sustainability and value of plant resources in and around the Park are being studied. PEK activities are being discussed in greater detail by Gary Martin in this Symposium, and are only summarised here.

PEK brings together Dusun-speaking collectors, Sabah Parks personnel, Malaysian scientists, and foreign specialists who are assessing plant resources around Mount Kinabalu. The infrastructure to carry out research within Kinabalu Park is being augmented, allowing Malaysian professors and students the opportunity to carry out ethnobotanical and phytochemical studies. The research focuses on the value and sustainability of non-timber forest products, including studies on biochemical prospecting for medicinal plants, assessment of the impact of harvesting on selected commercial species such as rattan palms, and the role of non-marketed useful plants in the health care and diet of local communities.

PEK collectors are helping carry out an ethno-floristic inventory in villages and forested areas that range from 400 m up to the high slopes of Mount Kinabalu. The collectors have a number of informants, older members of the community who know the Dusun names and uses of plants, but who are not necessarily fluent in Malay. The informants provide information about the Dusun classification and uses of plants, and they accompany the collectors on trips to the field when possible.

Herbarium specimens are being prepared and a computerised database for ethnobotanical information has been designed and programmed. In each of eight communities around Kinabalu an average of 40 – 50 voucher collections of useful plants are being made each month. More than 3,500 collections have already been obtained. The results of the ethno-floristic survey will be analyzed, with an emphasis on identifying plant resources that are vulnerable to over-harvesting and have under-exploited commercial potential. Some of these plants, including several species of rattan palms, will be selected for more intensive analysis of how they could be used in a sustainable way, potentially providing economic alternatives to local communities. A phytochemical laboratory is planned, which will allow initial screening of the chemical components and biological activity of selected plants.

The collectors are directly supervised by the Park Ethnobotanist, Luiza Majuakim, and by park rangers and naturalists, who ensure that collections are of good quality, ethnobotanical and locality data are accurately recorded, and collecting activities within Park boundaries are adequately monitored. They visit the collectors at regular intervals of at least once a month, pick up collections and deliver bottled gas (for specimen drying) and other supplies. The collectors and some park naturalists have attended the two training courses held thus far.

The ethnobotanical study will also have a significant GIS component. This application will facilitate integrating the taxonomic, ethnobotanical, and phytochemical databases into Park activities and monitoring aspects such as ecotourism development and illegal harvesting inside Park boundaries. It will also provide a means for accurate assessment of the distribution of useful plants and estimation of the proportion of the flora utilised by the Dusun peoples. One of the ultimate objectives of the ethnobotanical study is an analysis of Dusun botanical classification, following the approach outlined by Berlin (1992). Analyses of plant uses will in part involve GIS methodology, in which similarities in uses and naming of plants in different areas will be examined. Determination of the correspondence between Dusun and scientific categories will be required before the data can be fully integrated into the GIS and linked to the botanical inventory.

RELEVANCE TO *FLORA MALESIANA*

In his closing remarks for this Symposium Professor Pieter Baas read a statement (which he does not support) by van Steenis (1948) from the introduction to *Flora Malesiana*, concerning "The undesirability of compiling, at this stage, local floras in Malaysia." Van Steenis noted that "Only temporary profit may be gained from making local floras, and both valuable time and money are wasted by the enormous duplication which is unavoidable when the goal of a flora of a plant-geographical unit is to be reached along this tortuous road."

The value of a Flora covering a very large region to the preparation of local floras within that region is obvious, but the usefulness of local floras in the development of a work of the scope of *Flora Malesiana* may not be immediately apparent. Like Pieter Baas, we disagree with van Steenis concerning the undesirability of local floristic projects and can suggest several reasons why local floras are relevant and desirable within a region such as that of *Flora Malesiana*.

With respect to the Kinabalu project in particular, the following points may be made. (1) Specimens reviewed for the project provide a basis for understanding what areas need additional collecting, and provide rationale for collecting activities in under-collected areas. (2) The databases and GIS developed for local floras provide a nucleus around which more extensive databases and GIS can be developed. Already there is innate compatibility between the Kinabalu and Brunei databases, and other projects are on the horizon for further database expansion. (3) Local floras provide an immediate framework for development of hypotheses concerning the origin, evolution, and biogeography of particular species. (4) Testing of evolutionary hypotheses and intensive investigation of selected taxa result in improved understanding of component parts of the general flora. (5) Preparation of local floras provides a training ground for young systematists, both native and foreign, to gain valuable taxonomic research experience.

ACKNOWLEDGEMENTS

The project is currently supported by U. S. National Science Foundation grant DEB-9400888 to Michigan State University, and John D. and Catherine T. MacArthur Foundation grants to Sabah Parks and Universiti Malaysia Sarawak. Our research on Mount Kinabalu has been amply facilitated by Sabah Parks, and we greatly appreciate the cooperation of Datuk Lamri Ali, Sabah Parks Director, and the Kinabalu Park staff.

REFERENCES

Barkman, T. J., Repin, R. & Beaman, R. S. (1997). Extreme ultramfic habitat patches as silands on Mt. Kinabalu. In: Dransfield, J., Coode, M. J. E. & Simpson, D. A., F. M. 3 Symp. Vol.: 000 – 000.

Beach, J. H., Ozminski, S. J. & Boufford, D. E. (1993). An Internet botanical specimen data server. *Taxon* 42: 627 – 629.

Beaman, J. H. & Beaman, R. S. (1990). Diversity and distribution patterns in the flora of Mount Kinabalu. Pp. 147 – 160 *in* Baas, P. *et al. The Plant Diversity of Malesia*. Kluwer Academic Publishers, Dordrecht, Netherlands.

—— & —— (1993). The gymnosperms of Mount Kinabalu. *Contr. Univ. Michigan Herb.* 19: 307 – 340.

—— & Regalado, Jr., J. C. (1989). Development and management of a microcomputer specimen-oriented database for the flora of Mount Kinabalu. *Taxon* 38: 27 – 42.

——, Aman, R. H., Nais, J., Sinit, G. & Biun, A. (1996). Mount Kinabalu place names in Dusun and their meaning. In: Phillipps A. & Wong, K. M. (eds.), *Kinabalu: Summit of Borneo*. Ed. 2. Sabah Society Monograph.

Beaman, R. S., Thomas, A. G. & Beaman, J. H. (1997). A geographical information system (GIS) for Mount Kinabalu. In: Dransfield, J., Coode, M. J. E. & Simpson, D. A. F.M. 3 Symp. Vol.: 000 – 000.

Berlin, B. (1992). *Ethnobiological Classification. Principles of Categorization of Plants and Animals in Traditional Societies*. Princeton Univ. Press, Princeton, N. J.

Brooks, D. R. (1985). Historical ecology: a new approach to studying the evolution of ecological associations. Ann. Missouri Bot. Gard. 72: 660 – 680.

Legendre, P. (1993). Spatial autocorrelation: trouble or new paradigm? *Ecology* 74: 1659 – 1673.

Morain, S. A. (1993). Emerging technology for biological data collection and analysis. *Ann. Missouri Bot. Gard.* 80: 309 – 316.

Nelson, G. & Platnick, N. (1978). The perils of plesiomorphy: widespread taxa, dispersal, and phenetic biogeography. *Syst. Zool.* 27: 474 – 477.

Page, R. D. M. (1988). Quantitative cladistic biogeography: constructing and comparing area cladograms. *Syst. Zool.* 37: 254 – 270.

Parris, B. S., Beaman, R. S. & Beaman, J. H. (1992). *The Plants of Mount Kinabalu, 1. Ferns and Fern Allies*. 165 pp. + 5 pl. Royal Botanic Gardens, Kew.

van Steenis, C. G. G. J. (1948). *Flora Malesiana* I, 4: vii.

Wiley, E. O. (1988). Parsimony analysis and vicariance biogeography. *Syst. Zool.* 37: 271 – 290.

Wood, J. J., Beaman, R. S. & Beaman, J. H. (1993). *The Plants of Mount Kinabalu: 2. Orchids*. Royal Botanic Gardens, Kew. xii + 411 pp. + 84 pl.

Diversity in shoot architecture in *Pothos* (*Araceae*: *Pothoideae*): Observations towards a new infrageneric classification

PETER C. BOYCE[1] & A. HAY[2]

Abstract. The c. 70 species of *Pothos* (subtropical and tropical forest climbers centred on Malesia) show growth patterns and inflorescence presentation that have taxonomic significance at specific and generic levels in tribe *Pothoeae* (*Pothos*, *Pedicellarum* possibly congeneric and *Pothoidium*). In subgen *Allopothos* inflorescence presentation now suggests that the *P. insignis* group should be separated from, not allied with, the *P. rumphii* group. The latter seems closer to Bornean species previously thought to be isolated (e.g. *P. kinabaluensis*, *P. atropurpuracens*) and possibly the Australian *P. brassii*. The taxonomically isolated *P. falcifolius* (New Guinea) varies greatly in its in inflorescence presentation. The *P. luzonensis* group approaches *Pedicellarum* — both highly modified in inflorescence structure and sharing characters formerly used to separate the latter. In subgen *Pothos* the architecture appears more uniform. The *P. scandens* group flowers synchronously on short shoots in the axils of most leaves of a flowering branch; in the *P. papuanus* group inflorescences are fewer and irregularly positioned. In *Pothoidium* spadices are often arranged pseudopaniculately and mature sequentially but the vegetative architecture is similar to that in subg. *Pothos*, with which it also shares distinctive leaf morphology.

INTRODUCTION

Pothos L. is a genus of approximately 70 species of subtropical and tropical forest climbers centred in Malesia (Hay et al. 1995) belonging to *Araceae* subfamily *Pothoideae* (sensu Mayo, Bogner & Boyce, in press). Shared leaf, flower and fruit characters suggest that *Pothos* is most closely related to *Pedicellarum* M. Hotta (1976) and *Pothoidium* Schott (1857a,b). Indeed, *Pothos* and *Pedicellarum* may be congeneric. Until a full revision of *Pothos* is completed we are adopting a modified informal version of Engler's infrageneric classification of *Pothos* (Engler 1905). While we follow Engler's sections (as subgenera) we reject his serial concepts, instead recognizing informal species groups.

SHOOT ARCHITECTURE

In common with many *Araceae* lianes (see Schimper 1903: 193), species of tribe *Pothoeae* display differentiation of stem function manifesting as various types of shoot architecture (see Blanc 1977a, b, 1978, 1980). All *Pothos* species investigated have at least five types of shoot architecture (Fig. 1). On germination (Fig. 1, A) a

[1] Herbarium, Royal Botanic Gardens, Kew, Richmond, Surrey, TW9 3AE, U.K.
[2] Royal Botanic Gardens, Mrs Macquarie's Road, Sydney, NSW 2000, Australia.

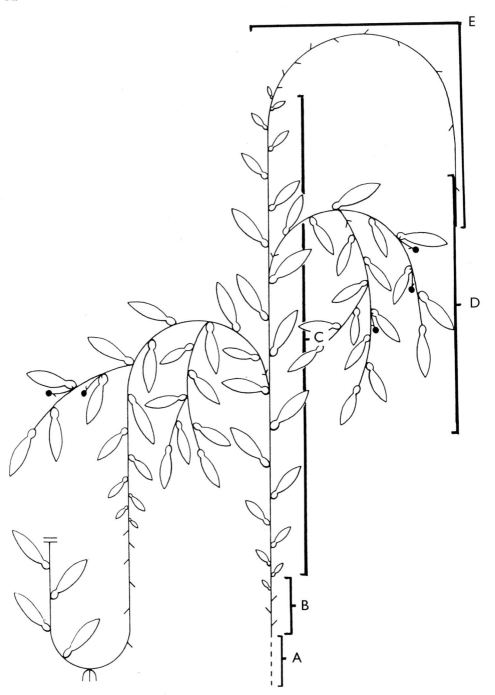

FIG. 1. Diagrammatic representation of growth habit of *Pothos* subgen. *Pothos*. **A**, eocaul phase; **B**, leafy juvenile phase; **C**, leafy sterile mature phase; **D**, sympodial fertile shoot; **E**, flagelliform foraging shoot. Drawn by Emmanuel Papadopoulos.

monopodial leafless (minute cataphylls present) thread-like shade-seeking (skototropic; see Strong & Ray 1975) creeper (eocaul) is produced. Once this shoot begins to climb (Fig. 1, B) a monopodial, leafy (shingling or not; see Madison 1977 and Boyce & Poulsen 1994), juvenile phase develops; this in turn (Fig. 1, C) leads to a monopodial, leafy, sterile mature phase; (Fig. 1, D) sympodial fertile shoots arise from the sterile mature phase. Both mature sterile and fertile shoots can give rise (Fig. 1, E) to a terminal extension consisting of a monopodial leafless flagelliform foraging shoot. A sixth shoot type occurs in some species of subgen. *Pothos* and in *Pothoidium* and is discussed later in more detail.

Subgenus *Allopothos*

Species of subgen. *Allopothos* display the above five types of shoot architecture. All species investigated pass through an eocaul phase (Fig. 2, A) that grows along the forest floor until a suitable climbing surface is encountered. Once a climbing surface is reached all but one (*P. brassii* B.L. Burtt) species investigated produce a distinctive shingle climber (Fig. 2, B) bearing ovate to lanceolate, almost sessile, distichously arranged leaves overlapping in the manner of roof tiles. This phase is closely attached to the substrate by roots arising from the nodes and occasionally produces geotropic feeder roots that reach the ground. The shingle phase continues for approximately 2 m when, abruptly, adult leaves are produced with long petioles and lanceolate to elliptic laminae (Fig. 2, C). This transition marks the beginning of the third type of shoot architecture (sterile mature). Many species of subgen. *Allopothos* remain at this third phase for a considerable period of time. The sterile mature shoots continue growth and often branch repeatedly from the lower parts, especially in the *P. luzonensis* (Presl) Schott group. Sterile mature shoots are often scandent, rooting at distant nodes and occasionally form hammocks (extensively branched aerial stem masses) in the canopy. Mature sterile shoots occasionally transform terminally into a foraging flagellum with a few leaf-like cataphylls and reduced foliage-leaves at first but soon becoming naked with slightly prominent nodes up to 10 cm distant. This stem forages for several metres until a suitable climbing surface is encountered and does not branch unless damaged apically when it then reiterates from the nearest healthy node. The foraging stem occasionally roots sparsely from the nodes but always roots prolifically on resumption of the climbing habit. Occasionally fertile mature shoots transform terminally into a foraging flagellum. Eventually, sterile mature stems branch from older, occasionally leafless, parts, giving rise to sympodial fertile shoots (Fig. 2, D). These fertile shoots, although initially produced from the mature sterile stem, later give rise (Fig. 2, E) to new fertile (enrichment) shoots that are produced from post fertile portions, usually the older mid-portions, the whole structure eventually forming a much-branched system. Each plant may bear many such systems and each is capable of producing a succession of inflorescences. The length of these individual branches and the degree of leaf development varies between species' groups. Extremes of branch length are exemplified by species of the *P. insignis* Engl. group (Boyce & Poulsen 1994) and the *P. luzonensis* group. Extremes of leaf morphological development are spanned by species of the *P. rumphii* (Presl) Schott and *P. luzonensis* groups.

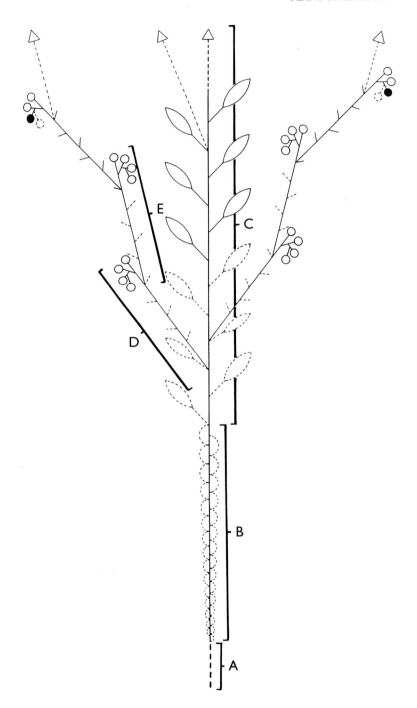

FIG. 2. Diagrammatic representation of growth habit of *Pothos* subgen. *Allopothos*. **A**, eocaul phase; **B**, leafy juvenile phase; **C**, leafy sterile mature phase; **D**, sympodial fertile shoot; **E**, enrichment shoot. Drawn by Emmanuel Papadopoulos.

Subgenus *Pothos*

Most species of subgen. *Pothos* appear to have the five-phase growth architecture described for subgen. *Allopothos*. A possible sixth phase, a modified version of the sterile mature growth, has been observed in four species of *Pothos* and in *Pothoidium lobbianum* and is discussed later. The eocaul is not known for all species in the subgenus but for those known, (*P. scandens* L., *P. repens* (Lour.) Druce (syn. *P. loureirii* Hooker & Arn.), *P. longipes* Schott (see Hay 1995), and *P. chinensis* (Raf.) Merr., this phase consists of a skototropic flagellar shoot that, on reaching a suitable climbing surface, alters into a shingle climber with closely arranged or nearly overlapping leaves of much the same shape as those of the mature growth phases. The shingle phase appears to be monopodial but also produces simple branching systems by reiteration (sensu Hallé, Oldemann and Tomlinson 1978). This phase continues until the production of the sterile mature growth phase. The transition between the seedling and sterile mature phases is gradual with leaves becoming progressively larger. The mature sterile phase is a densely leafy, later naked, appressed monopodial climber rooting copiously from the nodes. The sterile mature growth phase does not branch terminally unless damaged when it then reiterates from a node usually some distance back from the damaged apex. However, this phase does occasionally transform terminally into a foraging shoot. At some point the sterile mature phase begins to produce fertile mature shoots from lateral buds. These fertile shoots are sympodial, of varying determinate lengths and often branch to several orders from lateral buds. They are generally moderately leafy, later becoming naked below. Fertile mature shoots eventually produce much-abbreviated lateral shoots consisting of a minute prophyll and one to several cataphylls and terminate in one to several more-or-less synchronously-produced inflorescences. Occasionally fertile mature shoots are transformed terminally into a foraging flagellum.

In *P. scandens*, *P. macrocephalus* Scort. ex Hook.f., *P. repens*, *P. chinensis* and *Pothoidium lobbianum* Schott, the sterile mature growth phase occasionally produces exceptionally robust lateral branches in which the leaves, while of conventional shape and size, are tightly imbricated and not spreading. The factors triggering this modified growth phase are not known. It might represent another form of foraging shoot, a means for the plant to reiterate from lower buds when the top growth carried begins to exceed the capability of the functional root mass. To date this sixth shoot type has not been reported from subgen. *Allopothos*.

SHOOT ARCHITECTURE AND SPECIES GROUPINGS

Shoot architecture in *Pothos* can be used to make species groupings that appear to form a classification framework. To date, seven groups have been distinguished, representing c. 25 % of described species. The groups, their salient shoot architecture and the Malesian and [extra-Malesian] species thus far included are listed below.

Subgenus *Allopothos*

***P. insignis* group** (*P. borneensis* Furtado, *P. insignis*).

Inflorescences carried on greatly elongated robust orthotropic fertile shoots clothed with large inflated cataphylls and arising from lower, often leafless, parts of sterile mature stems.

This is an architecturally isolated group with no clear links to other groups. The fertile shoots are morphologically similar to the sixth shoot phase noted for subgen. *Pothos* and *Pothoidium*. *Pothos borneensis* is possibly conspecific with *P. insignis*.

P. rumphii group (*P. atropurpurascens* M. Hotta, *P. barberianus* Schott, [*P. brassii*], *P. brevistylus* Engl., *P. hosei* Rendl., *P. kinabaluensis* Furtado, *P. kingii* Hook.f., *P. lancifolius* Hook.f., *P. lorispatha* Ridl., *P. oxyphyllus* Miq., *P. rumphii*).

Inflorescences carried on plagiotropic leafy, very rarely on cataphyllary shoots, arising from upper parts or tips of sterile mature stems.

This group is not yet well defined. Even without critical investigation it is clear that a number of subordinate (separate?) groups can be formed using inflorescence morphology. Addition of more data will undoubtedly result in further division, e.g. a *P. barberianus* group, a *P. rumphii* group.

P. polystachyus group (*P. polystachyus* Engl., *P. cuspidatus* Alderw.)

Inflorescences carried on a plagiotropic or pendent, sympodial, much branched cataphyllary shoots.

To date only two (New Guinean) species have been assigned to this group (Hay 1995). However, cursory examination of several Bornean species (e.g. *P. leptostachyus* Schott) suggests that a similar, but less developed, branching inflorescence is present.

P. luzonensis group ([*P. armatus* Fischer], *P. beccarianus* Engl., *P. luzonensis*, *P. peninsularis* Alderw., [*P. remotiflorus* Hook.f]).

Inflorescences carried on much abbreviated stems clothed in minute cataphylls and arising from the mid-parts, rarely tips, of sterile mature stems.

Considering the characters cited here, *Pedicellarum* belongs in the *P. luzonensis* group. The floral differences between *Pedicellarum* and the *P. luzonensis* group (latrorse anther dehiscence, fused perigon elements and the presence of a 'receptacle' in *Pedicellarum*) are not strong.

The position of *P. falcifolius* Engl. & Krause and *Pothos motleyanus* Schott is debatable. Inflorescence type (comparatively scattered flowers on a more-or-less gracile spadix) support inclusion in the *P. luzonensis* group. However, the great plasticity of inflorescence presentation position combined with the curious inflorescence (flowers sunken into pits) is odd. More data are required.

Subgenus *Pothos*

P. scandens group ([*P. balansae* Engl., *P. chinensis* (Raf.) Merr.], *P. scandens* L.).

All species flower synchronously on short shoots in the axils of most leaves of a flowering branch.

A rather uniform group with an essentially Indo-Malesian distribution. Many species of the *P. scandens* group, especially those from Indo-China, are closely related. All species have rather small inflorescences. Entire fertile branches seem to function as a compound inflorescence.

P. papuanus group ([*P. grandis* Buchet ex P.C. Boyce & D.V. Nguyen], [*P. kerrii* Buchet], *P. macrocephalus, P. papuanus* Becc. ex Engl.).

Inflorescences few and irregularly positioned.

A geographically diverse assemblage with species from northern Vietnam (*P. grandis*) to Australasia (*P. papuanus*). All species have scattered, rather large, inflorescences which seem to function as separate 'flowers'.

P. repens group (*P. repens*).

Fertile shoots borne from terminal or several distal leaf axils on long plagiotropic or pendent sterile mature shoots.

Pothoidium has functionally unisexual spadices arranged pseudopaniculately and maturing sequentially but the vegetative architecture is similar to that in the *P. repens* group, with which it also shares distinctive leaf morphology. The structure of the pseudopaniculate fertile shoots is not clear. It is possible that they are developed by the reduction of all leaves to minute prophylls and cataphylls and that the entire pseudopanical is in essence a compound ('paniculate') fertile shoot. *Pothos repens* and *Pothoidium* are vegetatively identical. Depauperate flowering specimens of *Pothoidium* greatly resemble robust plants of *P. repens*. Despite these similarities with the *P. repens* group, *Pothoidium* appears to be generically distinct since it is readily separable by inflorescence characters. Interpretation of *Pothoidium* inflorescences suggests the species is functionally dioecious. 'Male' spadices have flowers with prominently visible anthers and an apparently sterile ovary. 'Female' spadices have flowers with a large unilocular fertile ovary and no stamens. Flowers of *Pothos* species are always bisexual with a trilocular ovary.

SUMMARY

Subgenus *Allopothos* consists of several species groups which appear to reflect some degree of evolutionary pattern. The *P. luzonensis* group is of particular interest since it forms a 'bridge' between *Pothos* and the presently generic *Pedicellarum*. The floral characters used to maintain *Pedicellarum* are not strong. All occur in various species of the *P. luzonensis* group, although not all in one species. It seems reasonable to suggest that either:

 i. *Pedicellarum* should be merged into a newly defined *Pothos* or
ii. The species comprising the *P. luzonensis* group should be removed from *Pothos* and merged into an expanded *Pedicellarum*. If the latter course is followed, the generic name would become *Goniurus* Presl (1851) by priority.

Subgenus *Pothos* seems more closely knit than subgen. *Allopothos*. The relationship between *Pothoidium* and *Pothos* must be investigated from this point. It is tempting to regard *Pothoidium* as a derived offshoot of subgen. *Pothos* in which functional dioecy has been favoured. The major obstacle to this elegant theory is the geographical disjunction between *Pothoidium* (Indonesia (Moluccas, Sulawesi, Sumatra), Philippines, and Taiwan) and the putatively related species in subgen. *Pothos* (*P. repens*, northern Indo-China, SW China).

REFERENCES

Blanc, P. (1977a). Contribution à l'étude des Aracées. I. Remarques sur la croissance monopodiale. Rev. gén. Bot. 84: 115 – 126.

—— (1977b). Contribution à l'étude des Aracées. II. Remarques sur la croissance sympodiale chez l'*Anthurium scandens* Engl., le *Philodendron fenzlii* Engl. et le *Philodendron speciosum* Schott. Rev. gén. Bot. 84: 319 – 331.

—— (1978). Aspects de la ramification chez des Aracées tropicales. Thèse pour l'obtention du Diplôme de Docteur de 3è. cycle à l'Université Pierre et Marie Curie, Paris.

—— (1980). Observations sur les flagelles des Aracées. Adansonia, sér. 2, 20 (3): 325 – 338.

Boyce, P. C. & Poulsen, A. D. (1994). Notes on *Pothos insignis* (*Araceae*: *Pothoideae*). Kew Bulletin. 49(3): 523 – 528.

Engler, A. (1905). *Pothos*. In Engler, A. (ed.), Das Pflanzenreich 21(IV.23B): 21 – 44. Leipzig.

Hallé, F., Oldemann, R. A. A. & Tomlinson, P. B. (1978). Tropical Trees and Forests; An Architectural Analysis. Springer-Verlag. Berlin.

Hay, A. (1995). The genus *Pothos* L. (*Araceae*: *Pothoeae*) in New Guinea, Solomon Islands and Australia. Blumea 40: 397 – 419.

—— , Bogner, J., Boyce, P. C., Hetterscheid, W. L. A., Jacobsen, N & Murata, J. (1995). Checklist & Botanical Bibliography of the Aroids of Malesia, Australia and the tropical western Pacific. Blumea Supplement 8, 210 pp.

Hotta, M. (1976). Notes on Bornean Plants III. *Pedicellarum* and *Heteroaridarum*, two new genera of the aroids. Acta Phytotax. Geobot. 27: 61 – 65.

Madison, M. T. (1977). A revision of *Monstera (Araceae)*. Contrib. Gray Herb. 207: 1 – 101.

Mayo, S. J., Bogner, J. & Boyce, P. C. (in press). The Genera of *Araceae*. Royal Botanic Gardens Kew.

Presl, C.B. (1851, "1849"). Epimeliae Botanicae. 244.

Schimper, A. F. W. (1903). Guilds in Plant-geography upon a physiological basis, part 2, ch. 2, pp. 192 – 206. English translation by W. R. Fisher revised and edited by P. Groom & I. B. Balfour. Oxford, Clarendon Press.

Schott, H. W. (1857a). *Aroideae* Fasc. 6. Gerold et Fil. Vienna.

—— (1857b). *Aroideae* Skizzirt. Oesterr. Bot. Wochenbl. 7: 69 – 70.

Strong, D. R. & Ray, T. S. (1975). Host tree location behavior of a tropical vine (*Monstera gigantea*) by skototropism. Science 190: 804 – 806.

The Genus *Selaginella* (*Selaginellaceae*) in Malesia

JOSEPHINE M. CAMUS[1]

Summary. The systematics of the Malesian species of *Selaginella* has been largely neglected since Alston's preliminary floristic accounts in the 1930s and 1940s. Subgeneric classifications have historically recognised a range of subgenera, sections and series. Palaeobotanical evidence supports the division of *Selaginella* into two genera. A number of morphological characters require investigation and assessment in Malesian species to evaluate species delimitation and phylogenetic relationships. Biogeographical analysis is complicated by the age of the genus, the geological and climatic history of the Malesian region, and the lack of ecological data for the species.

CLASSIFICATION

Small, herbaceous, heterosporous plants with ligulate, anisophyllous microphylls in Malesia belong to the genus *Selaginella* (*Selaginellaceae*). The family is the only taxonomic rank agreed on for these plants after 150 years of discussion. Where these plants fit into the plant kingdom still is not clear. The lycopsid lineage of the extant genera *Lycopodium* s.l. (i.e. *Huperzia, Phylloglossum, Lycopodium* s.s and *Lycopodiella*), *Selaginella* and *Isoetes* was recognised by Dumortier as early as 1829, though this was not generally accepted for many years (Pichi Sermolli 1973), and has been variously recognised as the order Lycopodiales (e.g. 'Klasse' in Engler & Prantl 1898 – 1902), the class Lycopodiatae (e.g. Kramer & Green 1990) or Lycopsida (e.g. DiMichele & Skog 1992) or the division Lycophyta (e.g. Bateman, 1996) with the family *Selaginellaceae* consequently being upgraded to the order Selaginellales. The relationships of extant and extinct lycopsids to each other have not yet been fully analysed (Bateman, 1996). Some molecular data is now accruing to separate the lycopsids from the ferns and reinforce their placement above the bryophytes and below the other vascular plants (Raubeson & Jansen 1992; Stein *et al.* 1992; Kolukisaoglu *et al.* 1995; Stevenson & Loconte 1996), but analysis of other data does not resolve relationships within the lycopsids (Manhart 1995).

The most comprehensive treatment of Malesian *Selaginella* remains Van Alderwerelt's (1915) monograph on Malayan fern allies, and its succeeding supplements (1915a, 1916, 1917, 1918, 1920, 1922). Van Alderwerelt arranged c. 300 taxa in a complex scheme (Table 1) of subgenera based on leaf arrangement, sections based on the number of steles in the stem, subsections based on sporophyll morphology, and groups (i.e. series) based on habit and pubescence. This system , although using the same major characters, was very different from that of Hieronymous (1902), which Van Alderwerelt (1915, p. 58) considered a '*natural arrangement ... but too intricate and too difficult for practical use*'. Van Alderwerelt

[1] Department of Botany, The Natural History Museum, Cromwell Road, London SW7 5BD. U.K.

TABLE 1. Subgeneric arrangement of *Selaginella* proposed by Van Alderwerelt, 1915.

Subgen.	*Isophyllum* (not in Malesia)		
	Homeophyllum (not in Malesia)		
	Heterophyllum		
Sect.		*Monostelicae*	
Subsect.			*Homeostachys*
Group			*Rosulantes*
			Decumbentes
			Radicantes
			Ascendentes
			Caulescentes
			Pubescentes
Subsect.			*Heterostachys*
Group			*Intertextae*
			Bisulcatae
			Suberosae
			Brachystachyae
Sect.		*Pleiostelicae*	

admitted that his own (loc. cit.) arrangement of subgenera, sections and series passed '*more or less gradually into each other*'.

Alston (1934, 1935a, 1935b, 1937, 1940) avoided the issue of subgeneric classification in his series of papers on *Selaginella* of the Malesian region, but did recognise four subgenera in his paper on Lycopodiinae with Walton (1938). Jermy (1986a) proposed five subgenera with some association with different ecological habitats: subgenus *Selaginella* comprises two species, one of which grows on base-rich mires and the other on bogs; subgenus *Tetragonostachys* comprises some 50 species which are mostly plants characteristic of seasonally dry areas; one of the species in subgenus *Ericetorum* grows in proteaceous heaths, the other two species are annuals. The species in subgenera *Stachygynandrum* and *Heterostachys* are generally terrestrial plants of lowland to mid-montane primary forest (Jermy 1990).

Current cytological data for species world-wide, with base numbers of $x = 7, 8, 9,$ 10, 11, and 12 (Tryon & Tryon 1982; Jermy 1990; Takamiya 1993; Praptosuwiryo & Darnaedi 1995) do not help to elucidate these subgeneric divisions. Jermy *et al.* (1967) found an interesting correlation between base numbers and coning habit: species with a base number of $x = 9$ exhibit sporadic coning (i.e. '*cones are produced haphazardly and any branch system can continue growth at intervals throughout the season*'), whereas in species with a base number of $x = 10$ there is a finite growth pattern in the lateral branch systems and cones are produced simultaneously at the tips of branchlets. Jermy (loc. cit.) further noted that in the tropics the former base number is characteristic of species of more open habitats, and the latter characteristic of species of dense forest.

There are good grounds for treating the isophyllous and anisophyllous groups as separate genera (Thomas & Quansah 1991). All the Malesian species are anisophyllous and would be assigned to the genus *Stachygynandrum*, encompassing Jermy's (1986a) subgenera *Stachygynandrum* and *Heterostachys*. Re-evaluation of historical concepts and characters such as stelar structure, use of current cell biology techniques and employment of cladistic analyses will be valuable in resolving subgeneric classification.

CHARACTERS

All species of *Selaginella* currently known from the Malesian region are anisophyllous. This means that there are three different types of leaves to be studied: the small median leaves on the upper surface of the stem, the larger lateral leaves, and the modified lateral leaves found at the stem dichotomies. Leaf shape, features of the margin, and details of auricles are important characters. Wong (1983) published some critical observations on Peninsular Malaysian *Selaginella* species in which he not only summarised leaf shape, but also assessed branching patterns. He also tackled the very difficult task of producing a key to non-fertile specimens. Many collections, including recent ones, comprise only a fragment of the plant making it impossible to determine the important features of habit and growth form from herbarium specimens.

The strobili or cones of fertile specimens provide either one or, more usually, two types of sporophylls. Three arrangements of sporophylls are found in the strobili: isophyllous (rare in Malesian species, e.g. *S. plana* (Desv.) Hieron.), anisophyllous in the same arrangement as the microphylls (the commonest arrangement, e.g. *S. rothertii* Alderw.), anisophyllous and resupinate (rare in Malesian species, e.g. *S. alutacia* Spring).

A number of authors in recent years have highlighted various features for taxonomic assessment. Amongst these, Horner & Arnott (1963) and Quansah (1988) described sporangial distribution patterns and found them useful at the series and species level; Bienfait & Waterkeyn (1974) published a brief survey of silicified callose deposits in the leaves and found the shape and distribution of these useful in determining species; Dahlen (1982) evaluated ligule shape, amongst other characters, and concluded that it may be a useful taxonomic character; Minaki (1984) studied macrospore morphology and found a relationship between this and homophylly and heterophylly of the sporophyte, but no correlation with habit or chromosome number; Tryon & Lugardon (1990) examined ornamentation of both megaspores and microspores, and found that the surface pattern of the former had a higher correlation with the five subgenera proposed by Jermy (1986a) than that of the latter; Mukhopadhyay & Sen (1981) described a curious structure projecting from the adaxial side of some sporophylls which Quansah & Thomas (1985) named a sporophyll-pteryx, but this feature is thought by Jermy (1986b) to be common to all species with a resupinate strobilus.

There is much still to be learnt and consolidated for this genus.

DISTRIBUTION

Biogeographical analysis of *Selaginella* in the Malesian region will not be easy. *Selaginella* has a long fossil record. The isophyllous and anisophyllous forms are known as far back as the Lower Carboniferous, some 300 million years ago (Thomas & Quansah 1991), and long before the Pangea supercontinent began to form. Kramer (1993) discusses long-range dispersal and the differences in distribution in central American regions of species of the homosporous genus *Lycopodium* with small light spores with that of species of heterosporous *Selaginella* with heavier megaspores. However, Tryon & Tryon (1982), Dahlen (1990) and Jermy (pers. comm.) have found evidence for simultaneous dispersal of mega- and microspores, and the first authors note that *Selaginella* has reached isolated oceanic islands. Webster (1992) reviews spore dispersal mechanisms and reproductive biology of this genus. Jermy (pers. comm.) has observed that invariably megaspores will be dispersed with microspores attached to their sculptured perispore. The natural occurrence of hybrids has not yet been assessed.

Data on the geological and climatic history, distribution of flora and fauna, and vegetation types in the Malesian region are accruing (Whitmore 1987; Baas *et al.* 1990; Michaux 1994), but there is still much to be discovered.

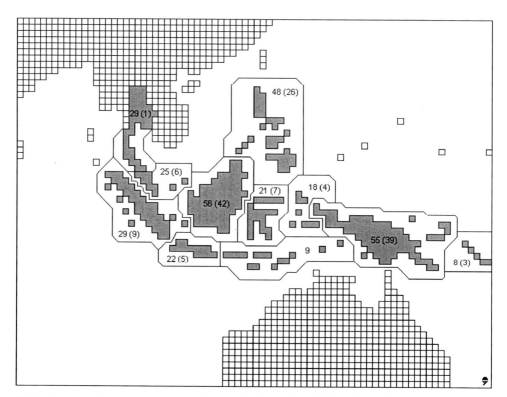

MAP 1. Number of species and, in parentheses, endemic species of *Selaginella* per island group in Malesia. (Preliminary data based on Alston, 1934, 1935a, 1935b, 1937, 1940; Tan & Jermy, 1981; Parris et al., 1984; Kato, 1988; Parris et al., 1992; Reed, 1966; BM herbarium specimens).

Little is yet known about the ecology of Malesian *Selaginella* species. Most are terrestrial in mesic, lowland to mid-montane primary and secondary forests, often in the more open areas near rivers. *Selaginella* is frequent in more open glades in Gunung Mulu National Park, Sarawak, where a number of species are found, e.g. *S. bluuensis* Alderw., *S. brooksii* Hieron., *S. conferta* T. Moore and *S. dielsii* Hieron. (Parris *et al.* 1984), but are less common at higher altitudes of mountain forest in spite of increased light. They have been collected at altitudes of c. 2000 m on Mount Kinabalu, Borneo, and c. 3200 m in western New Guinea. Some species are also found growing epiphytically, usually at a low height on tree boles, e.g. *S. involvens* (Sw.) Spring in Sarawak (Parris *et al.* 1984), although Tryon and Tryon (1982) report that *S. angustiramea* F. Muell. & Baker sometimes grows on branches 25 m above the ground. Others may be obligate epiliths, such as the limestone species *S. aristata* Spring (Kato 1988), *S. antimonanensis* B.C. Tan & Jermy and *S. pricei* B.C. Tan & Jermy from the Philippines (Tan & Jermy 1981), or *S. chaii* Jermy from Sarawak (Jermy 1986b). Too many collectors, past and present, record inadequate habitat data.

Distribution records for the species are incomplete. Van Alderwerelt (1915 *et seq.*) described many new species of *Selaginella*. Very few of the species that he recognised were known from more than one island in Malesia. His figures give 48 taxa (including ten varieties and two forms) from Sumatra and 71 taxa (including seven

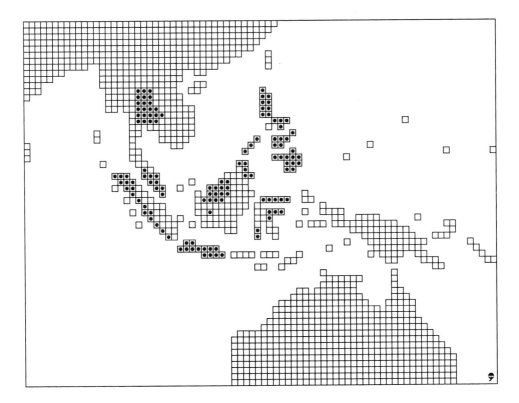

MAP 2. Approximate distribution of *Selaginella involvens* (Sw.) Spring in Malesia.

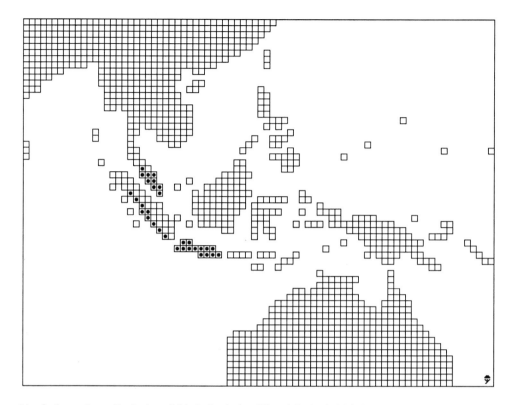

Map 3. Approximate distribution of *Selaginella stipulata* (Blume) Spring in Malesia.

varieties and one form) from the Philippines. Current estimates for these areas are 29 and 48 respectively. The lower numbers reflect the number of specific epithets reduced to synonymy by Alston (Reed 1966). Numbers of species and endemics for different parts of Malesia are given in Map 1. The correlation between species richness and endemism for this genus is not very high in Malesia. Mount Kinabalu, Sabah, has 14 species, two of which are endemic (Parris *et al.* 1992). The nearby Gunung Mulu National Park in Sarawak has 20 species, two of these endemic (Parris *et al.* 1984). Kato (1988) reported 18 species from Ambon and Seram, four of these being endemic. These figures suggest that the estimate of 42 endemic species given for Borneo in Map 1 reflects the number of species known from one or a few collections rather than the true number of endemics.

Preliminary data have been entered in WORLDMAP (Williams 1994), computer software developed for the fast, interactive assessment of priority areas for conserving biodiversity. Data from a phylogenetic analysis of *Selaginella* will be added in due course, enabling biodiversity hotspots to be identified. Species show a number of distribution patterns. Some species are widespread, such as *S. involvens* (Sw.) Spring (Map 2); others show a western distribution, e.g. *S. stipulata* (Blume) Spring (Map 3), or an eastern distribution, e.g. *S. caudata* (Desv.) Spring (Map 4). Others, such as *S. aristata* Spring (Map 5), show distributions that may reflect

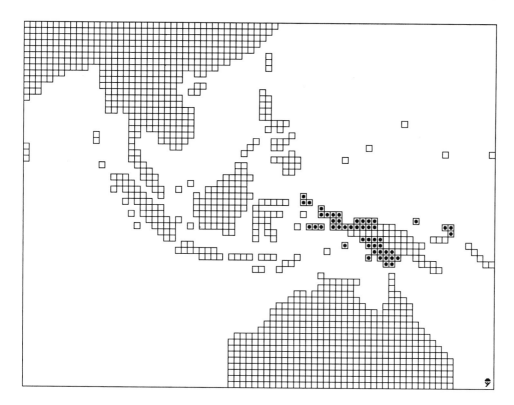

MAP 4. Approximate distribution of *Selaginella caudata* (Desv.) Spring in Malesia.

climatic factors such as seasonal dryness. Some regions, such as Kalimantan, have obviously low collecting indices.

The high number of 55 species in New Guinea is in dramatic contrast with a figure of about ten native species in Australia (Andrews 1990; Jermy & Holmes in press). However, New Guinea has a complicated geological history together with high altitudes and a great diversity of vegetation types (Whitmore 1984). Further botanical collections are needed from much of New Guinea.

The Philippines also have a large number of species of *Selaginella*. This group of islands lies at the edge of the Sunda shelf and is therefore expected to have derived much of its biota from Borneo and continental Indochina. However, Balgooy's (1987) plant geographical analysis of Sulawesi showed an affinity between that island and the Philippines, Lesser Sunda Islands plus Java, and the Moluccas. Humphries (1990) analysed moth and butterfly data with the interesting result of the Philippines, Java and Bali being biogeographically more closely related to Sulawesi/Moluccas/Western New Guinea in preference to other area relationships. Vane Wright (1990) gives a consensus of six biogeographic regions for butterflies and mammals within the Philippines and recognises four of these regions as together comprising a collective area of endemism. This complex situation may well apply to *Selaginella* where current data gives Mindanao as most species-rich and southeast Luzon as the area with most endemics.

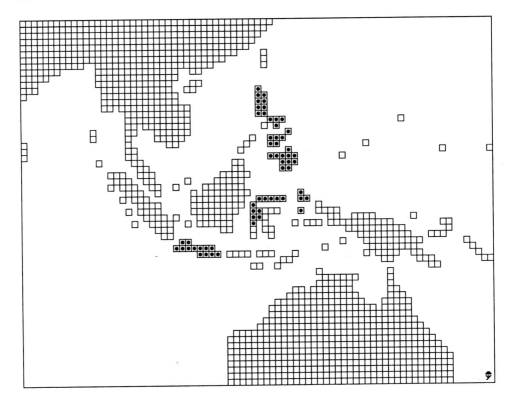

MAP 5. Approximate distribution of *Selaginella aristata* Spring in Malesia.

CONCLUSION

The systematics and biogeography of *Selaginella* in Malesia will not be explained by herbarium studies alone and may only be resolved when many more data have been acquired as the result of a research programme encompassing such aspects as phylogenetic analysis, cell biology and field research. Some preliminary analysis in conjunction with the biodiversity software of WORLDMAP should indicate areas where substantial field collecting, linked to country studies under the Convention of Biological Diversity and the development of a Protected Areas network, should be implemented.

ACKNOWLEDGEMENTS

I thank Clive Jermy for valuable discussions and substantial comments on an early draft of this paper, and an anonymous referee and editors for suggestions of further improvements.

REFERENCES

Alderwerelt van Rosenburgh, C. R. W. K. van. (1915). Malayan fern allies. xvi + 261 + 1 pp. The department of agriculture, industry and commerce, Batavia.

—— (1915a). New or interesting Malayan ferns 7. Bull. Jard. Bot. Buitenzorg série 2, 20: 1 – 28.

—— (1916). New or interesting Malayan ferns 8. Bull. Jard. Bot. Buitenzorg série 2, 23: 1 – 27.

—— (1917). New or interesting Malayan ferns 9. Bull. Jard. Bot. Buitenzorg série 2, 24: 1 – 8.

—— (1918). New or interesting Malayan ferns 10. Bull. Jard. Bot. Buitenzorg série 2, 28: 1 – 66.

—— (1920). New or interesting Malayan ferns 11. Bull. Jard. Bot. Buitenzorg série 3, 2: 129 – 186.

—— (1922). New or interesting Malayan ferns 12. Bull. Jard. Bot. Buitenzorg série 3, 5: 179 – 240.

Alston, A. H. G. (1934). The genus *Selaginella* in the Malay Peninsula. Gard. Bull. Straits Settlem. 8: 41 – 62.

—— (1935a). The Philippine species of *Selaginella*. Philipp. J. Sci. 58: 359 – 383.

—— (1935b). The *Selaginellae* of the Malay Islands: 1. Java and the Lesser Sunda Islands. Bull. Jard. bot. Buitenzorg série 3, 13: 432 – 442.

—— (1937). The *Selaginellae* of the Malay Islands: II. Sumatra. Bull. Jard. Bot. Buitenzorg série 3, 14: 175 – 186.

—— (1940). The *Selaginellae* of the Malay Islands: III. Celebes and the Moluccas. Bull. Jard. bot. Buitenzorg série 3, 16: 343 – 350.

Andrews, S. B. (1990). Ferns of Queensland. xx + 427 pp. Queensland Department of Primary Industries, Brisbane.

Baas, P., Kalkman, K. & Geesink, R.(1990). (eds) 'The plant diversity of Malesia. xii + 420 pp. Kluwer Academic Publishers, Dordrecht.

Balgooy, M. M. J. van. (1987). A plant geographical analysis of Sulawesi. In: T. C. Whitmore (ed.), Biogeographical analysis of the Malay Archipelago, pp.94 – 102. Oxford monographs on biogeography No. 4. Clarendon Press, Oxford.

Bateman, R. M. (1996). An overview of lycophyte pylogeny. In: J. M. Camus, M. Gibby, & R. J. Johns (eds), Pteridology in perspective, pp. 405 – 415. Royal Botanic Gardens, Kew.

Bienfait, A. & Waterkeyn, L. (1974). Contribution à l'étude systématique des *Selaginella*; spécificité des formation callosiques foliaires observées en fluorescence. Bull. Jard. Bot. État. 44: 295 – 302.

Dahlen, M. A. (1982). A taxonomic reassessment of the Hong Kong fern allies with special reference to *Selaginella* Beauv. 238 pp. Ph.D. thesis, University of Hong Kong (unpublished).

Dahlen, M. A. (1990). Komplementäre Oberflächenstrukturen der äusseren Sporenwand bei *Selaginella* (Engl. Summ.). Farnblätter 22: 20 – 27.

DiMichele, W. A. & Skog, J. E. (1992) The Lycopsida: a symposium. Ann. Missouri Bot. Gard. 79: 447 – 449.

Dumortier, B.-C. (1829). Analyse des familles des plantes. 104 pp. Tournay, Casterman.

Engler, A. & Prantl, K.(1898 – 1902). (eds), Die natürlichen Pflanzenfamilien 1,4. pp. vi + 808, Engelmann, Leipzig.

Hieronymous, G. (1902). *Selaginellaceae.* In: A. Engler & K. Prantl (eds), Die natürlichen Pflanzenfamilien 1, pp. 621 – 715. Wilhelm Engelman, Leipzig.

Horner, H. T. Jr, & Arnott, H. J. (1963). Sporangial arrangement in North American species of *Selaginella.* Bot. Gaz. (Crawforsdville) 124: 371 – 383.

Humphries, C. J. (1990). The importance of Wallacea to biogeographical thinking. In: W. J. Knight, & J. D. Holloway (eds), Insects and the rain forest of South East Asia (Wallacea), pp. 7 – 18. The Royal Entomological Society, London.

Jermy, A. C. (1986a). Subgeneric names in *Selaginella.* Fern Gaz. 13: 117 – 118.

—— (1986b). Two new *Selaginella* species from Gunung Mulu National Park, Sarawak. Kew Bull. 41: 547 – 559.

—— (1990). *Selaginellaceae.* In: K. U. Kramer & P. S. Green (eds), The families and genera of vascular plants 1. Pteridophytes and gymnosperms, pp. 39 – 45. Springer-Verlag, Berlin.

—— & Holmes, J. (in press). *Selaginellaceae.* In: A. E. Orchard (ed.), Flora of Australia 48: Ferns, Gymnosperms and Allied Groups. ABRS/CSIRO, Melbourne.

—— , Jones, K., & Colden, C. (1967). Cytomorphological variation in *Selaginella.* J. Linn. Soc. (Bot.) 60: 147 – 158.

Kato, M. (1988). Taxonomic studies of pteridophytes of Ambon and Seram (Moluccas) collected by Indonesian-Japanese Botanical expeditions 1. Fern-allies. Acta Phytotax. Geobot. 39: 133 – 146.

Kolukisaoglu, H. Ü., Marx, S., Wiegmann, C., Hanelt, S. & Schneider-Poetsch, H. A. W. (1995). Divergence of the phytochrome gene family predates angiosperm evolution and suggests that *Selaginella* and *Equisetum* arose prior to *Psilotum.* J. Mol. Evol. 41: 329 – 337.

Kramer, K. U. & Green, P. S., (1990). (eds), The families and genera of vascular plants 1. Pteridophytes and gymnosperms. pp. xiii + 404. Springer-Verlag, Berlin.

Kramer, K. U. (1993). Distribution patterns in major pteridophyte taxa relative to those of angiosperms. J. Biogeog. 20: 287 – 291.

Manhart, J. R. (1995). Cloroplast 16S rDNA sequences and phylogenetic relationships of fern allies and ferns. Amer. Fern J. 85: 182 – 192.

Michaux, B. (1994). Land movements and animal distributions in east Wallacea (eastern Indonesia, Papua New Guinea and Melanesia). Paleogeogr., Paleoclimatol., Paleoecol. 112: 323 – 343.

Minaki, M. (1984). Macrospore morphology and taxonomy of *Selaginella* (Selaginellaceae). Pollen et Spores 26: 421 – 480.

Mukhopadhyay, R. & Sen, U. (1981). The occurrence of a laminal flap in *Selaginella.* Fern Gaz. 12: 180 – 181.

Parris, B. S., Jermy, A.C., Camus, J. M. & Paul, A. M. (1984). Pteridophyta. In: A. C. Jermy (ed.), Studies on the flora of Gunung Mulu National Park Sarawak, pp. 145 – 233. Forest Department, Sarawak.

—— , Beaman, R. S. & Beaman, J. H. (1992). The plants of Mount Kinabalu 1 Ferns and fern allies. 165 pp. Royal Botanic Gardens, Kew.

Pichi Sermolli, R. E. G.(1973). Historical review of the higher classification of the Filicopsida. In: A. C. Jermy, J. A. Crabbe & B. A. Thomas (eds), The phylogeny and classification of the ferns. Bot. J. Linn. Soc. 67, Suppl. 1: xiv + 284.

Praptosuwiryo, N. & Darnaedi, D. (1995). Survai kromosom tumbuhan paku liar di Kebun Raya Bogor. Bull. Kebun Raya Ind. 8: 53 – 58.

Quansah, N. (1988). Sporangial distribution patterns in the strobili of African and Madagascan *Selaginella*. Ann. Bot. (London) 61: 243 – 247.

—— & Thomas, B. A. (1985). 'Sporophyll-pteryx' in African and American *Selaginella*. Fern Gaz. 13: 49 – 52.

Raubeson, L. A. & Jansen, R. K. (1992). Chloroplast DNA evidence on the ancient evolutionary split in vascular land plants. Science 255: 1697 – 1699.

Reed, C. F. (1966). Index Selaginellarum. Mem. Soc. Brot. 18: 1 – 287.

Stein, D. B., Conant, D. S., Ahearn, M. E., Jordan, E. T., Kirch, S. A., Hasebe, M., Iwatsuki, K. & Tan, M. K. (1992). Structural rearrangements of the chloroplast genome provide an important phylogenetic link in ferns. Proc. Natl. Acad. Sci. USA 89: 1856 – 1860.

Stevenson, D. W. & Loconte, H. (1996) Ordinal and familial relationships of pteridophyte genera. In: J. M. Camus, M. Gibby & R. J. Johns (eds), Pteridology in perspective, pp. 435 – 465. Royal Botanic Gardens, Kew.

Takamiya, M. (1993). Comparative karyomorphology and interrelationships of *Selaginella* in Japan. J. Plant Res. 106: 149 – 166.

Tan, B. C. & Jermy, A. C. (1981). Two new species of *Selaginella* from the Philippines. Fern Gaz. 12: 169 – 173.

Thomas, B. A. & Quansah, N. (1991). The paleobotanical case for dividing *Selaginella*. Fern Gaz. 14: 59 – 64.

Tryon, A. F. & Lugardon, B. (1990). Spores of the Pteridophyta. 648 pp. Springer-Verlag, New York.

Tryon, R. M. & Tryon, A. F. (1982). Ferns and allied plants with special reference to tropical America. xiv + 857 pp. Springer-Verlag, New York.

Vane Wright, R. I. (1990). The Philippines — key to the biogeography of Wallacea? In: W. J. Knight & J. D. Holloway (eds), Insects and the rain forest of South East Asia (Wallacea), pp. 19 – 34. The Royal Entomological Society, London.

Walton, J. & Alston, A. H. G. (1938). Lycopodiinae. In: F. Verdoon (ed.) Manual of pteridology, pp. 500 – 506. Martinus Nijhoff, The Hague.

Webster, T. R. (1992). Developmental problems in *Selaginella* (*Selaginellaceae*) in an evolutionary context. Ann. Missouri Bot. Gard. 79: 632 – 647.

Whitmore, T. C. (1984). A vegetation map of Malesia at 1:5 million. J. Biogeog. 11: 461 – 471.

—— (1987). (ed.) Biogeographical evolution of the Malay Archipelago. Oxford monographs on biogeography No. 4. x + 147 pp. Clarendon Press, Oxford.

Williams, P. (1994). WORLDMAP priority areas for biodiversity Version 3. (Computer software.)

Wong, K. M. (1983). Critical observations on Peninsular Malaysian *Selaginella*. Gard. Bull. Straits Settlem. 35: 107 – 135.

A preliminary investigation of leaf venation and cuticle features to characterise taxa within *Cordyline* (*Agavaceae* s.l.)

JOHN G. CONRAN[1]

Summary. The taxonomy of the largely Malesian and Australasian genus *Cordyline* (*Agavaceae* s.l.) is difficult to resolve using traditional techniques, as much of the herbarium material is sterile or lacking either flowers or fruits, both of which are important for identification. In addition, many of the taxa are very similar in the sterile state, and there is considerable confusion over the relationships and number of taxa, especially within the *C. fruticosa* complex, where it has been widely cultivated throughout Asia and the Pacific, and numerous horticultural forms have been developed. Leaf architecture and cuticle studies have been used in studies of various dicotyledons to augment taxonomic investigations, and although their applications in the monocotyledons are relatively limited, they show promise in the resolution of the taxonomy of such problematic and often sterile-collected Malesian taxa such as the *Agavaceae* and *Smilacaceae*. Preliminary studies of these leaf-based features in *Cordyline*, revealed major species differences, especially in cuticular sculpturing and stomatal patterns. These characters also appear to be related to leaf shape, and will be useful in an ongoing phylogenetic study of the genus. This is resulting in the development of a DELTA database for the genus, incorporating the venation and cuticle data, from which a complete picture of the variation within the genus should emerge, hopefully enabling rapid identification of otherwise sterile material.

INTRODUCTION

Cordyline L. (*Agavaceae* s.l.) is a widespread genus of arborescent monocotyledons extending from the Moluccas through Asia, Malesia, Australasia, the Pacific and South America, with centres of diversity in Australia and New Zealand (Baker 1875; Moore & Edgar 1970; Forster & Pedley 1986). There are currently between 15 and 20 species recognised, with many of the numerous taxa described from Malesia and the Pacific Islands considered to be variants of the widely cultivated and distributed *C. fruticosa* (Fosberg 1985). Similarly, the New Zealand *C. pumilio* has been thought to be a possible neotenous form of *C. australis* (Tomlinson & Fisher 1971), *C. kaspar* from the Three Kings Is. is possibly conspecific with *C. obtecta* from Norfolk I. (Green 1994) and there are a number of reported hybrids between the New Zealand species (Moore & Edgar 1970; Beever 1981). The position and affinities of the New Caledonian taxon with free tepals (previously placed in the genus *Cohnia*) also requires study.

The genus is traditionally associated with *Dracaena* and *Pleomele* in the *Agavaceae* (Cronquist 1981; 1988; Brummitt 1992), although Dahlgren *et al.* (1985) placed *Cordyline* within the *Asteliaceae*. The former placements represent widespread

---

[1]Department of Botany, The University of Adelaide, SA 5005, Australia.

71

pantropical associations, whereas the *Asteliaceae* are a primarily Gondwana-centred group, with extension into Asia and the Pacific (Conran 1995).

Current molecular data, in addition to overturning most of the relationships within the *Lilianae* suggest that *Cordyline* belongs with none of these families (Duvall *et al.* 1993; Eguiarte *et al.* 1994; Bogler & Simpson 1995; Chase *et al.* 1995), and is instead part of an expanded *Lomandraceae* (Rudall *et al.* 1996). The genus appears to have a long fossil history, with pollen recorded from Lower-Miocene New Zealand (Couper 1953), and there are macrofossil taxa with close affinities to *Cordyline* from Eocene South Australia (Conran & Christophel in prep.) and the Oligocene of Îles Kerguélen in the Southern Ocean (Conran in press). Cranwell (1962) suggested that the New Zealand *Cordyline* species represented an Oligocene influx from more tropical regions.

One of the problems with many of the herbarium collections is that, given the ease of identification of sterile material to generic level, the specimens often lack flowers or fruits (or both) making definitive identifications often difficult. This occurs especially within widely variable taxa related to *C. fruticosa*, given this species' wide distribution and the extensive development of ethnobotanical cultivars throughout Asia and the Pacific. There can also be problems with collections from different stages in the plant's development where both the size and form of the leaves can vary (Tomlinson & Fisher 1971). Such variation has given rise, for example, to the suggestion that *C. pumilio* represents a neotenous form of *C. australis*.

The usefulness of cuticular features in plant taxonomy has been developed by a number of authors including Stebbins & Khush (1961), Tomlinson (1974), Wilkinson (1979), and Baranova (1992). Leaf venation architecture is also useful for taxonomic studies (Hickey 1973; Hickey 1979; Hickey & Taylor 1989; Klücking 1995), and has also been used for the identification of Australian rainforest trees (Christophel & Hyland 1993). Although the venation architecture of the monocots has not been widely used in classification studies, such features have been employed in studies of the dicot-like net-veined monocots, both for extant taxa (Conover 1983; Inamdar *et al.* 1983; Conran 1985; Conran 1987; Conran 1989), and to examine the relationships of a fossil monocot with these extant groups by Conran *et al.* (1994). Cuticular characters have been utilised more widely in classificatory and evolutionary studies in the monocots generally by Dahlgren & Clifford (1982), Dahlgren & Rasmussen (1983), Dahlgren *et al.* (1985), and for the sedges (Bruhl 1995), "orchids" (Williams 1979; Rasmussen 1981; Yukawa *et al.* 1992) and the "lilies" sens. lat. (including *Cordyline*) by Gopal & Raza (1992), and net-veined taxa (Conover 1991; Conran *et al.* 1994).

During preliminary investigations of the venation and cuticular patterns in *Cordyline* to characterise the relationships of the possibly related fossil taxa (Conran & Christophel in prep., Conran in press), it was observed that individual *Cordyline* species seemed to have characteristic combinations of venation and cuticle features which allow each species to be identified from a small leaf portion. If these species-specific cuticle characters can be extended to determine most of the genus, the potential for accurate biodiversity surveys in the absence of the usual diagnostic characters such as flowers and fruits will be greatly enhanced.

The advent of the DELTA taxonomic database (Dallwitz *et al.* 1993) has seen the development of detailed character sets for numerous plant taxa from which both taxonomic descriptions and identificatory keys can be produced. The types of venation and cuticular data which the present study represents are ideally suited for the development of a DELTA matrix from which not only a detailed character set and taxon descriptions can be produced, but also accurate identification mechanisms for non-fertile *Cordyline* species. In the longer term, these data will provide the basis for a character set to be used in phylogenetic studies of the genus, and its relationships with other allegedly related taxa such as *Dracaena* and *Pleomele* (Dracaenaceae) and *Astelia, Milligania* and *Collespermum.(Asteliaceae)*.

Accordingly, the aims of the present study were to develop a DELTA character set using features of leaf architecture and cuticles by which the different species of *Cordyline* might be identified.

MATERIALS AND METHODS

Leaves of 13 species and a natural hybrid were obtained from the collections at AD, L and the living accessions at the Botanic Gardens of Adelaide (Table 1).

TABLE 1. *Cordyline* species and specimens used for the present study. Unless indicated otherwise, specimens are held in the leaf voucher collection at ADU

Cordyline specimens	Source
C. australis (G. Forst.) Endl.	J. Conran s.n. cult. Modbury North, Adelaide
	J. Conran s.n. cult. Adelaide Bot. Gard.
C. australis × *C. banksii*	R. Chinnock s.n., cult. Adelaide ex New Zealand
C. banksii J. D. Hook.	AD 97601458
	AD 97250311
C. indivisa (G. Forst.) Steud.	AD 97250329
	AD 97601469
C. kaspar W. R. B. Oliver	CHR G4624
	CHR G15155
	CHR G15249
C. pumilio J. D. Hook.	AD 97316172
	AD 97601477
C. obtecta (Graham) Baker	J. Conran s.n. cult. Adelaide Bot. Gard.
C. stricta (Sims) Endl.	AD 98661687
	J. Conran s.n. cult. Adelaide Bot. Gard.
C. rubra Otto & A. Dietr.	J. Conran 721 Samford SE Queensland
	J. Conran 722 Mt Glorious SE Queensland
C. murchisoniae F. Muell.	L 459061
	L 93531145 (as *C. haageana*)
	Christophel s.n. ADU
C. fruticosa (L.) A.Chev.	Queensland AD 96108540
	Oahu, Hawaii AD 98546185
	Hawaii AD 98350194
	New Guinea L 140137
	Queensland AD 97602797 as *C. hedychioides*
C. fruticosa cv 'rubra'	J. Conran s.n. cult. North Adelaide
C. petiolaris (Domin) Pedley	J. Conran s.n. cult. Adelaide Bot. Gard.
C. manners-suttoniae F. Muell.	J. Conran s.n. cult. Mt Coot-tha Bot. Gard.

Where possible, at least two leaves for each species in the first instance were taken from different collections. These were pressed, dried and x-rayed for three minutes in a Hewlett-Packard Faxitron series x-ray machine at 30 kV using Kodak AA-11 film under vacuum. The plates were then exposed as contact negatives onto Ilford No. 5 photographic paper. Leaf venation architecture was examined after the manner of Christophel & Hyland (1993).

A portion of the lamina from each leaf c. 5 × 5 mm was removed and the cuticle prepared following the methods of Christophel & Lys (1986), stained with crystal violet and mounted in phenol-glycerine jelly. The prepared cuticles were examined by Nomarski Differential Interference Contrast microscopy using a Zeiss microscope, and photographed using Kodak TMAX ASA 100 film.

Data were recorded for cuticular and vegetative morphological characters to develop a DELTA data matrix (Dallwitz *et al.* 1993) using the data editor developed by Gouda (1994). The leaf morphology and venation character set was derived, in part, from the characters outlined by Conran *et al.* (1994), Conover (1983; 1991), as well as those of Hill (1980), Blackburn (1980) and Christophel & Lys (1986). Additional stomatal features were based on those of Baranova (1992) Wilkinson (1979) and from a cuticular character set derived for the *Lauraceae* by Christophel *et al.* (in press).

RESULTS

The investigation of the leaf and cuticle features revealed 49 characters (12 leaf and 37 cuticle) which show promise for the development of a DELTA data matrix (Table 2).

TABLE 2. List of the characters used in the DELTA data matrix

---

#1. adaxial intercostal cell shape (surface view): 1. isodiametric; 2. elongated; 3. strongly elongated.

#2. upper epidermis cell wall shape: 1. angular ; 2. rounded; 3. undulate; 4. sinuous; 5. irregular rounded.

#3. adaxial sinuosity: 1. straight; 2. curved; 3. l=h; 4. l<h; 5. l<2h.

#4. adaxial cell size: 1. small (<12μm long); 2. medium small (12 – 19 μm); 3. medium (20 – 34 μm); 4. medium large (35 – 42 μm); 5. large (43 – 65 μm); 6. very large (>65 μm).

#5. lower vein cell periclinal wall sculpturing: 1. smooth; 2. granulate; 3. striate; 4. reticulate; 5: 1-pitted; 6. several pitted; 7. with a raised mesa-like centre.

#6. upper vein cells: 1. same as intercostal cells; 2. different from intercostal cells.

#7. upper vein cells shape: 1. isodiametric; 2. elongate; 3. very elongate.

#8. upper vein cell size: 1. small (<12μm long); 2. medium small (12 – 19 μm); 3. medium (20 – 34 μm); 4. medium large (35 – 42 μm); 5. large (43 – 65 μm); 6. very large (>65 μm).

#9. upper vein cell periclinal wall sculpturing (outer surface): 1. smooth; 2. granulate; 3. striate; 4. reticulate; 5: 1-pitted; 6. several pitted; 7. with a raised mesa-like centre.

#10. upper stomata distributed : 1. uniformly (incl. veins and areoles if visible); 2. in patches (areolate).

#11. stomate frequency on upper epidermis: 1. absent (0); 2. very rare (1 – 2 per field of view on × 20); 3. rare (2 – 15); 4. common (15 – 30); 5. abundant (>30).

#12. upper guard cell size: 1. small (<12µm long); 2. medium small (12 – 19 µm); 3. medium (20 – 34 µm); 4. medium large (35 – 42 µm); 5. large (43 – 65 µm); 6. very large (>65 µm).

#13. number of subsidiary cells associated with upper stomata: 1. 4; 2. 5; 3. >5.

#14. upper epidermal subsidiary cell shape (surface view): 1. rectangular; 2. triangular; 3. elongate; 4. curved; 5. very elongate.

#15. upper subsidiary cell periclinal wall sculpture: 1. smooth; 2. granulate; 3. striate; 4. reticulate; 5: 1-pitted; 6. several pitted; 7. with a raised mesa-like centre.

#16. upper guard cells position: 1. sunken; 2. not sunken.

#17. upper thickening on guard cells: 1. absent; 2. fine poral; 3. heavy poral.

#18. abaxial intercostal cell shape (surface view): 1. isodiametric (1:1; 2. elongated (2:1); 3. strongly elongated (3:1).

#19. lower epidermis cell wall shape: 1. angular ; 2. rounded; 3. undulate; 4. sinuous; 5. irregular rounded.

#20. abaxial sinuosity: 1. straight; 2. slightly curved; 3. l=h; 4. l<h; 5. l<2h.

#21. abaxial cell size: 1. small (<12µm long); 2. medium small (12 – 19 µm); 3. medium (20 – 34 µm); 4. medium large (35 – 42 µm); 5. large (43 – 65 µm); 6. very large (>65 µm).

#22. lower epidermis anticlinal cell sculpturing: 1. smooth; 2. beaded.

#23. lower epidermal periclinal wall sculpturing: 1. smooth; 2. granulate; 3. striate; 4. reticulate; 5: 1-pitted; 6. several pitted; 7. with a raised mesa-like centre.

#24. lower vein cells: 1. same as intercostal cells; 2. different from intercostal cells.

#25. lower epidermis anticlinal vein cell sculpturing (surface view): 1. smooth; 2. beaded.

#26. lower vein cell shape: 1. isodiametric (1:1); 2. elongate (2:1); 3. very elongate (3:1).

#27. lower vein cell size: 1. small (<12µm long); 2. medium small (12 – 19 µm); 3. medium (20 – 34 µm); 4. medium large (35 – 42 µm); 5. large (43 – 65 µm); 6. very large (>65 µm).

#28. lower vein periclinal cell wall sculpturing: 1. smooth; 2. granulate; 3. striate; 4. reticulate; 5: 1-pitted; 6. several pitted; 7. with a raised mesa-like centre.

#29. lower stomata distributed: 1. uniformly (over veins and areoles if venous cells are distinguishable); 2. stomata distributed in patches (in areoles only).

#30. stomate frequency on lower epidermis: 1. absent (0); 2. very rare (1 – 2 per field of view on × 20); 3. rare (2 – 15); 4. common (15 – 30); 5. abundant (>30).

#31. lower guard cell size: 1. small (<12µm long); 2. medium small (12 – 19 µm); 3. medium (20 – 34 µm); 4. medium large (35 – 42 µm); 5. large (43 – 65 µm); 6. very large (>65 µm).

#32. number of cells associated with lower stomata: 1. 4; 2. 5; 3. >5.

#33. subsidiary cell anticlinal wall: 1. smooth; 2. beaded; 3. undulate

#34. subsidiary cell periclinal wall: 1. smooth; 2. granulate; 3. striate; 4. reticulate; 5: 1-pitted; 6. several pitted; 7. with a raised mesa-like centre.

#35. lower guard cells position: 1. sunken; 2. not sunken.

#36. lower thickening on guard cells: 1. absent; 2. fine poral; 3. heavy poral.

#37. petiole: 1. absent; 2. present.

#38. petiole length (cm).

#39. petiole cross-section: 1. flat; 2. curved; 3. rolled, conduplicate.

#40. no. of primary veins at widest point.

#41. leaf blade length (cm).

#42. leaf blade width (mm).

#43. leaf blade length-width ratio.

#44. leaf shape index.

#45. leaf basal angle (degrees).

#46. leaf apex angle (degrees).

#47. leaf position of widest point from base of blade (cm).

#48. no. of secondary veins at widest point of blade.

#49. secondary vein angle (degrees).

Leaf Morphology

The leaves of *Cordyline* spp. tend to fall into two major categories: those with clearly defined petioles, and those without. The majority of the New Zealand taxa fall into the latter category, tending to have linear to ensiform leaves which gradually taper into a coalesced mass of veins basally before widening into the leaf base where it attaches to the stem (Fig. 1). In contrast, many of the Australasian taxa possess clearly defined petioles, and the lamina may either end ± abruptly above the petiole as in *C. manners-suttoniae* (Fig. 2) or may taper more gradually into a long thin petiole, as in *C. rubra* (Fig. 3). The nature of the petiole may also be of taxonomic significance, with its shape in cross-section (flat; curved or almost conduplicate) also useful in separating taxa (Tomlinson & Fisher 1971; Forster & Pedley 1986).

The venation is pinnate to parallelodromous (sensu Hickey (1973; 1979) and Conover (1983)) in all species. This pattern is also close to that described as arcuate by Klücking (1995), although his system is more difficult to interpret. In the petiolate taxa, the veins often arise for a considerable distance along the lamina, with wider vein angles and more veins in cross section at the mid-point of the leaf (Fig. 2 – 3). Venation in the non-petiolate taxa tends to arise nearer the basal region and to run the entire length of the leaf, converging near the apex (Fig. 1). The numbers of secondary veins and vein orders are also variable, with narrow-leaved taxa either compressing the venation or as in *C. pumilio* reducing both vein number and vein order levels. The degree of distinctness and size of the areoles created by the highest order cross venation also varies between species in both the leaf types.

Cuticle

Whereas leaf shape and, to a lesser extent, venation are often similar between different *Cordyline* spp., the patterns of cuticular sculpturing, cell shape (both costal and intercostal), stomatal features (including subsidiary cells) and the variation of these between adaxial and abaxial laminas appear to be highly diagnostic, providing the greatest source of morphological characters by which the identity and possible relationships of taxa within *Cordyline* might be explored. All species examined were amphistomatic, but there were often major differences between stomata from different surfaces of the same leaf. Each species appeared to have a more-or-less unique combination of features by which it could be recognised, and there were also suites of features which were common to groups of apparently related taxa.

In *C. australis* the upper (Fig. 4) and lower (Fig. 5) cuticles are similar and have ± isodiametric cells each of which has a generally single large pit (adaxial) or several pit (abaxial) on the outer surface, and the costal cells were similar to the intercostal cells on both surfaces. There was no clear structural distinction between the intercostal and subsidiary cells.

Cordyline banksii had radically different cuticles, with the adaxial (Fig. 6) consisting of generally elongate, unornamented and undifferentiated cells and few stomata, whereas the abaxial cuticle (Fig. 7) had a heavily ornamented intercostal region, with numerous stomata, and both the epidermal cells and subsidiary cells

FIG. 1. X-rayed leaf of *Cordyline kaspar* (CHR G15249) showing typical venation and structure for the non-petiolate group of *Cordyline* spp. Scale bar = 20 mm.

FIGS. 2 – 3. Leaf x-rays of Fig. 2: *C. manners-suttoniae* (Conran s.n., ADU); and Fig. 3 *C. rubra* (Conran 722, ADU), petiole shortened, as examples of the different forms of petiolate leaf. Scale bar = 20 mm.

with massive raised papillae or undulate ridges obscuring the surface. However, the abaxial costal cells were similar to both the adaxial costal and intercostal cells. The stomata were also almost completely obscured by the raised undulate thickenings of the subsidiary cell periclinal walls (Fig. 31).

Similarly, *C. indivisa* showed a marked contrast between the upper (Fig. 8) and lower (Fig. 9) surfaces. Here the upper cells were more isodiametric and unornamented and the stomata were largely covered by outgrowths of the inner subsidiary cells margins (Fig. 28) whereas the lower cells had several, often partly coalesced pits on the outer surface, both on the intercostal and costal cells, and the inner subsidiary cell margins were raised and greatly thickened, forming a four-sided box around each stomate (Fig. 30).

Cordyline kaspar was also ornamented with several pits per cell, but they remained discrete, and were found on both leaf surfaces (Fig. 10 – 11). Also, the ornamentation was far less heavy as than *C. banksii* and *C. indivisa*.

Cordyline obtecta (Fig. 12 – 13) and *C. pumilio* (Fig. 14 – 15) were similar in their cuticles, and both shared the relatively uncommon feature of having some stomata with five subsidiary cells rather than the usual four. Nevertheless, in the former there was a distinction between the ± isodiametric adaxial intercostal and elongate costal cells, but the abaxial cuticle cells were all similar. This contrasts with *C. pumilio* where the abaxial lamina cells and subsidiary cells were much more elongate than their adaxial counterparts. Also, in the latter there was slight beading of the anticlinal walls, and a shallow ridge around the stomata formed from the raised inner edges of the subsidiary cells (Fig. 29).

Despite being the most widely distributed *Cordyline* species (or species complex), the cuticles of the various accessions and cultivars of *C. fruticosa* examined so far had relatively unornamented cuticles with only some faint and shallow outer surface pitting (Fig. 16 – 17). There were some slight differences in degree of intercostal cell size and shape between the New Guinea and Hawaiian collections, but this requires further investigation of the complex over its range before characters which might delimit populations can be determined with certainty.

In contrast, the morphologically similar *C. petiolaris* had a similar upper cuticle to *C. fruticosa* (Fig. 18) but had a highly distinctive lower cuticle (Fig. 19), with the central portion of the intercostal cells raised up and mesa-like in appearance (Fig. 32). In addition, the subsidiary cells tended to have somewhat irregular or slightly undulate margins.

Cordyline manners-suttoniae showed an unornamented upper cuticle with some differentiation between the intercostal and vein cells (Fig. 20), but this was much more marked on the lower surface, where the intercostal cells were ± transversely striate to the leaf long axis, appearing almost wrinkled and only slightly 1-pitted, whereas the vein cells were more elongate, lacked striae, and were one- to several-pitted (Fig. 21).

In *C. stricta* both the upper (Fig. 22) and lower (Fig. 23) cuticles were somewhat granular on their periclinal walls. Whereas in the adaxial cuticle the vein cells were much narrower and more elongate than the intercostal cells, the lower cuticular cells were all relatively similar.

The adaxial cuticle of *C. murchisoniae* consisted of rather unspecialised ± isodiametric cells (Fig. 24). Stomata were extremely uncommon with some

FIGS. 4 – 9. Leaf cuticles of various *Cordyline* spp. viewed under Nomarski differential interference contrast. Scale bars all = 50 µm. Fig. 4: *C. australis* (Conran s.n., ADU) adaxial; Fig. 5: same, abaxial. Fig. 6: *C. banksii* (AD 97601458) adaxial; Fig. 7: same abaxial. Fig. 8: *C. indivisa* (AD 97250329) adaxial; Fig. 9: same, abaxial.

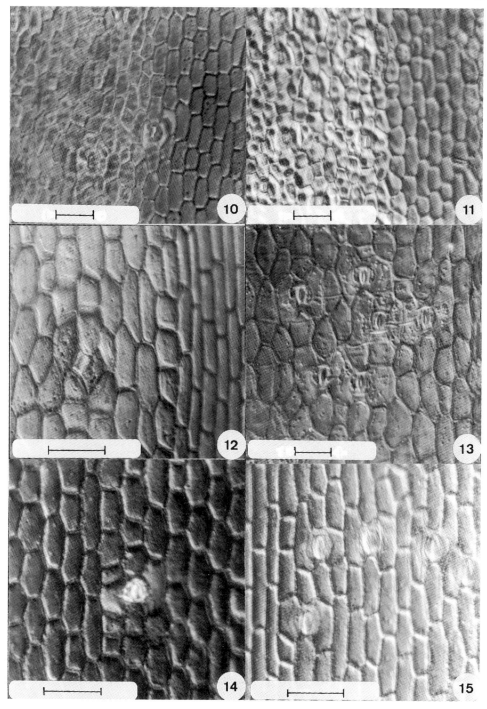

FIGS. 10 – 15. Leaf cuticles of various *Cordyline* spp. viewed under Nomarski differential interference contrast. Scale bars all = 50 μm. Fig. 10: *C. kaspar* (CHR G15429) adaxial; Fig. 11: same, abaxial. Fig. 12: *C. obtecta* (Conran s.n., ADU) adaxial; Fig. 13: same abaxial. Fig. 14: *C. pumilio* (AD 97601477) adaxial; Fig. 15: same, abaxial.

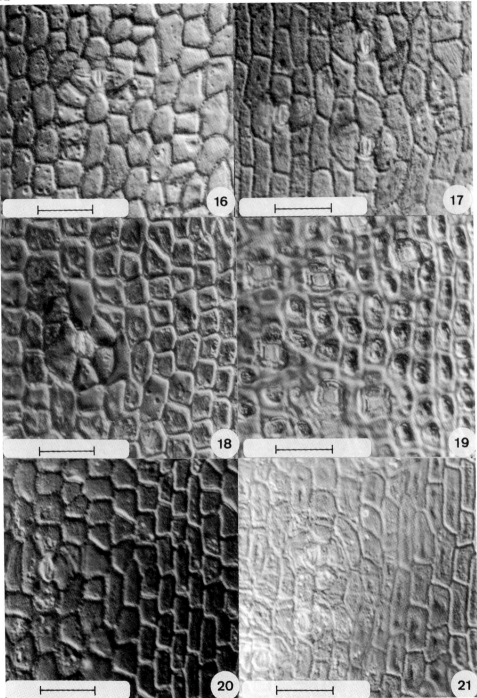

FIGS. 16 – 21. Leaf cuticles of various *Cordyline* spp. viewed under Nomarski differential interference contrast. Scale bars all = 50 μm. Fig. 16: *C. fruticosa* (L 140137) adaxial; Fig. 17: same, abaxial. Fig. 18: *C. petiolaris* (Conran s.n., ADU) adaxial; Fig. 19: same abaxial. Fig. 20: *C. manner-suttoniae* (Conran s.n., ADU) adaxial; Fig. 21: same, abaxial.

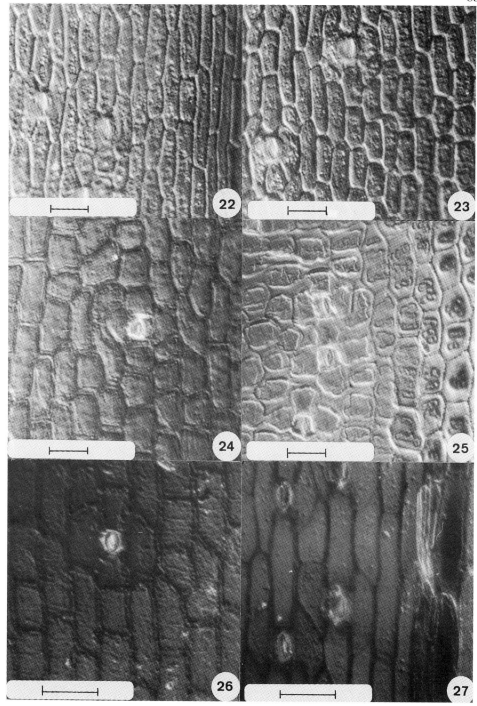

Figs. 22 – 27. Leaf cuticles of various *Cordyline* spp. viewed under Nomarski differential interference contrast. Scale bars all = 50 μm. Fig. 22: *C. stricta* (Conran s.n. ADU) adaxial; Fig. 23: same, abaxial. Fig. 24: *C. murchisoniae* (Christophel s.n., ADU) adaxial; Fig. 25: same abaxial. Fig. 26: *C. rubra* (Conran 722, ADU) adaxial; Fig. 27: same, abaxial.

collections apparently lacking any on their upper surface, but when present the subsidiary cells resembled those of the lamina and vein cells were not obvious. In comparison, the abaxial surface was heavily ornamented (Fig. 25) with the intercostal cells bearing similar raised central regions to *C. petiolaris* and the vein cells bearing multiple, almost tuberculate, papillae in pit-like depressions on the outer surface of the periclinal wall. The irregularly to almost triangularly-shaped abaxial subsidiary cells were also ornamented, with both raised regions bearing striae radiating out from the stomate (Fig. 33) and their margins were finely undulate.

Cordyline rubra cuticles were unusual in that the upper cuticle (Fig. 26) was more heavily ornamented than the lower (Fig. 27). There were fine granulations and occasional transverse striae on the adaxial lamina and subsidiary cells, all of which were ± isodiametric to slightly elongated. In comparison, the abaxial cuticle was mostly unornamented except for the vein cells which were deeply 1-pitted on the outer surface and all of the lower epidermal cells were much more elongate.

DISCUSSION

The shape of the leaves in *Cordyline* spp. has been an important feature in their classification (Baker 1875), the delimitation of species boundaries and the development of keys for identification (Moore & Edgar 1970; Forster & Pedley 1986). Although as Tomlinson & Fisher (1971) observed, there is wide variation within taxa, both in the development and maturation of the plant (e.g. *C. australis*) and within the widespread *C. fruticosa/terminalis* species complex generally. When these data are combined with the cuticular features examined, there appear to be clear suites of characters by which the different species might be identified in the sterile state.

There are also apparent species groups, with most of the New Zealand taxa more similar to each other than to the Australasian taxa and with many of the species sharing the same or similar patterns of papillate ornamentation to some degree on one or both cuticles. These features can also be used to support evidence of hybridisation, as the putative *C. australis* × *C. banksii* specimen was found to have a similar upper cuticle to *C. australis*, but the lower cuticle was more heavily ornamented than that species, but not as much as any of the specimens of *C. banksii* which were examined. The data also support the retention of *C. pumilio* as a distinct species. Tomlinson & Fisher (1971) may be right in the assertion that it evolved as a possibly neotenous form of something like *C. australis* whose seedlings it most closely resembles, but in cuticle structure, it is not very close to that species at present. Similarly, Green's (1994) query about the possible conspecificity of *C. kaspar* with *C. obtecta* was not supported from the data currently available, as both taxa have radically different cuticles.

Within the *C. fruticosa* s.l. complex many of the taxa formerly included within this polymorphic species such as *C. manners-suttoniae*, *C. petiolaris* and *C. murchisoniae* are distinctive on the basis of cuticular anatomy, even though they can be difficult to distinguish in the sterile state. Given the wide distribution of *C. fruticosa* from India through Asia to Australia, and as a result of human activity throughout the Pacific, a thorough investigation of the variation within the present

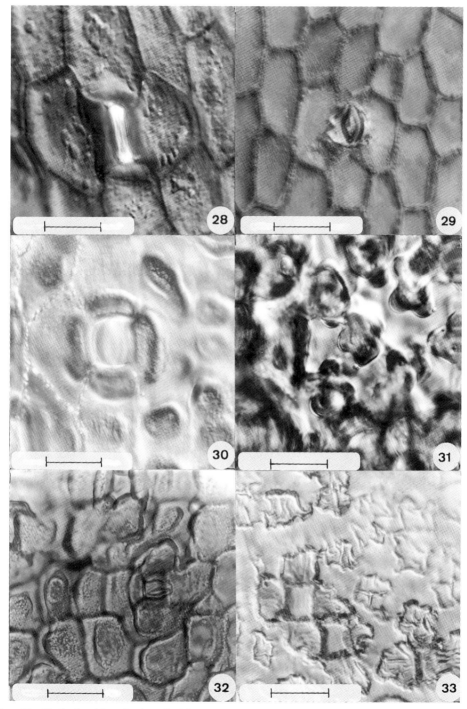

FIGS. 28 – 33. Leaf cuticles of various *Cordyline* spp. viewed under Nomarski differential interference contrast showing stomatal detail. Scale bars all = 20 μm. Fig. 28: *C. indivisa* (AD 97601469) adaxial; Fig. 29: *C. pumilio* (AD 97601477) adaxial. Fig. 30: *C. indivisa* (AD 97601469) abaxial; Fig. 31: *C. banksii* (AD 97601458) abaxial. Fig. 32: *C. petiolaris* (Conran s.n., ADU) abaxial; Fig. 33: *C. murchisoniae* (Christophel s.n., ADU), abaxial.

circumscription *C. fruticosa*, and especially the status of previously recognised taxa now included in synonymy such as *C. haageana*, is needed.

Similarly, the relationships between the cuticular features of *Cordyline*, the New Caledonian taxa of *Cordyline* with free tepals (sometimes placed in the genus *Cohnia*), other supposedly related genera such as *Dracaena, Pleomele* (*Dracaenaceae*) and the members of the *Asteliaceae* are needed. In the light of fragmentation of these groups arising from molecular studies of the *Asparagales* (Chase *et al.* 1995), it is most important to see whether there are any morphological or anatomical features by which the relationships of these taxa, both to each other and to the remainder of the *Lilianae* can be established. The leaves of *Dracaena* and *Pleomele* are not only similar in many gross morphological features such as shape and even aspects of the venation architecture (Klücking 1995), but preliminary investigations of several species from each genus show amphistomatic leaves with cuticles similar at least to the less specialised (ornamented) *Cordyline* spp.

This study represents a preliminary account of the types of characteristics which might be used in the development of a DELTA data matrix from which a generic treatment could be developed. There are still a number of taxa which have not been investigated, including *C. congesta* from northern Australia and the New Caledonian, Mascarene and South American species. The study needs to be expanded to sample within each taxon more thoroughly to establish the degree of variability. Moreover, the usefulness of the characters needs to be evaluated by multivariate numerical methods before taxa can be circumscribed and phylogenetic models tested. Morphometric cuticular and leaf characteristics such as those used by Conover (1991) need to be further expanded, as well as the utilisation of leaf anatomical features such as those used in studies of other monocots such as *Smilax* (Yates & Duncan 1970) and the *Iridaceae: Tigrideae* (Rudall 1991). Nevertheless, the use of leaf morphology and cuticular anatomy in the taxonomy of *Cordyline* spp. appears at this stage to be an extremely promising tool, and one which will hopefully enable the accurate identification of taxa in the sterile state.

ACKNOWLEDGMENTS

The Directors of AD, L and the Adelaide Botanic Gardens are thanked for access to their *Cordyline* collections, together with the Queensland Dept of Forestry for permission to collect material of *C. rubra* from south-eastern Queensland. Dr Murray Parsons (CHR) is thanked for material of *C. kaspar*, as is Dr Bob Chinnock (AD) for the *C. australis* × *C. banksii* material. The Botany Department at The University of Adelaide is thanked for the provision of facilities for this research, Dr David Christophel for advice and the use of his x-ray machine, Ms Heidi Wittesch, Ms Linda Allen and Miss Julie Dowd for specimen preparation, and Mr Brian Rowland for some of the photographic development. Dr Christophel and Ms Leigh Garde Conran are thanked for comments on the manuscript.

REFERENCES

Baker, J. G. (1875). Revision of the genera and species of *Asparagaceae*. J. Linn. Soc. Bot. 14: 508 – 632.

Baranova, M. (1992). Principles of comparative stomatographic studies of flowering plants. Bot. Rev. 58: 49 – 99.

Beever, R. E. (1981). Self-incompatibility in *Cordyline kaspar* (*Agavaceae*). N.Z. J. Bot. 19: 13 – 16.

Blackburn, D. T. (1980). A generalized distance metric for the analysis of variable taxa. Bot. Gaz. 141: 325 – 335.

Bogler, D. J. & Simpson, B. B. (1995). A chloroplast study of the *Agavaceae*. Syst. Bot. 20(2): 191 – 205.

Bruhl, J. J. (1995). Sedge genera of the world: relationships and a new classification of the *Cyperaceae*. Aust. Syst. Bot. 8(2): 125 – 305.

Brummitt, R. K. (1992). Vascular Plant Families and Genera. Royal Botanic Gardens, Kew.

Rudall, P., Chase, M. W. & Conran, J. G. (1996). New circumscriptions and a new family of asparagoid lilies: genera formerly included in *Anthericaceae*. Kew Bull. 51: 667 – 680.

Chase, M. W., Stevenson, D. W., Wilkin, P. & Rudall, P. J. (1995). Molecular systematics: a combined analysis. In: P. J. Rudall, P. Cribb, D. F. Cutler & C. J. Humphries (eds), Monocotyledons: Systematics and Evolution, London, Royal Botanic Gardens, Kew.

Christophel, D. C. & Hyland, B. (1993). Leaf Atlas of Australian Tropical Rainforest Trees. CSIRO, Canberra.

Christophel, D. C., Kerrigan, R. & Rowett, A. I. (in press). The use of cuticular features in the taxonomy of the *Lauraceae*. Ann. Missouri Bot. Gard. 83(3).

Christophel, D. C. & Lys, S. D. (1986). Mummified leaves of two new species of *Myrtaceae* from the Eocene of Victoria, Australia. Aust. J. Bot. 34: 649 – 62.

Conover, M. H. (1983). The vegetative anatomy of the reticulate-veined Liliiflorae. Telopea 2: 401 – 412.

—— (1991). Epidermal patterns of the reticulate-veined Liliiflorae and their parallel-veined allies. Bot. J. Linn. Soc. 107: 295 – 312.

Conran, J. G. (1985). The taxonomic affinities of the genus *Drymophila* (*Liliaceae* s.l.). Ph.D. Thesis., University of Queensland, St Lucia.

—— (1987). A phenetic study of the relationships of the genus *Drymophila* R.Br. within the reticulate-veined Liliiflorae. Aust. J. Bot. 35: 283 – 300.

—— (1989). Cladistic analyses of some net-veined Liliiflorae. Pl. Syst. Evol. 168: 123 – 141.

—— (1995). Family distributions in the Liliiflorae and their biogeographic implications. J. Biogeog. 22: 1023 – 1034.

—— (in press). *Paracordyline kerguelensis*: an Oligocene monocotyledon macrofossil from the Kerguélen Islands. Alcheringa.

—— & Christophel, D. C. (in prep.). *Paracordyline aureonemoralis*: an Eocene monocotyledon macrofossil from Adelaide, South Australia. Alcheringa.

——, Christophel, D. C. & Scriven, L. J. (1994). *Petermanniopsis angleseaënsis*: An Australian fossil net-veined monocotyledon from Eocene Victoria. Int. J. Pl. Sci. 155: 816 – 827.

Couper, R.A. (1953). Upper Mesozoic and Cainozoic spores and pollen grains from New Zealand. Pal. Bull. N.Z. Geol. Surv. Pal. 22: 1 – 77.

Cranwell, L.M. (1962). Endemism and isolation in the Three Kings Islands, New Zealand — with notes on pollen and spore types of the endemics. Rec. Auckland Inst. & Mus. 5: 215 – 232.

Cronquist, A. (1981). An Integrated System of Classification of Flowering Plants. 1262 pp. Columbia University Press, New York.

—— (1988). The Evolution and Classification of Flowering Plants. 555 pp. New York Botanic Gardens., New York.

Dahlgren, R. M. T. & Clifford, H. T. (1982). The Monocotyledons: A Comparative Study. Springer Verlag, Berlin.

——, —— & Yeo, P. F. (1985). The Families of Monocotyledons. Academic Press, London.

—— & Rasmussen, F. N. (1983). Monocotyledon evolution characters and phylogenetic estimation. In: M. K. Heckt, B. Wallace & G. T. Prance (eds), Evolutionary Biology Vol. 16, pp. 255 – 395, Plenum Publ. Corp., New York.

Dallwitz, M. J., Paine, T. A. & Zurcher, E. J. (1993). User's Guide to the DELTA System: A General System for Processing Taxonomic Descriptions. Draft of 4th Ed. CSIRO Division of Entomology, Canberra.

Duvall, M. R., Chase, M. W., Clark, W. D., Kress, W. J., Hills, H. G., Eguiarte, L. E., Smith, J. F., Gaut, B. S., Zimmer, E. A. & Learn, G. H., Jr. (1993). Phylogenetic hypotheses for the monocotyledons constructed from rbcL sequence data. Ann. Missouri Bot. Gard. 80: 607 – 619.

Eguiarte, L.E., Duvall, M. R., Learn, G. H., Jr. & Clegg, M. T. (1994). The systematic status of the *Agavaceae* and *Nolinaceae* and related asparagales in the monocotyledons: Analysis based on the *rbc*L gene sequence. Bol. Soc. Bot. Mexico 54: 35 – 56.

Forster, P. I. & Pedley, L. (1986). *Agavaceae*. In: A. S. George (ed.), Flora of Australia Vol. 46, pp. 71 – 87, AGPS, Canberra.

Fosberg, F. R. (1985). *Cordyline fruticosa* (L.) Chevalier (*Agavaceae*). Baileya 22(4): 180 – 181.

Gopal, B. V. & Raza, S. H. (1992). Stomatal structure as an aid to the taxonomy of *Liliaceae*. Asian J. Pl. Sci. 4(2): 51 – 56.

Gouda, E. J. (1994). Taxasoft Ver. 1.0. E. J. Gouda, Utrecht.

Green, P. (1994). *Agavaceae*. In: A. Wilson (ed.). Flora of Australia, pp. 522 – 526, AGPS, Canberra.

Hickey, L. (1973). Classification of the architecture of dicotyledonous leaves. Amer. J. Bot. 60: 17 – 33.

Hickey, L. J. (1979). A revised classification of the architecture of dicotyledonous leaves. In: C. R. Metcalfe & L. Chalk (eds), Anatomy of the Dicotyledons Vol. 1, pp. 25 – 29, Clarendon Press, Oxford.

—— & Taylor, D. W. (1989). Reexamination of leaf architectural characters of basal angiosperms and their sister-groups: implications for the origin and relationships of angiosperms. Amer. J. Bot. 89 (suppl.): 245 – 6.

Hill, R. S. (1980). A numerical taxonomic approach to the study of Angiosperm leaves. Bot. Gaz. 141: 213 – 229.

Inamdar, J. A., Shenoy, K. N. & Rao, N. V. (1983). Leaf architecture of some monocotyledons with reticulate venation. Ann. Bot. 52: 725 – 736.

Klücking, E. P. (1995). Leaf Venation Patterns Vol. 7: The Classification of Leaf Venation Patterns. 96 pp. J. Cramer, Berlin.

Moore, L. B. & Edgar, E. (1970). Flora of New Zealand Vol. 2. 354 pp. Government Printer, Wellington.

Rasmussen, H. (1981). The diversity of stomatal development on *Orchidaceae* subfamily Orchidoideae. Bot. J. Linn. Soc. 82: 381 – 393.

Rudall, P. (1991). Leaf anatomy in the Tigridieae (*Iridaceae*). Pl. Syst. Evol. 175: 1 – 10.

Stebbins, G. L. & Khush, G. S. (1961). Variation in the organisation of the stomatal complex in the leaf epidermis of monocotyledons and its bearing on their phylogeny. Amer. J. Bot. 48: 51 – 59.

Tomlinson, P. B. (1974). The development of the stomatal complex as a taxonomic character in the monocotyledons. Taxon 23: 109 – 128.

—— & Fisher, J. B. (1971). Morphological studies in *Cordyline* (*Agavaceae*) I: Introduction and general morphology. J. Arnold Arbor. 52: 459 – 478.

Wilkinson, H. P. (1979). The plant surface (mainly leaf). In: C. R. Metcalfe & L. Chalk (eds), Anatomy of the Dicotyledons 2nd Edition, pp. 97 – 165, Clarendon Press, Oxford.

Williams, N. H. (1979). Subsidiary cells in the *Orchidaceae*: their general distribution with special reference to development in the *Oncidieae*. Bot. J. Linn. Soc. 78: 41 – 66.

Yates, I. E. & Duncan, W. H. (1970). Comparative studies of *Smilax* section *Smilax* of the southwestern United States. Rhodora 72: 289 – 312.

Yukawa, T., Ando, T., Karasawa, K. & Hashimoto, K. (1992). Existence of two stomatal shapes in the genus *Dendrobium* (*Orchidaceae*) and its systematic significance. Amer. J. Bot. 79(8): 946 – 952.

A strategy for nature conservation to 2000 A.D.

EARL OF CRANBROOK (*CHAIRMAN OF ENGLISH NATURE*)

In this, my invited Keynote address, I have melded together my personal experience of working in the Malesian region with insights based on my present position as the chairman of the statutory nature conservation agency for England. I am convinced that lessons learnt in British circumstances are equally relevant to South-East Asian countries and to your own interests in the natural environment of this very special region of our fragile earth.

Points that I shall cover include a working definition of 'nature conservation' in the English context, which I hope will appeal to you. I shall look at the place of regulation and a government agency in a national strategy for nature conservation. I shall also examine the role of the individual — especially the land owner or manager, and other citizens and residents of our countries. I shall relate our activities to supranational and international obligations and I shall end with an introduction to a new approach, developed within English Nature (E.N.), which we call 'Natural Areas', one which I commend to you for consideration in the context of other geographical regions. This, I shall propose, provides useful guidance how we should proceed towards a sustainable 2000.

ECOLOGICAL DEVELOPMENT

The ecological development of modern Europe began when a global temperature rise of a few degrees Celsius ended the last Ice Age. Some 12,000 years ago, retreating glaciers re-exposed the English land surface to progressive recolonisation by plants and animals from southerly refugia. The melting ice raised world sea-levels and, at the opposite end of the Eurasian landmass, the Sunda shelf was inundated to re-create island Malesia. There, associated climatic changes restored a warm, humid environment and presumably established weather patterns familiar today.

Humans were represented in both regions at that time by stone age (Mesolithic in Europe) societies which subsisted by hunting and gathering. Although plausibly blamed for catastrophic declines or extinctions of some large mammal species in specific locations (e.g. on Mediterranean islands), their overall impact was slight. The Neolithic revolution reached Europe during the eighth millennium BP. Thenceforward, human intervention has progressively modified the pristine environment.

English Nature, Northminster House, Peterborough PE1 1UA, U.K.

In England, archaeology has shown that around Avebury — the greatest Neolithic henge of England — clearance of the primaeval forest was well advanced by 4000 BC. The stored nutrients of antiquity were depleted by the pioneer cultivations and natural woodland has never returned to the driest chalk soils of this downland landscape (Malone 1989). Near Peterborough, the removal of fenland peat has exposed orderly agricultural landscapes of ditched roads, hedged fields, habitations and stock buildings dating from the late Neolithic (~2000 BC) through the succeeding Bronze Age (Pryor 1991). Preserved in another peat bed, at Blakeway Farm, Somerset, an ancient wooden trackway (C_{14}-dated to ~2500 BC) incorporated hazel stems that had clearly been cut from coppiced stools. Here, the excavators deduced that Neolithic land-use included the management of woodland for such special products (Godwin 1981). Under present peat deposits at Thorne Moors, South Yorkshire, excavation has found evidence of managed park woodland dating from late Neolithic through the Bronze Age, and intermittent clearance by fire of pine-dominated successional (secondary) growth (Buckland, 1993). On Salisbury Plain, ancient earthworks in great diversity testify to several millenia of land-use by successive cultures. Their cumulative impact converted original forest to nutrient-poor downland habitat, largely abandoned since the 6th century AD but nowadays prized for its nature conservation value (Entwistle 1995). These instances illustrate how in England, for the past 5000 years or more, human intervention has made significant and enduring impacts on the natural environment, including some irreversible changes.

In western Malesia, the Neolithic revolution is associated with people of Austronesian type. Archaeology has shown that such people reached Borneo, the heartland of island Malesia, around 2500 BC, apparently bringing with them the cultivation of rice and domestic stock (fowls, pigs and dogs) (Bellwood & Cranbrook 1994). Buddhist and Hindu cultural monuments of the 6th - 15th centuries AD, found in parts of Malaysia and Indonesia, imply organised societies and local populations of some density. Where these were dependent on settled agriculture, they altered the natural wildlife no less profoundly than their contemporaries in England. Some of these artificial habitats prove to be valuable for nature conservation: for example, wetlands such as rice paddies or the settling pools of open-cast tin mines have increased areas favourable for certain migratory bird species (Medway 1976). The more general agricultural system of slash-and-burn shifting cultivation has evidently been prevalent from Neolithic times to the present day. Although dramatic in its immediate effects, there has been dispute over the degree of long-term ecological damage attributable to this farming method. I only know that, in 1977, in rain forest of the Melinau valley (now Mulu National Park), Sarawak, the local people showed me vegetational changes persisting some 150 years (5 generations) after the last known clearance for farming.

Through the history of England, mankind's progressive assault on nature has continued unabated. Population growth, industrialisation and urban development of the 19th century intensified pressures. In the 20th century, the agrochemical revolution redoubled the impacts and the side-effects of modern lifestyle have compounded the injury. In England today, as through much of Europe, truly natural climax plant communities are probably confined to remote rocky outcrops or coastal saltmarsh. Elsewhere the surviving nature conservation resource is 'semi-natural habitat', i.e. communities of native wild plants and associated animals, varying in composition in response to climate, soils and topography, occupying sites more or less modified by man and frequently maintained by human intervention at an arrested successional stage. Even these crucial habitats have become fragmented, especially in the lowlands. Many sites are small and isolated, vulnerable to changes of use on the land that surrounds them, to topical or ambient pollution and to chance effects. Many original animal species have disappeared and introduced aliens have altered natural ecology.

In the Melinau valley of Sarawak, after a lapse of 150 years, no owner was any longer recognised. In England, for centuries, proprietal rights have extended over the entire land surface, and indeed over the sea-bed to national limits. Where the state (or the Crown) holds land, it does so within the same legal framework as the citizen. There are some small sites where ownership is in doubt, but no major tracts of English land are statutorily unoccupied. Through the ~1200 years of England as a nation-state, a legal framework has evolved to control of the use of land or its products (including natural products such as timber, wild game or fish). Laws establishing the royal hunting parks of the 12th century Norman kings are often cited as early conservation legislation. The surviving sites (e.g. New Forest) are indeed valuable natural assets today, as are some game reserves decreed by 19th century Malesian sultans for their own sport (e.g. Kerau and Endau-Rompin, Malaysia).

ENGLISH NATURE: WHO WE ARE AND WHAT WE DO

However, the legislation backing present British nature conservation practice is much more recent, being based on the 1949 Act which set up the Nature Conservancy (NC), giving it the duty to notify Sites of Special Scientific Interest (SSSIs), to manage land for nature reserves and to perform related functions, including research. By later Acts, in 1973 the research stations were transferred to the Natural Environment Research Council and a Nature Conservancy Council (NCC) created; in 1981 and 1985, the duties and powers relating to SSSIs and National Nature Reserves (NNRs) were changed; and in 1991 the NCC was divided into separate agencies covering England, Scotland and Wales, who continue to work together on Great Britain-wide issues through a Joint Nature Conservation Committee (JNCC). English Nature is now the statutory body responsible for nature conservation in England, with duties including the management of nature reserves, and giving advice on nature conservation issues to Government, to local authorities, the owners and occupiers of land and the general public.

The process of identifying and notifying SSSIs has continued through these changes, following published guidelines originally devised by NCC but now reviewed and updated as necessary through the JNCC. The special interest may be based on animals, plants or geological or physiographical features, or a mix of these. There are now (31 March, 1995) 3829 notified SSSIs in England, covering

894,600 ha, i.e. about 6.8% of the land area of England above mean low water mark. Ownership of about 40% (by number) is in the hands of public bodies such as the Forestry Commission, the Crown Estate or Ministry of Defence, or voluntary conservation bodies including the National Trust, Royal Society for the Protection of Birds, Woodland Trust or county wildlife trusts. The balance is owned or managed privately — involving some 23,000 individuals when occupiers of all kinds are enumerated including, for instance, the holders of common grazing rights on open land.

Across England, some of the best and most representative SSSIs have been declared as National Nature Reserves (NNRs). Most NNRs are owned (or controlled through lease or agreement) and directly managed by English Nature, but increasingly we use a section of the 1981 Act to approve other owners, such as wildlife trusts, to manage NNRs with appropriate safeguards. There are now 164 NNRs in England, covering 65,000 ha of SSSI. There is also one statutory Marine Nature Reserve, at Lundy Island, Devon.

THE OVERALL OBJECTIVE OF NATURE CONSERVATION

When the Nature Conservancy was first set up, the aim was to "conserve nature", seemingly on the assumption that, if only man's influence could be removed, nature in England would stay the way it was — in our memories, in historic paintings or in literature. Nature would be fine if only we'd leave it alone. Unfortunately, in the intervening years, there have been huge changes, in agricultural scale and practice, in development pressure, in emissions to air and water, in the expectations of people for the chance to pursue countryside activities and yet, still, to appreciate "wildness". It is now clear that a *laissez-faire* attitude will not safeguard the traditional character of England that we all value so much. We have learnt that the long-term safeguard of nature requires positive action which will ensure continuation of the right kind of management practices.

In any circumstances, as a first step, we need to establish what we mean by 'nature conservation'. In English Nature we have agreed that our task is to create a situation where the **characteristic biodiversity** (= wealth of wildlife) and natural features of England are **maintained and enhanced** across **traditional ranges, for all to enjoy.**

Let us look more closely at the terms 'characteristic biodiversity' and 'natural features'. One categorisation of biodiversity is based on habitat, the determining elements of which are soils, topography and climate, together supporting spontaneous natural assemblages of plants and communities of wild animals (vertebrate and invertebrate) dependent upon them. One example would be beech woodland on the Chiltern Hills. Elsewhere in southern England, the rounded chalk downlands are associated with characteristic, flower-rich short grasslands between patches of broad-leafed woodland. In the extreme southwest, the Lizard promontory is composed of very particular rocks, forming soils of unique composition, supporting a mix of plant species some of which are confined to that one part of England.

Water is all-important, not only as rainfall (England is wetter to the west, drier in the east) but because of ecologically significant variations in the level of the water-table, rates of drainage and carrying capacity of soils, water-quality and pH. England has a varied heritage of flowing freshwaters, including turbulent upland rivers, rich chalk streams and sluggish, meandering channels of the flat lowlands, all often important for landscape and physiographical values. There are also lakes and other standing waters notable for their nature conservation value, and many marshes and other wetlands. The fens of East Anglia are a special form of wetland habitat developed through centuries of use, and now supporting characteristic rare plant communities and some special insects including the Fen raft spider (one of only two known localities).

Finally, Britain is an island, with a long and indented shoreline. The interaction between land and sea, and the varied topography of the coast are reflected in the wealth of distinctive coastal habitats, some of which are the most natural and least disturbed to survive in England.

ACTION FOR BIODIVERSITY

Although 'biodiversity', as we understand it, is more than the sum of species, EN's statutory duties do include obligations towards species that are rare, vulnerable or threatened. The Lady's Slipper orchid is undoubtedly England's rarest plant. Since about 1935, there was only one plant of this species growing wild in England. For many years, its blooms were picked each year as they emerged, to conceal its existence. Later, NCC invested large sums guarding it during the flowering season in order to allow seed to be set. When I visited, it was a quasi-military exercise, with a diversionary party, a concealed path, a resident guardian camped in a camouflaged tent, equipped with mobile telephone, a network of trip-wires and burglar alarms, and a rabbit-proof cage over the plant itself. For years, seeds were collected but failed to germinate. The task of propagation was passed to Kew, where the new technology of embryo culture has proved successful. I am glad to say that, under English Nature's Species Recovery Programme, the parent plant in the wild is now surrounded by a thriving family of test tube babies!

The Species Recovery Programme consists of a set of partnership projects aiming to restore, maintain or enhance populations of plants or animals under threat and, ultimately, to ensure their long term survival as part of England's wildlife. In 1995, we supported 32 projects, from a budget of £320,000. One project is tackling the Large Blue butterfly. This insect actually became extinct in England despite NCC 's endeavours, before the complexities of its lifestyle were fully understood. This butterfly is totally dependent on a symbiotic relationship with one species of ant. The eggs are laid on thyme, a characteristic perennial herb of short chalk grassland. After hatching and undergoing a couple of moults, the small caterpillars are gathered up by the ants, carried into the nest where they complete their growth on a diet of the larvae of their unfortunate hosts, pupate and, ultimately, emerge as adult butterflies. But this is not the whole story. The exudates of the young butterfly larva will attract ants of two closely related species. Both will carry it to their nests but only one provides the conditions in which it has a good chance of survival to complete its life-cycle. Careful research has shown that vegetation cover determines which ant species dominates. The successful host of the butterfly is favoured by a very short, tight sward and is replaced by the other if the average height of the vegetation increases by only a few centimetres. Now that the biological details are known, the right conditions can be created by hard grazing (or close mowing). Under the SRP, Large Blues have been imported and released into carefully managed habitat, where successful propagation has been achieved.

Success has also been achieved in the re-introduction to England of the Red Kite. This large raptorial bird, formerly common over much of Britain, was exterminated in England during the last century. For several years, young stock have been collected from the wild in other European countries, acclimatised in spacious pens and released when full grown. Several pairs have established themselves, and this project has a very favourable prognostication.

The referee is still out in the case of the red squirrel. This attractive small mammal, Britain's only native squirrel, is threatened by an introduced North American species, the Grey Squirrel. The latter is slightly larger, and (like so many introductions the world over) has proved to out-compete the native form, progressively replacing it across England. Trial measures to reverse the unfortunate situation have included the development of one special hopper with a sprung floor across which Red Squirrels can get access to supplementary food but through which heavier Grey Squirrels will fall, and another, with a magnetically closed access door which only the stronger Grey Squirrels can open (to find poisoned bait). It also appears that Reds may be able to hold their own against Greys in the special habitat of monoculture pine plantations. Accordingly, in the Forestry Commission holdings in eastern England, we have built a special 2-ha enclosure where Reds will be raised and from which they will periodically be released. The outcome of this project is uncertain and, at its best, we recognise we are unlikely to restore the native species across its full original range, or to eradicate the introduction entirely.

ACTION FOR NATURAL FEATURES

Within the term 'natural features', we include geological features. For its size Britain has a remarkably varied geology including, for historical reasons, many sites of international importance. About 1250 English SSSIs have been selected for their nationally important geological or geomorphological interest. Their selection is based on the national Geological Conservation Review (GCR), an immense programme costing millions of pounds sterling, which has run since the late 1970s and is now nearing fulfilment with the publication of the set of chosen sites in a multi-volume series to be completed by 2000.

The array of geological SSSIs will together illustrate the whole history of geological succession in England. Many elements of this are based on fossils, some of which are spectacular, Recently, we were involved with the recovery of an articulated marine reptile (plesiosaur) skeleton of late Jurassic age (150 my) from the Oxford Clay, found in a brick-clay quarry near Peterborough.

EN is also concerned to safeguard natural features of the coast, such as the massive shingle bar at Chesil Beach, formed by coastal currents, maintained and kept in place by natural processes and such a striking feature of the Dorset coast. It is 29 km long, and without parallel in Britain.

River systems can also be important for the conservation of active landform features, carved by the flood and flow of water, as well as for the wildlife value of unpolluted natural waters. English rivers are extensively used for water supplies, the removal of liquid effluents and floodwaters, for navigation and recreation, and other purposes. EN has selected 25 rivers as the best of their type in England, to form a national series to be protected by SSSI status, with appropriate management.

MAINTAIN AND ENHANCE

This brief review, I hope, will have given you some idea of the biodiversity and landform heritage of England, which determines English Nature's basic tasks. So, what do we mean by 'maintain and enhance'?

Clearly, we have to promote positive action to replicate, or restore management practices which were responsible for the creation and maintenance of these features in the past. This may have been traditional agriculture now only perpetuated in rural life museums. Elsewhere, it has become impractical, and certainly uneconomic to farm in these ways. Our task then is to select, or mimic the essential elements. Very frequently the most useful tool is the grazing animal. In EN, we like to enter into contracts with local farmers, to provide grazing stock, to mutual benefit. But (like several voluntary conservation bodies) we also acquire animals of our own, often choosing the more primitive English breeds.

At other sites, we perpetuate traditional production, such as hay making, recreating each year the cycle of growth, maturity, harvest and regeneration to which the specialised plants or small animals are adapted.

THE TRADITIONAL RANGES

The traditional ranges set boundaries to our actions. We seek to ensure that what is being promoted fits the local "natural" circumstances — we are not trying to create a zoo or a botanic garden. For example, we don't want to cultivate heathlands on chalk downs, or peatlands in London parks. We may need to research the traditional ranges of organisms that have become rare, but we respect the natural variations in distribution that contribute to local and regional diversity.

FOR ALL TO ENJOY

What do we mean by 'for all to enjoy'? To succeed, we cannot let nature conservation be an elitist activity, for the privileged few. We must provide and promote opportunities for everyone, and accept that not everyone wants to see the same thing, or necessarily have the same priorities as us. There are places, or times when access has to be restricted: for instance, the breeding sites of protected birds in their nesting season. But it is EN's general aim to encourage people to experience the nature conservation resources of England, to enjoy the experience and to value it.

THE CHALLENGES OF NATURE CONSERVATION IN ENGLAND

The point is made, I hope, that the peculiarity of my country's natural heritage is that it is rarely wilderness or desert, but generally the product of centuries of interaction between man and nature. Almost every hectare of English countryside is grazed, tilled, cropped or managed in some way, or has been at some time. The effects of these activities over hundreds of years have created the natural environment which we in England value. This countryside rapidly loses its characteristic habitats and species if the practices which produced them cease or change. If we in EN are to succeed in maintaining and enhancing characteristic plant and animal communities and natural features of England, our actions must encourage beneficial land management practices.

Is this so different to the challenge for nature conservationists elsewhere or, indeed, for anyone concerned with the state of the natural environment anywhere on earth? In some parts of Malesia, even today, mankind still makes modest demands on the natural environment. In others, you have more or less recently and more or less rapidly reached or surpassed what I might call the 'English condition'. Is it not up to us all to sharpen understanding in our societies so that, at every level from state planning agencies and industrial corporate managers, to the small-holder and subsistence farmer or fisherman, land (and water) uses give value to the conservation of wildlife and natural features, and acknowledge the place of nature conservation in the aspirations for national achievement?

THE INTERNATIONAL AND SUPRANATIONAL CONTEXT

This objective fits well with the commitment made by Governments at the Earth Summit in Rio in 1992, which resulted in the signature of the Biodiversity Convention and agreement on Agenda 21, a comprehensive programme of action needed to achieve a more sustainable pattern of development for the next century, world-wide.

In UK we are now committed to local Agenda 21 across all administrative tiers. English Nature is especially involved with local government in drawing up local versions. The promotion of statutory Local Nature Reserves is one related activity. There are now 444 approved LNRs managed by local authorities in England, in consultation with EN. On these, management of the land for nature conservation is a priority, often linked closely with public enjoyment and appreciation. A key element of EN's urban programme is to encourage the provision of green space in and around towns and cities, offering nature conservation experience within easy reach.

EN is also involved in the national production of the UK Biodiversity Action Plan and the national Sustainability Action Plan. The latter includes the UK Round Table on Sustainable Development, on which I have been invited to sit.

Membership of the European Union (EU), a supra-national body based on inter-governmental treaties, has had a profound effect on environmental legislation in the UK. ASEAN member nations may reflect on the potential of that treaty-based body as a force for similar benefits. Especially important for nature conservation are the Wild Birds directive and the more recent Habitats and Species directive, which was implemented in UK law by regulation in October 1994. The directive indicates certain types of habitat and certain species of European importance which should be given special protection because of their vulnerability or rarity. In Britain, all terrestrial sites designated under these Regulations will already be SSSIs; marine sites will represent an extension of statutory nature conservation protection. The selection process is already in hand. When finally approved, these sites will contribute to the Natura 2000 series representing the best across the EU.

THEORY INTO ACTION

The challenge of translating good intentions into positive action is considerable. EN believes that, if we are to succeed in maintaining and enhancing England's diversity of wildlife and natural features, we must find more effective ways of achieving the positive management of special sites. If we do not succeed there will be continuing steady deterioration of the quality of our wildlife resource.

For instance, there are present risks that English upland heather moorland will degrade as a result of over-intensive grazing. Conversely, for lack of grazing, lowland heaths are being overwhelmed by scrub and woodland. We also need to address the challenge of habitat fragmentation and isolation, which threaten the survival of scarce or irregularly distributed species. The effects of land-use practices on areas adjacent to reserves or other special sites can also cause concern.

On a national scale, priorities for transport infrastructure battle with the preservation of open spaces and landscape features. Plant communities on poorly buffered soils are persistently threatened by acid emissions to air from industrial sources and the internal combustion engine, carried by atmospheric routes. In rivers, lakes or estuaries, standards of effluent quality that meet hygienic regulations can be inadequate to protect natural aquatic communities in the receiving waters. These problems are English, but they are also universal.

PARTNERSHIP

The statutory conservation agency cannot secure the necessary levels of positive management on its own. The way ahead, perceived by EN, is through working more closely with decision makers, and especially with landowners, land managers or occupiers, towards common goals and objectives which support wildlife. We aim to build on the sense of stewardship felt by landowners. We need to secure shared commitment.

On our part, we recognise that we must promote positive incentives to encourage the management of land in ways which support wildlife. We must foster the sense of stewardship more widely in society as a whole.

All who enjoy the benefits of a healthy natural environment need to appreciate their responsibilities for its maintenance. Through example, we hope to inspire others to work with us towards a sustainable natural environment. EN wants to share our decades of experience. This experience in managing nature reserves and special sites, underpinned by scientific research, puts us in a strong position to suggest means whereby nature conservation objectives can be achieved through land management. In their daily activities, staff are already sharing this knowledge with other land managers, and with land-use planners. With them, and with others concerned with the quality of the natural environment, we need to develop joint objectives for the countryside.

Some of English Nature's successful programmes may be instructive in showing how we have achieved shared objectives and delivery through partners:–

The **Wildlife Enhancement Scheme** is designed to provide flat-rate hectarage payments through a simple agreement process, to support traditional management practices which will retain wildlife interest. This is working well in Culm Grasslands, Pevensey Levels, etc.

Community Action for Wildlife has built upon the eagerness of people to get involved in practical activities to improve their local environment. For instance, churchyards are places where use as burial grounds, often for centuries, can preserve and protect wild plants and small animals. Appropriate management can enhance this value, and appeal to local sentiment. Our programme recognises the benefits of getting people involved and is deliberately not too purist.

Species Recovery Programme, on the other hand, has to be based on a good scientific understanding of the ecology of the species concerned. We have to know why it has declined, and what factors will have a good chance of overcoming the problems. In operation, it has moved the attitude of English Nature from the defensive, protectionary role so often prevalent in nature conservation, into one of positive action which, as I have told you, can increase the populations of some of our rarest and most vulnerable species.

NATURAL AREAS — A FRAMEWORK FOR ACTION

Success in finding more effective ways of working with partners to maintain an increasingly threatened wildlife resource, and enhance or restore the elements most at risk, depends on the development of a sound framework for action. The challenges of habitat management, habitat fragmentation and habitat isolation require a move from a site specific focus to a landscape scale. We must be able to focus on the matrix in which the special sites lie, as well as the special sites themselves. The natural dynamics of species involve local extinctions and recolonisation, and this process is constrained in a fragmented landscape. The restoration of linkages and connections between sites is now a vital nature conservation issue. Success will depend on wide acceptance of the need for action, and on practical involvement by people. It is therefore essential that any framework relates to people, and to the sense of place that they feel for the area where they live. This is the basis for EN's concept of Natural Areas.

England is a very varied country, reflected in differences in local geology and topography, traditions patterns of land use, natural wildlife and custom in such things as the style of houses or agricultural buildings. Such regional variation will be familiar to you in your own countries. For English examples, compare:
— North York Moors: upland heather moorland, traditionally grazed
— The Wash: a huge area of alluvial coastal mud

— Leicestershire: the agricultural heartland of rural England

Such variations have led us to adopt our Natural Areas approach.

It is an empirical observation, in England (as elsewhere) that tracts of land, unified by their underlying landforms, rocks and soils, and influenced by climate, altitude and aspect, not only display characteristic natural vegetation types and wildlife species but also support broadly similar land uses and settlement patterns. These factors together identify a series of areas across England, and in our coastal seas, characterised by different combinations of features. EN is working closely with the statutory landscape agency (Countryside Commission) and the agency for ancient monuments and historic buildings (English Heritage) to integrate the nature conservation character with landscape character, and produce a single map which all of us will use to develop our programmes.

Within each Natural Area we shall set up dialogues with those who live and work there, and those who own and manage the land, to agree the character of the area, and what makes it special; the shared objectives that all want to achieve, to ensure that this character is maintained and enhanced; and action programmes to deliver these objectives. Action will focus at all scales, from the individual special sites to the overall pattern of the area, and will look at ways of strengthening and re-establishing sustainable management of the important features, as well as restoring links between isolated or fragmented features.

The natural areas framework is intended to encourage the use of existing information in a more ecologically sound manner. Instead of focusing on a local authority basis, we can consider the whole extent of an ecological unit, such as the heathlands of Surrey, Sussex, Hampshire and Berkshire, and compare them with the distinct heathland complex of the New Forest and Dorset. We can consider the natural coastal process cell extending from Folkestone to Selsey Bill as a single unit, without the disrupting effects of changes of policy or emphasis which often occur at local authority boundaries.

Natural Areas also serve to translate national and international priorities into local action. Thus, while national priorities are identified in the Biodiversity Action Plan, or in the EU Habitats and Species Directive, the natural areas approach simplifies the review of England's priorities for species groups or vulnerable habitat types, giving a framework within which to relate national priorities to the particular areas, and then to set up plans for delivery.

The approach also correlates intelligence on evolving issues which may need attention. From information about the state of comparable Natural Areas across the country, for example in chalk areas or heathland areas, a picture can be built of existing challenges, incentive schemes may be developed, policy issues explored, or generic research conducted.

PRIME BIODIVERSITY AREAS

A map-based approach easily develops into a system to focus effort and to obtain maximum nature conservation advantage. It can show where to reverse fragmentation and isolation effects.

By the use of such a map, we can plot:

— Existing areas of high quality managed grassland (NNR).
— Opportunities to focus effort on areas that can link existing high quality areas of grassland, and also extend the area.
— Some high quality grassland in need of management.
— Already been able to secure some reversion of ex-arable land to grassland through long-term Set Aside and through Countryside Stewardship.
— Can identify areas of current arable land that could usefully be the subject of arable reversion to grassland.

The map can also provide a mechanism to study a range of countryside incentive schemes as part of a co-ordinated overall plan. It can thereby improve our ability to guide bodies such as the Ministry of Agriculture in targeting initiatives in the wider countryside, into ways which will contribute positively to the maintenance and enhancement of biodiversity.

Conclusions

English Nature's hopes for nature conservation into the next millennium will rely on public support and understanding, and on strong working partnerships with land owners and managers, planners and policy makers, and commerce and enterprise. If we achieve a clear shared vision of the wildlife and countryside that everyone values, and a shared commitment to take action, we can hope to ensure that our heritage of wildlife and natural features is passed on to future generations in a better state than we inherited. I believe that this vision is as valid in Malesian countries as I hope it will prove to be in the United Kingdom.

References

Bellwood, P. & Earl of Cranbrook (1994). Human Prehistory. In: Earl of Cranbrook & D. S. Edwards, A Tropical Rainforest, pp. 336 – 337. Royal Geographical Society/ Sun Tree Publishing, Singapore.

Buckland, P. C. (1993). Peatland archaeology: a conswrvation resource on the edge of extinction. Biodiversity and Conservation 2, 513 – 527.

Entwistle, R. (1995). Prehistoric and Romano-British settlement of Salisbury Plain. Sanctuary (1995), 26 – 27.

Godwin, H. (1981). The Archives of the Peat Bogs. Cambridge University Press.

Medway, Lord (1976). Migratory birds. In: Lord Medway & D. R. Wells, The Birds of the Malay Peninsula. Vol. V. London, Witherby.

Malone, Caroline (1989). Avebury. B. T. Batsford/English Heritage, London.

Pryor, F. (1991). Flag Fen: prehistoric fenland centre. B. T. Batsford/English Heritage, London.

On the *Curcurbitaceae* of Malesia

W. J. J. O. DE WILDE[1] & B. E. E. DUYFJES[1]

Summary. An enumeration of the genera of *Cucurbitaceae* extant in Malesia and practical keys to the genera are presented.

INTRODUCTION

The family *Cucurbitaceae* occurs worldwide (c. 130 genera, c. 900 species) mainly in tropical regions, with major centres of occurrence in Africa (drier regions), South and Central America, and continental East and Southeast Asia. However, Australia and Malesia are comparatively poor in species. A treatment of the family for Flora Malesiana is currently under way (De Wilde & Duyfjes in prep.)

Compared to other families there are many, quite distinct genera, and in Malesia most genera have only few or a single species. In Malesia there are 27 genera of which nine or ten are only in cultivation or known as locally established aliens.

Trichosanthes is the largest indigenous genus, possibly with some 20 species in Malesia (see Table 1). This is a complex genus and can be divided into several subgenera and sections. It is currently being revised by Rugayah at Bogor.

Table 1 shows the 27 genera in Malesia, divided into two subfamilies, subfam. *Cucurbitoideae* being divided into several tribes. *Cayaponia martiana* is naturalized on the Ijen Plateau in East Java while *Cyclanthera brachystachya* is found running wild in mountainous areas in Western Java. Taxonomic problems on the species level are still to be solved in *Trichosanthes*, *Zehneria*, and *Neoalsomitra* in Malesia.

During a five month stay in Indonesia during 1995, the collections in the Bogor Herbarium, including dry fruit and spirit collections, were reviewed and checked, and during field excursions photographs were taken of living plants. A selection of the latter, together with photographs of herbarium specimens representing all Malesian genera, were shown at the Third Flora Malesiana Symposium.

Taxonomic justification for the division of *Cucurbitaceae* in Southeast Asia was given by Jeffrey (1980 a,b). Recent Flora treatments of adjoining areas were published by Keraudren-Aymonin (1975) and Telford (1982).

Provisional practical keys to the Malesian genera, for male flowering specimens and for female flowering and fruiting specimens respectively, are presented below.

---

[1]Rijksherbarium/Hortus Botanicus, PO Box 9514, 2300 RA Leiden, Netherlands.

103

TABLE 1. Genera of *Cucurbitaceae* occurring in Malesia

	Number of species in Malesia
Subfam. *Cucurbitoideae*	
Tribe *Abobreae*	
1. *Cayaponia* Manso (introduced)	1
Tribe *Benincaseae*	
2. *Benincasa* Savi (cultivated or running wild)	1
3. *Citrullus* Schrad. (cult.)	1
4. *Coccinia* Wight & Arn. (also cult.)	1
5. *Diplocyclos* (Endl.) Post & Kuntze corr. Jeffrey	1
(*Ecballium* A. Rich.) (cult.)	(1)
6. *Lagenaria* Seringe (cult.)	1
7. *Luffa* Mill. (cult. and possibly partly indigenous)	2
Tribe *Cucurbiteae*	
8. *Cucurbita* L. (cult.)	1 (– 4)
Tribe *Cyclanthereae*	
9. *Cyclanthera* Schrad. (introduced)	1
Tribe *Joliffeae* (subtribe *Thladianthinae*)	
10. *Baijiania* Lu & Li	1
11. *Momordica* L. (also cult.)	3 (or 4)
12. *Siraitia* Merr.	1
13. *Thladiantha* Bunge	1
Tribe *Melothrieae* (several subtribes)	
14. *Cucumis* L. (cult.)	2
15. *Kedrostis* Medik.	1
16. *Muellerargia* Cogn.	1
17. *Mukia* Arn.	3
18. *Solena* Lour.	1
19. *Zehneria* Endl.	± 5
Tribe *Sicyoeae*	
20. *Sechium* P. Br. (cult. and running wild)	1
Tribe *Trichosantheae*	
21. *Gymnopetalum* Arn.	2
22. *Hodgsonia* Hook.f. & Th. (subtribe *Hodgsoniinae*)	1
23. *Trichosanthes* L. (some cult.)	15 – 20(– 25?)
Subfam. *Zanonioideae*	
Tribe *Zanonieae*	
24. *Alsomitra* (Bl.) Roem.	1 or 2
25. *Gynostemma* Bl. (subtribe *Gomphogyninae)*	1 (or 2)
26. *Neoalsomitra* Hutch.	± 3
27. *Zanonia* L.	1
total	± 60(– 70)

KEYS TO THE GENERA

Key to male flowering specimens

1 a. Male inflorescences paniculate (at least once branched) (partly ± racemose in *Cayaponia* and *Cyclanthera*) · 2
   b. Male flowers either solitary, or fascicled, or in racemes · · · · · · · · · · · · · · 9
2 a. Corolla diam. c. 1.5 cm or more (female flower and fruit not known). N Sumatra · **12. Siraitia**
   b. Flower c. 1 cm diam. or less · 3
3 a. Leaves compound · 4
   b. Leaves simple, entire or lobed · 5
4 a. Margin of leaflets entire · **26. Neoalsomitra**
   b. Margin dentate (teeth sometimes minute). Flowers minute, c. 3 mm diam. at anthesis · **25. Gynostemma**
5 a. Leaf margin dentate or denticulate. Perianth (calyx) 5-merous · · · · · · · · · 6
   b. Leaf margin entire. Calyx 3- or 4-merous. (Fruit capsular, seed winged) · · · 8
6 a. Flowers minute · 7
   b. Flowers nearly 1 cm diam. Locally escaped from cultivation, E Java · · · · · · · ·
      · **1. Cayaponia**
7 a. Dioecious. Leaves compound, or sometimes entire. Fruit berry-like, small, not soft-spiny · **25. Gynostemma**
   b. Monoecious. Leaves entire, coarsely 3-lobed. Fruit soft-spiny, asymmetrical, exploding. A weedy climber, locally naturalized in W Java · · **9. Cyclanthera**
8 a. Male flowers c. 1 cm diam. (Fruit large, subglobose, 20 – 30 cm diam.) · · · ·
      · **24. Alsomitra**
   b. Male flowers c. 0.5 cm diam. (Fruit elongated, much smaller) · · **27. Zanonia**
9 a. Flowers medium or large, diameter of corolla at anthesis(0.8 –)1 cm or more. Small or large climbers · 10
   b. Diameter of corolla less than 0.8 cm. Tendrils simple (forked in *Cyclanthera*; naturalized). Small or medium herbaceous climbers · · · · · · · · · · · · · 25
10 a. Male inflorescences spike-like (racemose) and manifestly peduncled · · · · 11
   b. Male flowers either solitary, or in few-flowered fascicles, or in racemes, sessile or with but short peduncle · 17
11 a. Corolla white or pinkish, lobes long-fimbriate · · · · · · · · · · · · · · · · · · 12
   b. Corolla white or (pale) yellow, lobes entire or dentate · · · · · · · · · · · · · 13
12 a. Perianth tube 7(– 8) cm long or less. Probract present, membranous or carnose. (Fruit a leathery many-seeded berry) · · · · · · · **23. Trichosanthes**
   b. Perianth tube (7 –)8 – 10 cm long. Probract absent, but a hard, nail-like protuberance manifestly present. (Fruit hard, with few large seeds) · · · · ·
      · **22. Hodgsonia**
13 a. Plant growing wild; of medium or small size. Tendrils simple or 2-fid. · · · ·14
   b. Cultivated, robust climbers or creepers. Tendrils 2 – 6-fid. Flowers yellow · · ·16
14 a. Leaves ovate, cordate, unlobed; margin (sub)entire or crenulate-dentate. Corolla yellow · 15
   b. Leaves 5-angular or shallowly or deeply lobed; margin often coarsely dentate. Tendrils simple. Corolla white · · · · · · · · · · · · · · · · **21. Gymnopetalum**

15 a. Leaf margin crenulate-dentate. Tendrils simple. Sumatra, Java. (Perianth
    within with scales towards the base) · · · · · · · · · · · · · · **13. Thladiantha**

    b. Leaf margin (sub)entire (minute spaced dents occasionally present). Tendrils
    2-fid. Borneo (Sabah, SE Kalimantan) · · · · · · · · · · · · · · · **10. Baijiania**

16 a. Corolla 1.5 cm diam. or more, pale or bright yellow. Probract present,
    distinctly glandular. Cultivated or apparently wild · · · · · · · · · · **7. Luffa**

    b. Corolla 1 – 1.5 cm diam., pale (greenish-)yellow. Probract absent. Cultivated · ·
    · **20. Sechium**

17 a. Flowers long-pedicelled (or peduncled), with one conspicuous bract. Corolla
    inside with 1 – 3 incurved scales at base · · · · · · · · · · · · · **11. Momordica**

    b. Large bract on pedicel absent. Inner scales in perianth absent · · · · · · · · 18

18 a. Corolla white · 19

    b. Corolla greenish-yellow or yellow · 21

19 a. Plant stout, leaves large (blade c. 10 cm or more); two glands at apex of petiole
    at the transition to the blade. Cultivated · · · · · · · · · · · · · · **6. Lagenaria**

    b. Plants (collected samples) usually more delicate. Glands absent or on
    different locations on the blade. Mostly not cultivated. · · · · · · · · · · · 20

20 a. Corolla (above calyx lobes) connate for halfway or more. Small glands on
    lower leaf surface in the axils of the basal nerves always present. Plant
    frequently coastal; sometimes cultivated · · · · · · · · · · · · · · · **4. Coccinia**

    b. Corolla (above calyx lobes) incised (nearly) to the base, lobes subentire or
    finely dentate. Glands on leaf blade scattered or absent, no glands in basal
    nerve axils · **21. Gymnopetalum**

21 a. Leaves deeply (to over halfway) lobed · 22

    b. Leaves (sub)entire, ovate or subcircular, or blade 5-angular, or shallowly lobed
    (sometimes more deeply lobed in cultivated *Benincasa* and *Cucurbita*) · · · 23

22 a. Climber to 6 m tall; not cultivated. Flowers fascicled or in very short racemes.
    Leaves palmi-lobed · **5. Diplocyclos**

    b. Trailer; cultivated or running wild. Flowers solitary. Leaves frequently ±
    pinnilobed · **3. Citrullus**

23 a. Male (and female) corolla large, c. 6 cm diam. or more. Flowers solitary (rarely
    few-fascicled in *Cucurbita*), long pedicelled. Tendrils 2 – 6-forked · · · · 24.

    b. Corolla smaller, c. 2 cm diam. or less. Flowers solitary or fascicled, short-
    pedicelled. (Anther connectives with sterile appendage). Tendrils simple
    · **14. Cucumis**

24 a. Probract (at node of stems) present. Corolla lobes (above sepal-lobes) largely
    free; sepal lobes shortish, frequently patent or reflexed. Stamens 3,
    completely free, with all three anthers free and 2-thecous. (Ovary densely
    long-haired) · **2. Benincasa**

    b. Probract absent. Corolla connate for c. half-way; sepal-lobes short or long,
    narrow or foliaceous at apex, erect, not reflexed. Stamens 3, only the
    filaments free, the anthers coherent into an elongate mass; two anthers 2-
    thecous and one anther 1-thecous. (Ovary short-haired or subglabrous) · ·
    · **8. Cucurbita**

25 a. Plants dioecious · 26

b. Plants monoecious. Probract absent or present. Leaves various of shape, petiole comparatively long. Filaments long or short, anthers erect, straight · · · · 27

26 a. Leaves (Malesian specimens) usually distinctly hastate or sagittate, with short petiole, 0.5 – 1(– 2) cm long. Probract present, minute. Male inflorescence a short densely flowered raceme with peduncle only 0.5 – 1 cm or less; pedicels distinct, bracteoles present. Stamens 3, with long filaments, anthers transverse, two anthers 2-thecous, one anther 1-thecous · · · · · · **18. Solena**

b. Petiole long (*Zehneria mucronata*) · · · · · · · · · · · · · · · · · · · **19. Zehneria**

27 a. Ovary (and fruit) ornamented with warts or soft spines · · · · · · · · · · · 28

b. Ovary (and fruit) smooth. Tendrils simple. Probract absent · · · · · · · · · · 29

28 a. Probract distinct. Tendrils simple. Male inflorescence long-peduncled; male flowers long-pedicelled · **16. Muellerargia**

b. Probract absent. Tendrils 2-fid. Male inflorescence a (sub)sessile spike (raceme) or condensed panicle; flowers short-pedicelled. A weedy garden escape, W Java · **9. Cyclanthera**

29 a. Male flowers short-pedicelled, arranged in elongate spikes with short or long peduncle. (Fruit cherry-like in size, colour and consistency). Western half of Java, E Borneo · **15. Kedrostis**

b. Male flowers (rather) long-pedicelled, arranged in rich or few-flowered fascicles or in short racemes (spikes) which are either sessile, or short or long-peduncled · 30

30 a. Male flowers in short (sub)sessile fascicles, or in short (condensed) sessile spikes appearing as fascicles; pedicels short · · · · · · · · · · · · · · **17. Mukia**

b. Male flowers either solitary, or (few) fascicled, or in distinctly peduncled (condensed) spikes; male pedicels comparatively much longer · **19. Zehneria**

Key to female flowering and fruiting specimens

1 a. Female inflorescences (or infructescences) few or many-flowered, essentially paniculate (i.e. at least once branched). Flowers greenish, small (less than 1 cm diam.). (Including *Cayaponia*, a garden escape in E Java) · · · · · · · 2

b. Female inflorescences consisting of a single flower or of a several-flowered fascicle or (stalked) or sessile short raceme; inflorescence associated with adjoining male inflorescence or not · 5

2 a. Leaves simple and entire (not lobed). Flowers c. 0.5 – 1 cm diam. Seed winged at two sides · 3

b. Leaves compound (rarely simple in *Gynostemma* and in juvenile stages) · · · 4

3 a. Tall liana. Fruit a large subglobose pendent capsule, 20 – 30 cm diam.; seed thinly butterfly-like winged, c. 15 cm · · · · · · · · · · · · · · · · · **24. Alsomitra**

b. Medium liana. Fruit a pendent cylindrical capsule, less than 10 cm long. Seed including wings c. 5 cm long · · · · · · · · · · · · · · · · · · · **27. Zanonia**

4 a. Medium-sized liana. Margin of leaflets entire. Flowers c. 0.5 cm diam. Fruit a cylindrical pendent capsule opening at apex, 4 – 5 cm long; seed winged at one side · **26. Neoalsomitra**

   b. Small (or medium) herbaceous climber. Margin of leaflets (leaves rarely entire) serrate or finely remotely dentate. Flowers less than 0.5 cm diam. Fruit small, berry-like, globose, with 2 or 3 verrucose seeds · **25. Gynostemma**

  5 a. Inflorescences situated at and towards the apex of the stems, forming panicle-like inflorescences with leaves reduced in size. Fruit a dry ellipsoid berry, c. 1.5 cm long, striped. A weedy climber or trailer, escaped from cultivation on the Idjen Plateau, E Java · **1. Cayaponia**

   b. Mature inflorescences situated lower down on the twigs, associated with normal-sized leaves · 6

  6 a. Flowers medium or large, expanded corolla more than (0.8 –)1 cm diam. Stout or medium creepers or climbers (*Momordica charantia* of delicate habit) · 7

   b. Flowers small, corolla diam. 0.5(– 0.8) cm or less. Mostly delicate, slender-stemmed plants, less than 3(– 5) m long. Fruit smallish, c. 4 cm or less · · 24

  7 a. Hypanthium elongated; corolla white, rarely pinkish, lobes long-fimbriate. Fruit smooth (glabrous or glabrescent), not ornamented. Tendrils usually 2 – 5-forked · 8

   b. Hypanthium shallow or elongated (*Gymnopetalum*); corolla white or yellow, lobes entire or crenulate, not long-fimbriate. Fruit various, angular or ribbed or not, ornamented or not · 9

  8 a. Hypanthium 8 cm long or less. Fruit generally red, smooth; seeds many, small, of various shape. Mostly medium-sized herbaceous or subwoody climbers. Probract mostly present, (sub)herbaceous. (*Trichosanthes anguina* cultivated). Leaves entire, or lobed, or compound · · · · **23. Trichosanthes**

   b. Hypanthium c. 8 cm long. Fruit dull greyish or reddish, large, globose, hard, containing few large, hard seeds. Tall liana. Probract absent, replaced by an ungulate coriaceous excrescence. Leaves 3 – 5-lobed · · **22. Hodgsonia**

  9 a. Plant cultivated, or temporarily maintaining itself in waste places · · · · · · ·10

   b. Plant growing wild, either in original or in disturbed vegetation · · · · · · · ·18

10 a. Corolla white ·11

   b. Flowers yellow or greenish yellow ·12

11 a. Plants stout. Tendrils mostly 2-fid. Two small glands at blade-base, at the transition to the petiole. Fruit large, whitish · · · · · · · · · · · · **6. Lagenaria**

   b. Plants more delicate, though sometimes forming dense mats. Tendrils simple. Small glands at blade-base, in the angles of the palmate basal nerves. Fruit elongate, red, 3 – 7 cm long. (Sometimes cultivated for the leaves, or for the fruit in India?; also growing wild) · · · · · · · · · · · · · · · · · · **4. Coccinia**

12 a. Delicate climber with slender stem and delicate, mostly much dissected leaves. Probract (at stemnode) absent. Fruit lengthwise irregularly furrowed and warted, with bitter taste. (*Momordica charantia*) · · · · · · · · · **11. Momordica**

   b. Plants stouter · 13

13 a. Probract absent. Corolla lobes (above sepals) ± free or to halfway united · · · 14

   b. Probract present. Corolla-lobes (almost) free · · · · · · · · · · · · · · · · · · 16

14 a. Medium-sized trailers. Corolla c. 2 cm diam.; lobes (above sepals) largely free. Tendrils simple · **14. Cucumis**

b. Plants robust. Tendrils 2 – 5-fid · 15
15 a. Flowers (expanded corolla) c. 6 cm diam. or more; corolla fused into a tube
        for c. half-way · **8. Cucurbita**
    b. Flowers (expanded corolla) c. 1.5 – 2 cm diam. Leaves subentire or shallowly
        5-lobed or -angular. Fruit pear-shaped, 10 – 15 cm, pendent, pale green,
        firmly fleshy, irregularly furrowed, 1-seeded · · · · · · · · · · · · **20. Sechium**
16 a. Flowers (expanded corolla) c. 1.5 – 2 cm diam. Leaves deeply ± pinnately
        lobed. Fruit large, subglobose · **3. Citrullus**
    b. Corolla large, c. 2 cm diam. or more · 17
17 a. Probract glandulous. Tendrils 2 – 5-fid. Fruit pulp (pericarp) coarsely spongy
        fibrous. (Male flowers in a peduncled raceme) · · · · · · · · · · · · **7. Luffa**
    b. Probract not glandulous, convex. Tendrils 2 – 3-fid. Fruit hard-shelled,
        whitish-waxy. (Male and female flowers solitary) · · · · · · · · · **2. Benincasa**
18 a. Flowers solitary, with on the pedicel (peduncle) a conspicuous bract clasping
        the flower or not. Fruit generally ornamented. Petiole with (0 or)1 – 5
        glands. Corolla white or pale yellow · · · · · · · · · · · · · · · **11. Momordica**
    b. Single bract sustaining female flower absent. No glands on petiole · · · · · 19
19 a. Corolla white · 20
    b. Corolla greenish-yellow or yellow · 21
20 a. Hypanthium elongated; corolla lobes (above calyx lobes) largely free. No
        glands present abaxially in the axils of the palmate basal nerves of the
        blade. Fruit either (subglobose or) ellipsoid or fusiform, ribbed or not.
        Male flowers solitary or in bracteate racemes · · · · · · · · **21. Gymnopetalum**
    b. Hypanthium not elongated; corolla fused for c. halfway. Blade abaxially with
        glands in the angles of the basal palmate nerves. Fruit elongate, not
        ribbed. (Male flowers solitary) · **4. Coccinia**
21 a. Leaves deeply palmately lobed. Flowers greenish-yellow. Fruit globose, 2 – 3
        cm diam., bright red with white blotchy bands. Seeds turbinate. Monoecious.
        (Accompanying male flowers solitary or in poor fascicles) · · · **5. Diplocyclos**
    b. Leaf blade cordate, (sub)entire (but margin may be serrate or toothed).
        Flowers yellow. Fruit not white blotched. Seeds flat. Dioecious. (Male
        flowers in panicles or racemes) · 22
22 a. Male inflorescences paniculate (branched). Female flowers and fruit not
        known. North Sumatra · **12. Siraitia**
    b. Male flowers in a bracteate raceme · 23
23 a. Fruits few, 1 or 2 on a long-peduncle. Borneo (Sabah, SE Kalimantan) · · · · ·
        · **10. Baijiania**
    b. Fruit single, short-stalked. Sumatra, Java · · · · · · · · · · · · · **13. Thladiantha**
24 a. Ovary and particularly the fruit warty or soft-spiny · · · · · · · · · · · · · · · 25
    b. Ovary and fruit smooth · 26
25 a. Probract present, curved around the stem. Flowers white. Fruit long and
        straight pedicelled, not exploding · · · · · · · · · · · · · · · · **16. Muellerargia**
    b. Probract absent. Flowers greenish-yellow. Fruit asymmetrical, exploding.
        Occasionally cultivated as a vegetable, escaped and running wild in
        montane areas, W Java. (*Cyclanthera brachystachya*) · · · · · · · **9. Cyclanthera**

26 a. Petiole short (c. 1 cm long), blade 6 – 20 cm long, generally with conspicuous hastate-sagittate base enclosing the main stem. Fruit ellipsoid-oblong, 4 – 5 cm long, minutely pubescent, glabrescent; seeds subglobose · · **18. Solena**

   b. Petiole proportionally longer, blade generally smaller, the base not clasping the main stem · 27

27 a. Climber to 5 m tall. Fruit cherry-like (of size, consistency, and colour) containing 1 or 2 globose seeds. (Male inflorescences an elongated raceme). Formerly found locally in the Western half of Java, also E Borneo · **15. Kedrostis**

   b. Small sized trailers. Fruit usually smallish, green, or whitish, red, or purple black, juicy or not; seeds many, flattened · 28

28 a. Female flower(s) or fruit together with male flowers in (sub)sessile fascicles; flower pedicels or fruit stipe short. Indumentum of petiole of coarse scabrid hairs · **17. Mukia**

   b. Female flower either 1(or 2), long pedicelled, together with 1 or 2 (rarely more) male flowers, fascicled, or female flower associated with a slenderly peduncled raceme-like male inflorescence. Monoecious, or plant dioecious in *Zehneria mucronata*, with female flowers or fruits 1-several in clusters, these subsessile or peduncled · **19. Zehneria**

ACKNOWLEDGEMENTS

The authors are grateful to WOTRO, The Hague, and Greshoff Rumphius Fonds, Leiden, The Netherlands, for subvention, and to the Indonesian authorities for allowing their stay and study in Indonesia.

REFERENCES

Jeffrey, C. (1980a). The *Cucurbitaceae* of Eastern Asia. 60 pp. Royal Botanic Gardens, Kew.
Jeffrey, C. (1980b). A review of *Cucurbitaceae*. Bot. J. Linn. Soc. 81: 233 – 247.
Keraudren-Aymonin, M. (1975). *Cucurbitaceae*. In: Aubréville, A. & Le Roy, J.-F. Flore du Cambodge, du Laos et du Viêt-Nam 15: 3 – 116. Paris.
Telford, I.R. (1982). Flora of Australia 8: 158 – 198.

Distribution of epiphytic ferns in Sumatran rain forest

E. Gardette[1]

Summary. A floristic and ecological study of epiphytic ferns in different forest types was carried out in Central Sumatra, Indonesia. Each potential host-tree was divided into six height sections. Sampling was done with a new climbing technique. Clear vertical and horizontal distribution patterns were discovered for epiphytes in lowland and foothill rain forest. Two categories of epiphytes, specialists and generalists, were found. There is no clear host specificity with epiphytic ferns. Discontinuity surfaces were clearly apparent between habitats in full sunlight and those in shade. Epiphytic ferns are distributed in 'ensembles' of different size and number. These ensembles are fragmented after disturbances such as logging in secondary rain forest and require very many years to recover their initial status.

INTRODUCTION

Although epiphytes are often very characteristic life-forms in the rain-forest canopies, their inaccessibility has resulted in few reports on their distribution over large areas (Johansson 1974, Cornelissen & Ter Steege 1989, Grazioli & Prosperi 1990, Tuomisto & Ruokolainen 1994 and Dzwonko & Kornas 1994).

Whereas numerous works have been published on systematics in the Malay Peninsula (Holttum 1954 & 1981, Piggot & Piggot 1988) there has been little documentation on the flora and pteridophytes of Sumatra. However, what information is available provides a basis for carrying out studies on the ecology and the distribution of ferns. Only few detailed studies have been done on pteridophyte ecology in the Malesiana region (Johns 1985, Parris 1983 and Tryon 1985, 1986 & 1989) and no authors have made investigations of these plants over large areas. Therefore the need for such studies is apparent.

A second reason was the opportunity to work in permanent sample plots originally set up by Laumonier in 1984, in Pasirmayang, Jambi province, Sumatra, Indonesia, where the information concerning the trees was already recorded and available: the identification, height and diameter measurements, crown projection, dynamic phases of trees (dbh > 10 cm) as well as the recording and identification of saplings. These data have been partly published (Laumonier 1991) and they are stored in a data base at SEAMEO-BIOTROP, Bogor, Indonesia. These permanent sample plots have been monitored every year since 1986.

A third and critical reason for carrying out this study was recent improvement in climbing techniques (Perry 1978, Dial & Carl Tobin 1994) allowing safer and more effective tree-climbing.

---

[1]Institut de la Carte Internationale de la Végétation, Laboratoire d'écologie terrestre,UMR 9964 (CNRS/UPS), 13, avenue du Colonel Roche, BP 4403, 31405 Toulouse CEDEX, France.

The distribution of epiphytic ferns was approached in different ways:
— to determine the vertical distribution by making an inventory of the different ecological niches occupied by the fern species on a host-tree
— to determine the horizontal distribution over large zones in order to identify possible patterns
— to seek possible relationships between epiphytic fern patterns and the forest mosaic.

If the vertical distribution has been well documented for vascular epiphytes (Johansson 1974) and other groups such as bryophytes (Cornelissen & Ter Steege 1989), similar studies in depth for pteridophytes have not yet been carried out, but have been pointed out by Oldeman (1974).

Our work was done in lowland rain forest in several sites. 2,500 trees were sampled.

STUDY AREA

Field work was carried out from September 1992 to August 1993, in a logging concession at Pasirmayang (Latitude 1°5' S and Longitude 102°10' E, altitude 100 m), near Muarabungo, in Jambi Province, Central Sumatra, Indonesia. A local climate recording station, located within the logging company's area, records that rainfall approximates 2500 mm/year with mean values. The wet season is from October to April. The drier season is from May to September with two months (July and August) with less than 100 mm/month. Average temperature ranges from 27° to 30°C and air humidity is over 80%.

The study was carried out on dipterocarp forests in several representative sites, which include different environments or habitats (valley, slope, ridge, swamp), scattered within the logging company's concession. This primary rain forest has a density of 670 trees (dbh>10 cm) per hectare; for further information about this forest type, see Laumonier (1991).

On 2000 trees in the primary rain forest, 29 epiphytic ferns were collected; at the same time the other vascular epiphytes were also sampled but these data are not included in this paper.

METHODOLOGY

The data were collected in three hectares of primary rain forest and two hectares of secondary rain forest which correspond to 2500 trees.

Following Johansson (1974) and Cornelissen & Ter Steege (1989), each tree was divided into six height sections (Fig. 1).

Several climbing techniques were used, with the contribution of caving, rock-climbing and 'arborist' skills (Perry 1978 and Dial & Carl Tobin 1994). A new arborist method, widely used in France by the employees of ONF (Office National des Forêts), was tested successfully in tropical rain forest for the first time.

The presence or absence on each tree of epiphytic ferns was recorded for the different height sections and the cover index of these epiphytes estimated so as to evaluate the amplitude of their horizontal distribution.

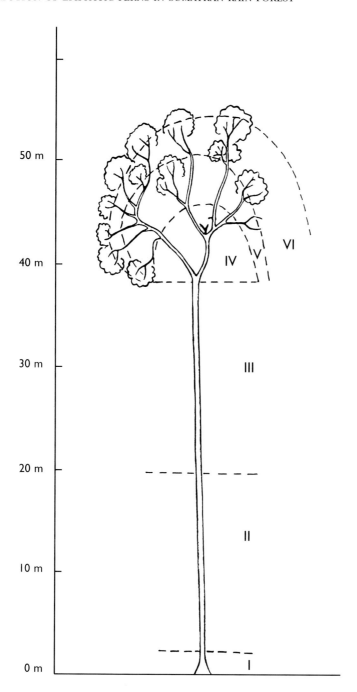

FIG. 1. Height sections on a rain forest canopy tree, after Johansson (1974) and Cornelissen & Ter Steege (1989). I = trunk base (0-3 m), II = lower trunk (from 3 m up to the middle of the trunk), III = upper trunk, IV = lower canopy (main branches), V = middle canopy (medium branches), VI = upper canopy (small branches and twigs).

As most of epiphytic ferns are clonal individuals, it was not possible to count each individual satisfactorily. The 'strand', a term used by Johansson (1974), appears to be difficult to use in the field. There was no distinction made between adult and juvenile ferns. Each species was collected and named by the author, using Holttum's *Ferns of Malaya* and several other of his publications (Holttum 1954, 1959, 1963 & 1981).

Vouchers were deposited in the Herbarium of the Royal Boranic Gardens, Kew (K) and in the BIOTROP-SEAMEO of Bogor, Indonesia. The identification of my specimens has been checked by Mr Peter Edwards of Kew.

RESULTS AND DISCUSSION

A. Vertical distribution in primary rain forest

28 epiphytic species were recorded on 254 host-trees which represent 12.7 % of the total number of potential hosts. This percentage is very similar to reports from other lowland rain-forests quoted in the literature (Richards 1952). In a 1-hectare block, this percentage can reach 16.3% which appears to be the upper limit for this kind of rain-forest. Values of over 20 % can be obtained only on small plot sizes which are not representative of the rain-forest heterogeneity.

Among these species, 21 were distributed in two to three height sections with a maximum of records in two height sections. These species are termed as 'specialists' as they have a small range of distribution.

TRUNK SPECIALIST SPECIES can be distinguished: *Asplenium phyllitidis, Hymenophyllum acanthoides, Elaphoglossum sp., Vittaria scolopendrina* (Fig. 2.).

CANOPY SPECIALIST SPECIES are more numerous, *Drynaria quercifolia, Platycerium ridleyi, Lecanopteris lomarioides, Huperzia phlegmaria, Pyrrosia longifolia, Vittaria ensiformis.* (Fig. 2.).

Trunk specialists occur only on the main trunk either at the buttresses or the trunk itself while canopy specialists are distributed mainly in the crown of medium or large trees. These specializations are certainly linked with environmental conditions, requirements for light and water with adaptation of the morphology, architecture and physiology of these species.

The others are distributed in four or more height sections. I consider these to be 'generalist' species and are less numerous (Fig. 3). Among them are found: *Asplenium nidus, Davallia solida, Drynaria sparsisora, Davallia triphylla, Pyrrosia angustata, Vittaria elongata, Selliguea heterocarpa.* However, *Davallia solida* occurs more frequently in a restricted area on the main and the medium branches and *Selliguea heterocarpa* occurs more frequently on the first fork.

If we total the number of records for each height section and the number of species, we find that the largest number of species occurs in the canopy, starting at the first ramification. There is a limit or discontinuity surface between the trunk species and the canopy species that can be related to the 'inversion surface' (Oldeman 1974 & 1990), situated at 23 m above the ground. This surface corresponds to the crossing point of the humidity and light gradients. Another discontinuity surface can be distinguished between the second and third height

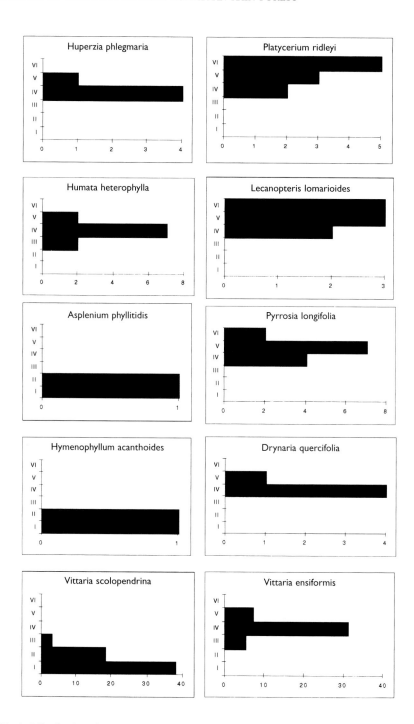

FIG. 2. Vertical distribution of specialist trunk and canopy species in a 3-hectare sample plot, Pasirmayang, in terms of occurrence in each height section.

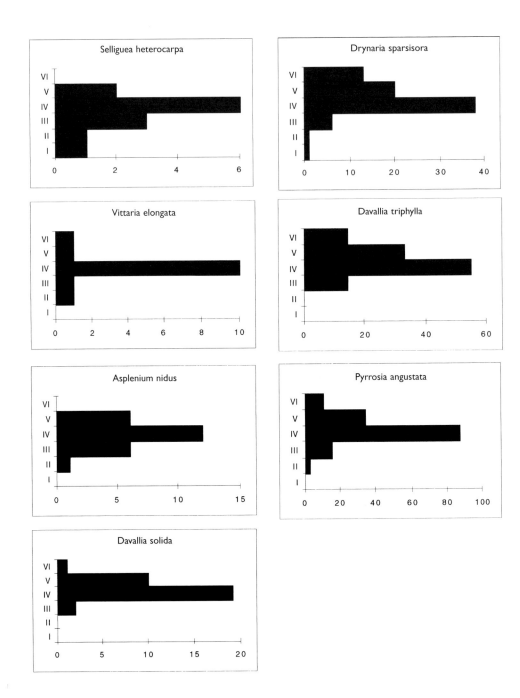

FIG. 3. Vertical distribution of generalist species in a 3-hectare sample plot, Pasirmayang, in terms of occurrence in each height section.

sections at 9 m above the ground which marked the distribution of very specialized trunk species (Fig. 5a & b).

B. Vertical distribution in secondary rain-forest

On a 2-hectare plot after 15 years of selective-logging, only 12 species were recorded: half of them specialist and half generalist. The number of records is very low in comparison with those of primary rain forest and cannot be solely attributed to the remoteness of the large trees which are usually logged by foresters (Table 1 & 2).

Some specialist species such as *Vittaria ensiformis* and *Vittaria scolopendrina* show the same distribution as they do in primary rain-forest. Some other species such as *Platycerium coronarium* and *Pyrrosia longifolia* appear to be more abundant, with a different vertical distribution (Fig. 4).

The discontinuity surface, separating trunk species from canopy species, is still well-marked, but the second surface is not distinct and could easily be linked to the effect of logging which disorganises the layer structure of these ecosystems (Fig. 5c & d) and therefore hampers the re-establishment of fern communities on the remaining trees.

In logged-over secondary rain-forest, where all the trees were felled 15 years ago, all the specialist species have disappeared. We can find only three generalist species, *Asplenium nidus, Drynaria sparsisora* and *Pyrrosia angustata* —the same species as those found in young rubber plantations.

Fig. 8 summarizes the effect of different disturbances on specialist and generalist species.

C. Horizontal distribution

The epiphytic fern communities have been mapped on a 3-hectare sample plot in Pasirmayang and a 1-hectare sample is illustrated in Fig. 6. On this map, only the crown projections of the trees over 30 cm diam. (dbh 1.30 m) are drawn and the trees are tagged in the plot with numbers. The crown projection of the trees between 10 and 30 cm has been drawn on another frame. In order to make the figure readable, I have shown the position and cover of epiphytic ferns first on the host-trees over 30 cm trunk diam. (thin line) and second on the host-trees between 10 and 30 cm diam. (dashed line). The numerous trees in this latter section are not drawn on this map. The positions of epiphytic communities have been marked either by the position of the trunk or by the outline of crown projection of host-trees. The crown projections without epiphytes are drawn with dotted lines.

Patterns are very distinctive from places where epiphytes are either very abundant or absent. Then 'ensembles' (Fig. 6: thick lines) can be drawn with very accurate limits. An ensemble is formed by the clustering of several host trees carrying communities of epiphytic ferns whose crowns overlap or whose distance from one another is < 2 m. They can be divided in three classes according to the surface covered (Fig. 7). The first class (4 – 25 m²) has ten ensembles. The second class (26 – 75 m²) has nine ensembles and the third class has 11 ensembles. These are average values from three hectares of primary rain-forest.

TABLE 1. Presence of pteridophyte species in each height section of host-trees, in a 3- hectare sample plot, Pasirmayang. Functional status (Fun. Sta.): S = specialist, G = generalist. Abundance (Abun.) expressed as a percentage.

Height sections	Fun. sta.	I	II	III	IV	V	VI	Total	Abun.
Aspleniaceae									
Asplenium glaucophyllum Alderw.	S				2			2	0.3
Asplenium nidus Linn.	G		1	6	12	6		25	4.1
Asplenium phyllitidis D.Don	S	1	1					2	0.3
Davalliaceae									
Davallia denticulata (Burm.) Mett.	S			2	4			6	1
Davallia solida (G.Forst.) Sw.	G			2	19	10	1	32	5.3
Davallia trifhylla Hook.	G			14	55	33	14	116	19.1
Humata repens (L.f.) Diels var Gar.	S				3			3	0.5
Humata heterophylla (Sm.) Desv.	S			2	7	2		11	1.8
Hymenophyllaceae									
Hymenophyllum acanthoides (Bosch.) Rosenst.	S	1	1					2	0.3
Lycopodiaceae									
Huperzia phlegmaria (L.) Rothm.	S				4	1		5	0.8
Huperzia squarrosa (Forst.f.) Trevis	S				2			2	0.3
Lomariopsidaceae									
Elaphoglossum sp	S			1				1	0.16
Nephrolepidaceae									
Nephrolepis biserrata (Sw.) Schott agg.	S				1			1	0.7

TABLE 1 continued

Height sections	Fun. sta.	I	II	III	IV	V	VI	Total	Abun.
Polypodiaceae									
Crypsinus stenophyllus (Blume) Holttum	S				1			1	0.16
Drynaria quercifolia (L.) J.Sm.	S				4	1		5	0.8
Drynaria sparsisora (Desv.) T.D.Moore	G	1		6	38	20	13	79	13
Goniophlebium verrucosum J.Sm.	S				1			1	0.16
Lecanopteris lomarioides (J.Sm.)Copel.	S				2	3	3	8	1.3
Platycerium coronarium (König ex O.F. Müll.) Desv.	S					1		1	0.16
Platycerium ridleyi Christ.	S				2	3	5	10	1.65
Pycnoloma metacoelum (Alderw.) C. Chr.	S				1			1	0.16
Pyrrosia lanceolata (L.) Farw.							1	1	0.16
Pyrrosia angustata (Sw.) Ching	G		3	15	87	34	10	149	24.6
Pyrrosia longifolia (Burm.f.) C.V.Morton	S				4	7	2	13	2.1
Selliguea heterocarpa Bl.	G	1	1	3	6	2		13	2.1
Vittariaceae									
Vittaria elongata Sw.	G		1	1	10	1	1	14	2.3
Vittaria ensiformis Sw.	S			5	31	7		43	7.1
Vittaria scolopendrina (Bory) Thwaites	S	38	18	3				59	9.7
Total of occurrence records		42	27	60	296	131	50	606	
Number of species		5	8	12	22	15	9	28	
Total of trunk and crown occurrence				129	477				

TABLE 2. Presence of each pteridophyte species in each height section of host-trees, in a 2-hectare selective-cutting sample plot, Pasirmayang. Functional status (Fun. sta.): S = specialist, G = generalist. Abundance (Abun.) expressed as a percentage.

Height sections	Fun. sta.	I	II	III	IV	V	VI	Total	Abun.
Aspleniaceae									
Asplenium glaucophyllum Alderw.	S		1		7			8	5.6
Asplenium nidus L. s.l.									
Asplenium phyllitidis D.Don									
Davalliaceae									
Davallia denticulata (Burm.f.) Mett.	S								
Davallia solida (G.Forst.) Sw.	S			1	6	2		9	6.4
Davallia triphylla Hook.	G			1	9	9	4	23	16.2
Humata repens (L.f.) Diels var Gar.									
Humata heterophylla (Sm.) Desv.									
Hymenophyllaceae									
Hymenophyllum acanthoides (Bosch.) Rosenst.	S	1						1	0.7
Lycopodiaceae									
Huperzia phlegmaria (L.) Rothm.									
Huperzia squarrosa (Forst.f.) Trevis.									
Lomariopsidaceae									
Elaphoglossum sp									
Nephrolepidaceae									
Nephrolepis biserrata (Sw.) Schott agg.	S				1			1	0.7

TABLE 2 continued

Height sections	Fun. sta.	I	II	III	IV	V	VI	Total	Abun.
Polypodiaceae									
Crypsinus stenophyllus (Blume) Holttum	S								
Drynaria quercifolia (L.) J.Sm.	S								
Drynaria sparsisora (Desv.) T. Moore	G	1	2	2	17	6	4	32	22.5
Goniophlebium verrucosum J.Sm.									
Lecanopteris lomarioides (J.Sm.) Copel.									
Platycerium coronarium (König ex O.F.Müll.) Desv.	S				2			2	1.4
Platycerium ridleyi Christ.	S								
Pycnoloma metacoelum (Alderw.) C. Chr.	S								
Pyrrosia lanceolata (L.) Farw.	S								
Pyrrosia angustata (Sw.) Ching	G			2	13	9	4	28	19.7
Pyrrosia longifolia (Burm.f.) C.V.Morton	S				7	4	3	14	9.9
Selliguea heterocarpa Blume									
Vittariaceae									
Vittaria elongata Sw.	G	2			2			4	2.8
Vittaria ensiformis Sw.	S				11			11	7.7
Vittaria scolopendrina (Bory) Thwaites	S	8	1					9	6.4
Total of occurrence records		12	4	6	75	30	15	142	
Number of species		4	3	4	10	5	4	12	
Total of trunk and crown occurrence				22	120				

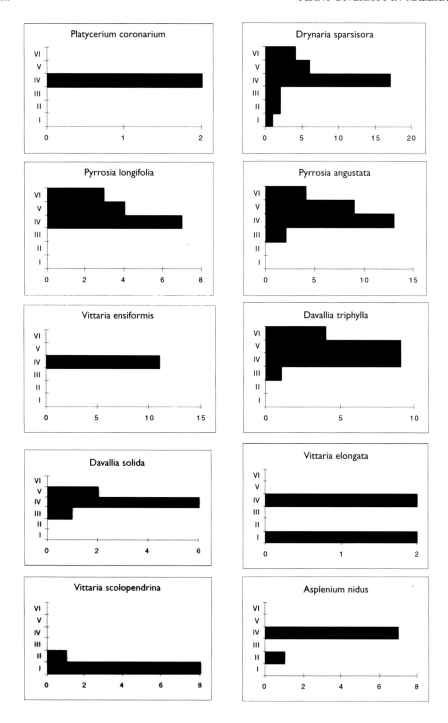

FIG. 4. Vertical distribution of epiphytic species in two hectares of selective-cutting rain forest, Pasirmayang, in terms of occurrence in each height section.

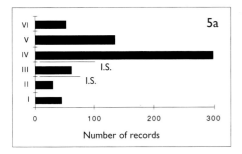
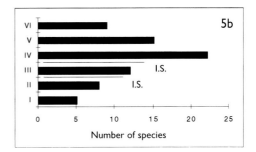

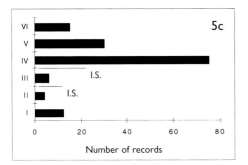
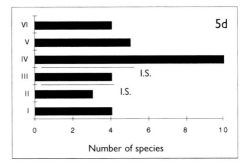

FIG. 5a & b. Total number of cumulated records and species of epiphytic ferns in each host-tree height section in 3-hectare of primary rain forest, Pasirmayang. Inversion surface (I.S.).

FIG. 5c & d. Total number of cumulated records and species in each host-tree height section in 2-hectares of selective-logged rain forest, Pasirmayang. Inversion surface (I.S.).

The first class is often comprised of occasional ensembles, with very poor communities and seems to have a random distribution in the plot. The second class represents ensembles in growth of communities starting to settle on one or two host-trees. The third class comprises mature units of well-developed communities and where individuals provide reproductive forms.

The surface covered by these ensembles represent 23% of the total surface and varies from 13% to 33% in different 1-hectare blocks. If we overlap the dynamic map (Laumonier 1991) with the epiphytic ensembles, we have a robust correlation between the location of these ensembles and the position of mature sylvigenetic phases which are dominated by the abundance of 'trees of the present' to the detriment of 'trees of the future'. Trees of the present correspond to either to trees that are reiterating their architectural model or to trees which fully display their architectural model without reiterating (Hallé, Oldeman & Tomlinson 1978); they allocate more resources to their growth in diameter than height. Trees of the future, on the other hand, grow faster in height than in diameter and have not yet begun to reiterate.

Independent from height or host specificity, epiphyte communities develop more easily in those zones of the forest in which there is an assemblage of trees of the present and trees of the future (60 – 40%), the former category being the largest colonized.

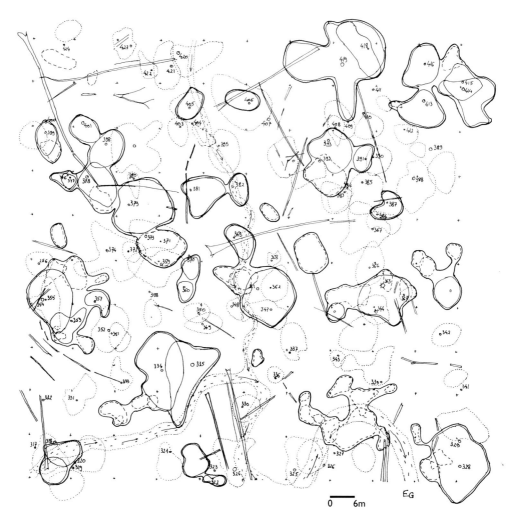

FIG. 6. Horizontal distribution of epiphytic ferns 'ensembles' and epiphytic ferns communities in a 1-hectare sample plot, Pasirmayang. Trees (diameter >30cm) without epiphytes with dotted line; host-trees (diameter 10 – 30 cm) with dashed line; host-trees (diameter >30 cm) with thin line; Epiphytic ferns 'ensembles' with thick lines.

CONCLUSIONS

These results confirm observations made by many authors (Richards 1952, Johansson 1974, Cornelissen & Ter Steege 1989, Whitmore 1990).

The distribution of epiphytes was shown to be restricted to a narrow zone in the host-tree for the majority of species, whereas few generalist species occur in a wide range.

In our first graphical approach, these communities form ensembles, closely related to particular phases of forest dynamics and in particular 'trees of the present' prove to be the most colonized.

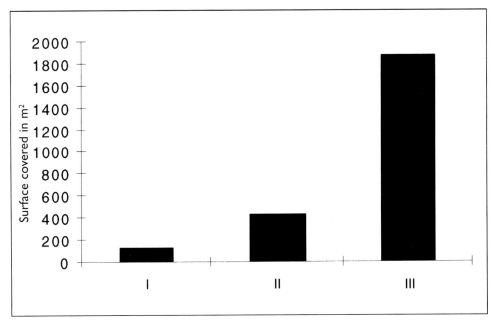

FIG. 7. Surface covered (m²) with each epiphytic ferns units in a 3-hectare sample plot with mean values, Pasirmayang, in terms of three size classes: Class I = 'ensembles' formed by 1 host-tree; Class II = 'ensembles' formed by 1 – 2 host-trees; Class III = 'ensembles' formed by 2 – 9 host-trees.

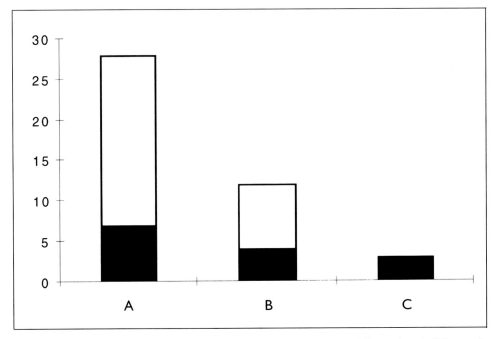

FIG. 8. Specialist and generalist species cumulated in each height section in different sites. A= Primary rain forest. B = Selective-logging secondary rain forest. C = Logged-over forest. Specialist species in white rectangles, generalist species in black rectangles.

There is no correlation between epiphytic ferns either with the species of a host tree or with the height of the host (even though some tree species or some tall trees are more frequently colonized by epiphytes). This has already been tested in Africa by Johansson (1974).

These comparative studies on primary and secondary rain forests demonstrate that, after 15 years of selective logging, only 13 species were recorded and it will take more than a sylvatic growth cycle for epiphytes to recover from such disturbance — maybe very many years will be necessary. We should never forget that a forest is a complex ecosystem with different elements and, even if the time period for the regeneration (turn-over) of tree species is reckoned to be 35 years (Whitmore 1990), other species may respond more slowly after disturbance before recovering their previous abundance.

The study of the distribution of epiphytic fern species can be a very useful tool for the management and conservation of tropical rain forests. It allows us to understand the structure and the functioning of these ecosystems better.

Acknowledgements

I thank the 'Fondation Ushuaîa' for their financial support and the University Paul Sabatier, Toulouse. The logistic support was carried out in co-operation with SEAMEO-BIOTROP of Bogor, Indonesia and the I.C.I.V., International Institute for the mapping of Vegetation of Toulouse, France. The logging company, PT IFA, from Barito Pacific Timber Group contributed to this study by offering accommodation.

I am grateful to Dr Y. Laumonier and Dr M. Torquebiau for the sharing of data in primary and secondary rain-forest plots, making such a study possible.

The herbarium samples were identified in collaboration with P. Edwards from the Herbarium of the Royal Botanic Gardens, Kew, England.

I thank also Dr Irène Umboh and Dr Upik Rosalina for their support. I am indebted to Pak Musa who gave me assistance during the fieldwork and several local people: Doni and Pak Hioto.

References

Cornelissen, J. H. C. & Ter Steege, H. (1989). Distribution and ecology of vascular epiphytes in lowland rain forest of Guyana. Journal of Tropical Ecology 5: 131 – 150.
—— & —— (1989). Distribution and ecology of epiphytic bryophytes and lichens in dry evergreen forest of Guyana. Biotropica 21(4): 331 – 339.
Dial, R. & Carl Tobin, S. (1994). Description of arborist methods for forest canopy access and movement. Selbyana 15(2): 24 – 37.
Dzwonko, Z. & Kornas, J. (1994). Patterns of Species Richness and Distribution of Pteridophytes in Rwanda (Central Africa) — A Numerical Approach. Journal of Biogeography 21(5): 491 – 501.
Grazioli, V & Prosperi, J. (1990). Les jardins suspendus de Guyane, Bromeliaceae et épiphytes à terreau. Mémoire de DEA. 67pp.

Hallé, F., Oldeman, R. A. A. & Tomlinson, P. B.(1978). Tropical trees and forests. An architectural analysis. Springer-Verlag, Berlin: XVII + 411 p.

Holttum, R. E. (1954) Ferns of Malaya. A revised Flora of Malaya. Vol. 2. Govt. Printing Office, Singapore, 1 – 643 pp.

—— (1959). Taxonomical revisions of *Gleicheniaceae.* series II Pteridophyta Vol. 1., published by Martinus Nijhoff, Dr W. Junk, The Hague, Boston, London, 1 – 36 pp.

—— (1963). Taxonomical revisions of *Cyatheaceae.* series II Pteridophyta Vol. 1., published by Martinus Nijhoff, Dr W. Junk, The Hague, Boston, London, 65 – 176 pp.

—— (1981). Taxonomical revisions of *Thelypteridaceae.* series II Pteridophyta Vol. 1., published by Martinus Nijhoff, Dr W. Junk, The Hague, Boston, London, 331 – 560 pp.

Johansson, D. (1974). Ecology of vascular epiphytes in west African rain forest, Acta phytogeographyca Suecica 59, 129 pp.

Johns, R. J. (1985). Altitudinal zonation of pteridophytes in Papuasia. Proc. R. Soc. Edinburgh, Sect. B., 86, 381 – 389.

Laumonier, Y. (1991). Végétation de Sumatra, écologie, flore, phytogéographie. *Thèse doctorat d'état.* I.C.I.V. Université de Toulouse, France, 350 pp.

Oldeman, R. A. A. (1974). L' architecture de la forêt guyanaise. In cahier scientifique de l'ORSTOM, 204 pp.

—— (1990). Forests, Elements of Sylvology. Springer-Verlag, Heidelberg, 624p.

Parris, B. S. (1983). A taxonomic revision of the genus *Grammitis* Swarts (*Grammitidaceae*, Filicales) in New Guinea. Blumea, 29: 13 – 222.

Perry, R. D. (1978). A method of access into the crowns of emergent and canopy trees. *Biotropica* 10(2): 152 – 157.

Piggot, A. G. J & Piggot, C. J. (1988). Ferns of Malaya in colour. Tropical Press, Kuala Lumpur. 458 pp.

Richards, P. W. (1952). The tropical rain forest. An ecological study. Cambridge University Press, Cambridge, 450 pp.

Tryon, R. (1985). Fern speciation and biogeography. Proc. R. Soc. Edinburgh, Sct. B. 86, 353 – 360.

—— (1986). The bioeography of species, with special reference to Ferns. Bot. Rev., 52(2): 118 – 156.

—— (1989). Pteridophytes. In: Lieth, H & Werger, M.J.A. (eds.). Tropical rain forest Ecosystems 14 B: 327-338. Elsevier, Amsterdam.

Tuomisto, H. & Ruokolainen, K. (1993). Distribution of Pteridophyta and Melastomataceae along an edaphic gradient in an Amazonian rain forest. J. Veg. Sc. 4: 25 – 34.

Whitmore, T. C. (1990). An introduction to tropical rainforest. Clarendon Press, Oxford, 226 pp.

Techniques used in the production of an electronic and hardcopy Flora for the Bukit Baka–Bukit Raya area of Kalimantan

J. K. Jarvie[1], Ermayanti & U. W. Mahyar[2]

An electronic and hardcopy Flora is being prepared for the Bukit Baka–Bukit Raya National Park and environs, focusing on trees and shrubs. Discussed here are the needs of our user groups and the three computer systems we have developed central to the task at hand. The first is a specimen database, currently holding about 7,000 records. One half of these records come from historical collections of woody groups from West Kalimantan, deposited in Herbarium Bogoriense. The other half is from the project's continuing collecting programme in and around the park. The second is a glossary database which, given the diversity of the target audience for our outputs, has to be both botanically accurate and generally understandable. The most important, in terms of tangible outputs, is a DELTA system used to produce a partly illustrated interactive key to the relevant genera and textual descriptions, in both English and Bahasa Indonesia. The first edition of the interactive key is available on the Internet from one of the Harvard servers. The hard-copy Flora, still in development, will be based around the interactive key to genera augmented by annotated keys to species. It will be in a spring folder book form to enable users to customize and update the Flora at will. We hope that the systems developed for this project will be flexible enough to assist future projects of a similar type, intending to maximise use of taxonomic data by translating it to an accessible and adaptable information base.

Introduction

Bukit Baka–Bukit Raya National Park (BBBR) is in the centre of the island of Borneo. Across it runs part of the border between West and Central Kalimantan. It covers approximately 181,000 ha and ranges in altitude from around 250 m to 2310 m, the latter being the summit of Bukit Raya which is the largest mountain in Kalimantan. The park is principally made up of hill Dipterocarp forest. At high elevations *Agathis* and Fagaceous forest predominate. Along the higher ridges cloud forest is encountered. The park is the focus for a number of projects and it is envisaged that BBBR will be among the best planned parks in Indonesia.

This paper discusses a project which is producing an inventory database and Flora for the park in a collaboration between Herbarium Bogoriense and the Arnold Arboretum. The philosophy behind the task is to use all generally available tools to turn taxonomic data into synthesized packages of information needed by, and hence made for, a diverse user-group in Indonesia. The comments given herein reflect an Indonesian perspective, but we hope that they will be of relevance in the wider region.

---

[1]The Arnold Arboretum of Harvard University, 22 Divinity Avenue, Cambridge, MA 02138, U.S.A.
[2]Herbarium Bogoriense, Jl. Juanda 22, Bogor – 16122, Indonesia

THE USER-GROUP

Is the question of target audience a consideration for writers of Floras? The perception of non-taxonomists is generally no. The information in Floras is locked in a technical language often bearing little more than a passing resemblance to English (Jarvie & Welzen 1994). Whereas some of this language is necessary to accurately communicate information about structures, their characters and the attributes of their characters, glossaries reflecting how the terms are used in the Flora are often lacking. The recent Australian Tropical Rain Forest Trees (Hyland & Whiffin 1993) is a notable exception to this wherein terms found in the computerised key are described, sometimes with comments about how their definition of a structure, aril is an example, differs from the possible definition of other authors. When a definition is required in a different language, especially non-European, this issue is complicated much further.

What is our target audience? Following consultation with groups including relevant national and regional government agencies, NGOs conducting field surveys, international collaborative aid projects, commercial groups and national universities a Flora would appear to be needed which ultimately has to meet the following criteria:

— available in Indonesian and English
— written in understandable language, with a dedicated glossary to clarify technical terms
— illustrated
— match the needs of each user group and contain the information they require

THE LIMITATIONS OF TRADITIONAL FORMATS AND HOW WE ATTEMPT TO GET AROUND THEM

One of the interesting points to come out of discussion with potential users was that very few were interested in species with the exception of those of major economic importance, the numbers of which are minor in relation to the size of the plant biota. If a particular species is renowned, it is most likely to be widely known in the field. A good example is ironwood, locally known as belian or ulin (*Eusideroxylon zwageri*: *Lauraceae*). This finding is in agreement with Kessler & Sidiyasa (1994), who have prepared their own Flora to trees in *Dipterocarpaceae* dominated plots in East Kalimantan. All users were particularly interested in tree groups, often those delimited by local classifications or commercial wood categories. Everyone wanted to know how to identify the trees, but could not get started. The project, however, was faced with the realization that in a series of six collecting trips we were not going to find and collect every species in the park. This was potentially a serious problem in consideration of a point made above, that a Flora was required which would match the needs of each user group and contain the information they require.

Questions arose of what could be contributed in a Flora as complete information, and what could be legitimately be provided while incomplete? A strategy evolved, rather than being decided a priori, which was to:

— set up information databases covering areas of information needed by different users
— design a flexible-format Flora which could be (de)constructed by users from a range of outputs from databases and other files, some of which could be very narrowly defined by users. The end-result would be a Filofax/Dayrunner format, an obvious success because of the flexibility available in a utilitarian tool

A final point to be addressed concerns keys. Without keys information cannot be efficiently accessed. Keys are accurate with ten taxa, unwieldy but usable with 20, effective only if well written to 30 and well into the realm of high probability of a poor outcome by 50. A decision had to be taken to determine the most detailed level of taxonomic currency the project could effectively use and communicate effectively. A Flora to the trees and shrubs was chosen. Based on the requirements specified above and problems with keys, the overall Flora will consist of:

— a computerized interactive key to the tree and shrub genera of BBBR
— text descriptions of the genera, generated from the database for the interactive key
— annotated keys to the species, manually generated
— illustrations of genera and species

THE TECHNIQUES

All databases are stored in PARADOX 4 for DOS format. The decision to use this package was personal preference.

Specimen database

The field structure for the specimen database reflects standard information pertaining to habit, habitat, collectors, local names and uses. As the data will be later transferred to a project-dedicated GIS and a proposed Indonesian biodiversity database, and because various subsets have to be provided to different users, the design (Table 1) is made with a maximized number of narrowly defined fields. The philosophy behind this is that when receiving data for incorporation into an alternative database, it is far easier to combine data fields than to divide them up.

Two databases are kept. One is for historical collection data from Herbarium Bogoriense of woody families found in BBBR. The other contains collections from the project's expeditions. The total number of collections is about 7000, with each database containing about 50% of the total.

Data input forms have been designed in Indonesian and English. The project's collections are entered in English, but regional projects and universities have asked for the system to be modified to allow an Indonesian language front end. A family and genus lookup table checks accuracy in spelling, and entry of the genus name causes the family to be filled in automatically. Herbarium label generation is accomplished through a WordPerfect merge file. This system is relatively flexible, although rudimentary. It is easy to teach and users find it easy to adapt label formats to their own tastes.

TABLE 1. Structure of specimen database

Field Name	Field Type	Field Name	Field Type
INST_CODE	A10	N_S	A1
CAT_#	N	E_W	A1
FAMILY	A35	GEO_SOURCE	A25
GENUS	A35	CONTINENT	A10
SPECIES	A35	COUNTRY	A25
SUBSPECIES	A25	REGION	A25
RANK	A25	ISLAND	A25
AUTHOR	A50	STATE	A50
MORPH1	A255	KABUPATEN	A50
MORPH2	A255	ALTITUDE	N
COLL_1	A25	LOCALITY	A255
COLL_2	A25	HABITAT	A255
COLL_3	A25	DATE_ENTRY	D
COLL_#	A25	DATE_AMEND	D
DATE	D	SITE_NAME	A50
MICROHAB	A255	CATALOGER	A15
LATDEGREE	N	REMARKS	A255
LATMINUTE	N	LOCAL_NAME	A75
LATSECOND	N	DETERMINER	A50
LONGDEGREE	N	#_SHEETS	N
LONGMINUTE	N	#-REPS	N
LONGSECOND	N		

Glossary database

The words needed for the glossary are taken from the DELTA (Dallwitz 1980; Dallwitz et al. 1993)CHARS file, discussed in the next section. We have fields for the English word, English definition, Indonesian word and Indonesian definition (Table 2). A recent dictionary (Rifai 1993) provided Indonesian words and therefore a basis for the definitions. Some botanical terms do not yet have Indonesian equivalents and the original words have been retained. This database contains fields indicating inclusion in the dictionary, and/or the glossary. Through WordPerfect merge functions, the database can be searched and an updated glossary or dictionary produced at any time. We think of a printed output as a static entity, whereas the database continually improves and develops.

TABLE 2. Structure of glossary database

Field Name	Meaning	Field Type
NUMBER	Code for word	N
ENG_WORD	English word	A255
ENG_DEF	English definition	A255
IND_WORD	Indonesian word	A255
IND_DEF	Indonesian definition	A255
DICT_INC	To be included in dictionary?	A1
GLOS_INC	To be included in glossary?	A1
DIFFICULT	Unsure about definition	A1

The DELTA related databases and systems

The CHARS file is central to the entire DELTA system. It was modified (Jarvie et al. 1994) from the automated descriptions of the families of flowering plants (Watson & Dallwitz 1991), adding field characters, deleting those we would not be using and modifying others. From the 244 remaining, 113 were selected from this table for the production of the interactive key and descriptions. Four subdirectories were set up, one for each of:

— the English language interactive key
— the Indonesian language interactive key
— the English language descriptions
— the Indonesian language descriptions.

Descriptions can be generated directly from the interactive key database. Edited hardcopy descriptions are produced in both of our target languages. Batch files have been written to take the output from PARADOX 4 databases, which are in the ITEMS format that DELTA requires, and call the algorithms necessary to produce English and Indonesian keys and descriptions.

Illustrations of the genera were taken from the literature. Species illustrations are made by project artists and will be inserted in both the interactive key and hardcopy products.

THE FUTURE

As we move toward the next millennium the issues of biodiversity and its conservation will become ever more urgent. Currently we practise the art of describing nature, the basis for many activities relating to natural resource management including conservation, in a way little changed for centuries. We cannot convince the wider community that Floras make a worthwhile product until, developed from them, are useful packets of information. Local Floras, which we would argue include regional Floras (it is all a matter of scale) can interact with each other, exporting and repatriating data and products useful to both targeted and general audiences. We hope the techniques used in the production of a Flora for the Bukit Baka-Bukit Raya area, from ascertaining market needs to developing the computerized systems for information dissemination while getting on with the collecting and describing, may in part show how Flora efforts could be improved.

ACKNOWLEDGEMENTS

Our thanks to Dr. Purnadjaja for checking all Indonesian translations and much constructive criticism, which has improved the quality of our work considerably.

REFERENCES

Dallwitz, M. J. (1980). A general system for coding taxonomic information. Taxon 29: 41 – 46.

——, Paine, T. A. and Zurcher, E. J. (1993). Users guide to the DELTA system, 4th edition. CSIRO Division of Entomology, Canberra.

Hyland, B. P. M. & Whiffin, T. (1993). Australian Tropical Rain Forest Trees, vol. 1 & 2. CSIRO, Australia.

Jarvie, J. K., & Welzen, P. van. (1994). What are tropical floras for in SE Asia? Taxon 43: 444 – 448.

——, Mahyar, U. W. & Ermayanti. (1994). A DELTA character list for family and generic descriptions of plants in English and Indonesian is now available. Trop. Biodiversity 2(1): 280 – 281.

Kessler, P. J. A & Sidiyasa, K. (1994). Trees of the Balikpapan – Samarinda area, East Kalimantan. The Tropenbos Foundation, Wageningingen, The Netherlands.

Rifai, M.A. (1993). Glosarium Biologi. Pusat Pembinaan dan Pengembangan Bahasa & Departemen Pendidikan dan Kebudayaan, Jakarta.

Watson, L. & Dallwitz, M. J. (1991). Automated descriptions of the families of flowering plants. Aust. Syst. Bot. 4: 681 – 695.

An update on the Flora of the Philippines project

Domingo A. Madulid[1] & Seymour H. Sohmer[2]

Summary. An update of the Flora of the Philippines project since its inception in 1990 is presented.

Almost five years have passed since the start of the Flora of the Philippines. Starting with just two staff in October 1990 there are now 15 full time staff members consisting of three field teams, administrative support staff, computer programmer/encoder, a collections manager and herbarium technicians. To date the Flora project has accomplished many significant and tangible results.

Field trips undertaken

A total of 65 field trips were undertaken by the three field teams. This covered botanically-rich areas throughout the country from the northernmost islands in Luzon (i.e. Batanes) to the southernmost tip of Mindanao down to the Sulu Archipelago. Selection of sites for collection was based on botanical richness of the site, history of collection in the area and relative peace and order condition. Table 1 summarizes the localities and the number of collections made from the start of the project to the present.

Specimens collected

From 1990 to 1995 a total of 17,049 specimens in ten duplicates have been collected. These consist mostly of vascular plants and ferns with some bryophytes, fungi and algae. The families of flowering plants which were frequently collected were *Leguminosae, Rubiaceae, Euphorbiaceae, Moraceae, Meliaceae, Palmae, Pandanaceae, Araceae, Orchidaceae, Rutaceae*, and *Gramineae*.

Noteworthy collections

Even at this stage when the Philippines Plant Inventory (PPI) specimens are still being collected in the field several new taxa were already detected from the collections. Some of these were given names by specialists while others are still being formally described. For example several new species of palms have been collected. The late Dr Benjamin C. Stone had segregated several specimens of

---

[1] Philippine National Herbarium
[2] Botanical Research Institute of Texas

pandans which he considered to be new taxa and was in the process of naming them but unfortunately he died before the names were published. Several specimens turned out to be new records for the Philippines or they represent the second recollection of the species since they were first described more than 70 years ago. Many of these were Merrill's and Elmer's species.

TABLE 1. Sites Visited by the PPI team and Number of Collections

Sites	No. of Specimens (in 10 sets)	Sites	No. of Specimens (in 10 sets)
Luzon		**Visayas**	
1. Batanes	141	1. Negros	340
2. Palanan	350	2. Samar, Mt Sohoton	404
3. Polillo Island	394	3. Mt Madia-as, Panay	220
4. Mt Nilisan and Mt Mayon	380	4. Mt Canlaon, Negros Or	189
5. Bicol National Park	435	5. Leyte and Biliran Is.	321
Quezon National Park		6. Mt Lokilokon, samar	325
6. Catanduanes Island	365	7. Mt Guiting-Guiting	877
7. Mindoro	439	8. Ormoc, Leyte	316
8. Mt Mariveles; Mt Masinloc	360	9. Cebu and Bohol	321
9. Mt Pulog, Benguet	72	10. Mt Baloy	418
10. Baler	120	11. Palawan	435
11. Mt Isarog	475	12. Mt Mantalingahan	310
12. Ifugao; Mt Mayaoyao	310	13. Coron Island	74
13. Ilocos Norte; Mt Arayat	314	14. Mt Capoas, Palawan	253
14. Mt. Makiling; Mt Samat	74	15. Culion & Busuanga	245
15. Casiguran	301	16. Ransang, Quezon, Palawan	225
16. Tagkawayan, Quezon	221	17. Brooke's Pt Palawan	508
17. Ilocos Norte, Benguet	89		
18. Tarlac	130		
19. Mt Pinatubo	150	**Mindanao**	
20. Subic Forest	187	1. Mt Apo	324
21. San Mariano, Isabela	209	2. Agusan Marsh	767
22. Biak-na-Bato N.P.; Angat	313	3. Dinagat Island	441
Watershed		4. Zamboanga	331
23. Penablanca, Claveria	296	5. Mt Matutum	312
24. Mt Isarog	27	6. Tawi-Tawi Is.	239
25. Mt Amuyao, Ifugao	192	7. Mt Timolan	175
26. Mt Palay-Palay N.P.	250	8. Mt Kitanglad	1785
Mt Natib, Bataan		9. Camiguin Is.	249
		10. Mt Malindang	249
		11. Siargao Is.	330
		12. So. Cotabato	147

DISTRIBUTION OF COLLECTIONS

As far as practicable ten duplicates were collected for each plant in the field. The duplicate sets are intended to be distributed to various collaborating herbaria, namely: US, A, BISH, L, K, BRIT, SING, BO, CUHP, and PNH. On 16 February, 1995 the first batch of duplicate specimens numbering 25,000 were sent out by PNH to BRIT and from there the individual sets were sent to US, A, L, K, and BISH. For its part the PNH will be sending duplicate sets to BO, SING and CUPH. Through this scheme it is intended that the collaborating institutions will also send specimens in exchange to PNH and BRIT thus increasing the collections of these herbaria.

COLLECTIONS MANAGEMENT PROJECT (CMP)

This component of the Flora Project is being undertaken with the kind support of the John D. and Catherine T. MacArthur Foundation. Running for three years now the three staff members hired by the project were able to process a total of 331,206 backlog specimens. Their activities included freezing, mounting, strapping, accessioning, sorting, distributing, and identification. A total of 10,759 backlog PNH specimen records were also entered into a computerized database. A summary of the activities of the CMP is presented in Table 2.

COMPUTERIZED DATABASE

In order to access relevant information about the specimens gathered in the field a computerized database was devised by our computer programmer/encoder. This database is modified from existing herbarium databases such as Labels 3 and Tropicos. To date we have entered the data for most of our field collections and they are sorted into family, genus, species, specimen, local name, locality, date of collection, etc.

TABLE 2. Summary of Activities and Accomplishments of the Collections Management Project

Activities	Number of Specimens
No. of specimens freeze-treated	183476
No. of specimens examined	76873
No. of specimens mounted	28050
No. of specimens strapped	15661
No. of specimens accessioned	7499
No. of specimens identified	8888
No. of specimens entered into dbase	10759
Total	331206

PERMANENT PLOTS

During the second phase of PPI a plan was conceived to construct three semi-permanent plots to serve as long-term ecological research areas where one can study the species richness and diversity of the various vegetation types. To date two plots have been established, one in Mt Kitanglad, Mindanao which is a submontane forest and another one in Mt Guiting-Guiting in Sibuyan Island, which is a forest over ultramafic soil. A report about the Mt Kitanglad plot was presented by J. Pipoly and D.A. Madulid during the Biodiversity Symposium held at Smithsonian Institution in May, 1995. This topic was also presented to the Flora Malesiana Symposium (Pipoly & Madulid 1997). The third plot will be established in another lowland primary forest in Mt Pulog, in the island of Luzon.

CONSORTIUM OF UNIVERSITIES

Through the efforts of BRIT and the PNH a consortium of universities in Texas and Manila were organized to encourage young Filipino and American students to undertake undergraduate and postgraduate training in taxonomy, systematics and related fields. Students under this program will be able to participate in the Flora Project by taking up revisions, monographic studies, systematic problems, etc. pertaining to Philippine plants. Appendix I is a list of U.S. and Philippine universities who formed this consortium.

PLANNING THE RESEARCH, WRITING AND DOCUMENTATION OF THE FLORA PROJECT

1. BRIT Meeting From 26 – 28 May, 1994 a workshop was held at BRIT to plan the research, writing and documentation of the Flora of the Philippines Project. The workshop was attended by P. Ashton (A), J. Beaman (Mich), K. Kalkman (L), P. Morat (P), D. Stevens (MO), W. Wagner (US), C. Lamoureux (Univ. of Hawaii), E. Fernando (CUPH), D.A. Madulid (PNH) and S.H. Sohmer (BRIT), G.C. Casal (PNM), J. Pipoly (MO), B. Lipscomb (BRIT), J. Ward and B. O'Kennon.

2. Manila Meeting In October 1994 the second workshop to plan the research, writing and documentation of the Flora of the Philippines Project was held in Manila, Philippines. The workshop was attended by S.H. Sohmer, G.S. Casal, D.A. Madulid, K. Kalkman, J. Beaman, C. Lamoureux, S. Kawano and by members of the Philippine Advisory Committee led by Senator Alvarez, J. Velasco and P.M. Zamora. The workshop was also attended by American graduate students in plant taxonomy as part of their learning/training exercise on flora making.

PHILIPPINE FLORA NEWSLETTER

In order that collaborating scientists, members of the Advisory Committees, foreign and local herbaria, and research institutions interested in the Philippine Flora project are kept abreast of the progress and achievements of the Flora project a semi-annual newsletter is published. The first issue came out in June 1991 and the 7th issue came out last December 1994. The eighth issue is now being prepared by June Cunningham, editor of the newsletter and was expected to be published by the end of July 1995.

FUTURE PLANS FOR THE FLORA PROJECT

1. Short Term

A Revised Enumeration of Philippine Flowering Plants. This project aims to revise Merrill's Enumeration based on our present knowledge of this identity, extent and occurrence of Philippine flowering plants. Each taxon will be fully annotated and it will deal with the recognition of taxa within the context of the entire Malesian region. We propose that this effort be formulated as a National Science Foundation proposal ready for submission in Autumn 1995. A Revised Enumeration will take three years to accomplish, would be relatively easy to perform, would be facilitated by the latest computer technology and software and would cost much less than a full Flora. Furthermore, the Revised Enumeration would serve as a sound basis for the actual writing of the Flora in the future.

Continuation of Plant Inventory. We propose that the Philippine Plant Inventory is continued for another three years after 1996 to bring the collecting activities of the present project through to the turn of the century. We believe that the urgency of continuing the collecting programme is dictated by the reality of the deforestation occurring in the Philippines. The continuance of the collecting program would be linked to the maintenance and the establishment of permanent plots as has already begun under the present phase of the Philippine Plant Inventory. We suggest that the relevance of this activity is one of the most important future aspects of the project.

2. Long Term

Production of the Flora. The actual writing of the Flora will probably take 10–15 years to finish and will consist of 5 – 7 volumes. It will be done partly by full-time in-house staff and by specialists who will contribute part of their time to the project. There will be an editorial board who will decide on the format and contents of the Flora and a chief editor who will be responsible for the general editing work and ensuring that the writers submit their contributed manuscripts on time.

CONTRIBUTORS TO THE FLORA OF THE PHILIPPINES

Botanists writing the Flora of the Philippines are of two categories: in-house staff and part time contributors. Earlier on, the late Dr. Benjamin Stone, S.H. Sohmer and D.A. Madulid started to "advertise" in international publications such as *Taxon*, etc. for botanists interested in contributing to the Flora project. As a result numerous persons from various institutions around the world responded to the call. To date there are already more than 70 contributing botanists in our list and we expect to list more names in the future.

REFERENCES

Pipoly, J & Madulid, D. A. (1997). The vegetation of a submontane moist forest on Mt Kinasalapi, Kitanglad range, Mindanao, Philippines. Flora Malesiana Symposium Volume: 000 – 000.

APPENDIX I

Summary of Graduate Training Available for Members of the Dallas-Fort Worth Consortium and the Flora of the Philippines Project.

Plant Systematics:
    Molecular Systematics: UNT
    Classical, floristics, biogeography, phylogenetic: BRIT
    Classical and molecular: UTA

Plant Ecology:
    Community Ecology: BRIT, TCU, UNT, UTA
    Physiological Ecology: TCU
    Aquatic Ecology: TCU, UNT
    Remote Sensing, GIS, Landscape ecology: BRIT, TCU, UNC
    Pollination Biology: BRIT, TCU, UNT

Plant physiology-developmental Biology, Morphogenesis: UNT, TCU, BRIT, TCU, UTA

Plant Anatomy/morphology: BRIT, TWU, UNT, TCU

Plant Conservation Biology: BRIT

Economic Botany, Ethnobotany: SMU, BRIT, UNT

Plant Population and Evolutionary Genetics: UNT

Plant Pathology: BRIT (rusts only), UTA

DFW Consortium Members:

    BRIT — Botanical Research Institute of Texas
    SMU — Southern Methodist University
    TCU — Texas Christian University
    TWU — Texas Wesleyan University
    UNT — University of North Texas
    UTA — University of Texas at Arlington

A Flora for Tamilnadu, South India

K. M. MATTHEW[1]

Summary. Phase I (1976 – 1983) covering a representative sector of the plains (600 field days, 30,000 collections) published in four volumes (The Flora of the Tamilnadu Carnatic, 1981, 1982, 1983 & 1988). Volumes 2 and 4 comprised 1,814 full-page illustrations. The work was summarised as An Excursion Flora of Central Tamilnadu, India (700 pp) in 1991, reprinted in 1995, with a vernacular version in 1993.

Phase II (1984 – 1996) covering the montane counterpart, the Palni Hills (300 field days, 15, 000 collections), will be published in the same format. Illustrations are due in 1996, and the text (3 parts) in 1997 to be followed by the Excursion Flora, English and the vernacular. Format: (A) Text: (a) Nomenclature (all protologues seen); synonymy to cover all (to 9) concerned regional Floras since Roxburgh; (b) Vernacular names; (c) Original descriptions from own materials; (d) Occurrence and field notes in two paras; (e) Distribution. (B) Illustrations: Habit with a dozen structural details, comparable, all original.

THE SCOPE

At the start of the project, the three previous *Floras* covering the region were half a century to over a century old, and needed urgent revision. *The Flora of British India* (Hooker, J.D. *et al.* 1872 – 1897), the national *Flora*, had little specific to the region. *The Flora of the Presidency of Madras* (Gamble & Fischer 1915 – 1936), probably the best provincial *Flora* in India, covered a vegetationally heterogenous area; besides, there were no illustrations. *The Flora of the South Indian Hill Stations* (Fyson 1932), covering three hill stations, needed thorough revisions of the keys, descriptions, nomenclature and illustrations. Finally my own *Exotic Flora of Kodaikanal* (Matthew 1969), besides the necessary revision, had to be integrated into a comprehensive general *Flora*.

So, a modern, illustrated, regional *Flora* was the immediate goal. Apart from its all-purpose objective, this *Flora* (Matthew 1981 – 88) would serve as the take-off base for applied work, especially conservation and biodiversity for which the extensive field work of some 1,000 days would be the indispensable resource base. The resulting collections would further serve the long-term goal of monography.

The *Excursion Flora* in English (Matthew 1991) was a second step, aimed at the dissemination of knowledge of plants to ordinary people.

A *vernacular version* (Matthew 1993) was a third step towards wider outreach to people without facility in English. The prices were almost nominal, thanks to hefty subsidies, less than a quarter of the commercial ones, and for the *Excursion Flora* a mere 10%. This low pricing that obviously kept commercial booksellers away, led to the discovery of a happy marketing procedure as an integral part of the production of the *Flora* itself. This entire methodology is presented as a model, especially for third world botany.

---

[1] The Rapinat Herbarium, St Joseph's College, Tiruchirapalli 620 002, INDIA

THE TECHNICAL PROGRAMME

The choice of the tract for exploration was crucially important: on the one hand, all significant vegetational types should be covered, and on the other, exploring of extensive tracts with monotonously similar vegetational types should be avoided. The latter was, besides, a necessity for a private foundation without State support, and depending entirely on time-bound funding.

A two-stage programme was devised to cover the *plains* and *hills* separately, the sum of which would adequately cover the Peninsula excluding the evergreen of the western side of the Western Ghats (Map 1).

Table 1 sets forth the details of the two parts of the project.

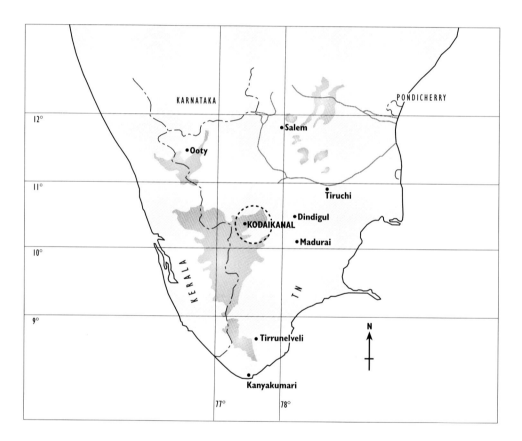

MAP 1. South India showing the area explored
  (a) The Tamilnadu Carnatic;
  (b) The Palni Hills

TABLE 1. The two parts of the work: plains and hills

Area	Duration	Number of		
		Field Days	Collections	Species
Plains (Carnatic) (27,794 sq. km.)	1976 – 83	631	30,722	2,037
Hills (Palnis) (2,068 sq. km.)	1984 – 94	323	14,987	2,478
TOTALS		954	45,709	3,015 (× 4,515 i.e. 1,500 species common)

1. THE TAMILNADU CARNATIC (plains) Map 2

A.1 The tract and field work

The middle sector of Tamilnadu State (27,794 sq. km, one-fifth of the area of the State), with the Ponnaiyar river in the north, the Cauvery-Kollidam rivers in the south and the Bay of Bengal in the east, was chosen: the seven hill ranges of the Pacchaimalais (425 sq. km.), the Kollimalais, with the Bodamalais (490 sq. km.), the Kalrayans (1,095 sq. km.), the Chitteris (640 sq. km.), the Servarayans (470 sq. km.), rising to over 1,600 m at Yercad, the (Tamilnadu) Melagiris (2,910 sq. km.) bordering on the Karnataka State, together with the typical scrub jungles at the foot-hills, the Picchavaram mangroves, the 50 km coastal and dune vegetation between Parangipettai and Cuddalore, the riparian vegetation at the lower reaches of the rivers Cauvery, Kollidam and Ponnaiyar, stationary fresh water vegetation (Veeranam Lake) and a sector of the plains vegetation. *The tract is a phytogeographical unit that qualitatively covers all the vegetational types of the Peninsula east of the Western Ghats.* Further, eight out of the nine tribal belts of the State falling within the tract is ethnobotanically significant.

A.2 Documentation

The Field Trip Report (Matthew 1981: 86 – 101) records the execution of the programme over the years under seven *hill* ranges and five special habitats in the *plains* under the following headings: (a) location; (b) camping facilities; (c) various routes with distances marked; (d) plants of special interest and (e) altitude of localities. All these details are available on two maps, with a detailed gazetteer of localities, especially the smaller ones not found in general purpose maps.

A day-by-day tabulated report of the collections (183 trips, 631 field days, 30,722 field numbers during 1976 – 83) is given in Matthew 1981: 45 – 72, 1983: 1971 – 72 and 1988: 900). The pages on "Field work in execution" (Matthew 1981: 73 – 85) are an original and practical scheme of how the field work was organised. Supplementary collections (evanescent corollas and delicate herbs in paper flimsies,

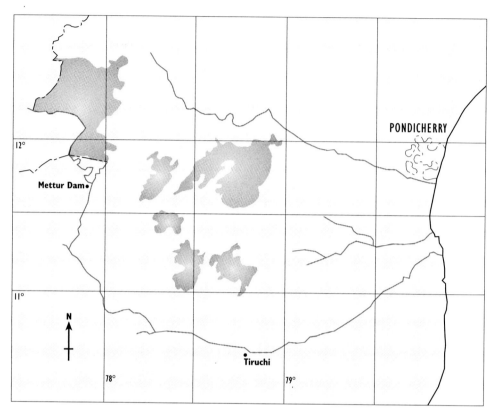

MAP 2. The Tamilnadu Carnatic

spirit collections, timber and bark, transparencies and photographs) received special attention. To have draughtspersons accompanying the field team was extremely useful. The elaborate methodology, despite serious handicaps (especially financial), actually succeeded because the team was highly motivated. The comprehensive specimen citation (Matthew 1981: 135 – 415) provides a modern inventory from the five regional herbaria: RHT 26,166; MH 2,250 (Nair *et al.* (1983–89)); FRC 74; PCM 419 and HIFP 474.

A.3 Illustrations
The objectives were unpretentious: to illustrate one species per genus. The job of the draughtsperson (never a student of botany!), working side by side with the research student, was to sketch the significant details according to a simple methodology (Matthew 1982: vi – vii), so that each plate would contain, besides the habit, about a dozen details, all to scale. We had a pleasant surprise: that in spite of the indifferent quality of the plates, the *Illustrations* (Matthew 1982) sold out faster than the other volumes. In fact our very idea of embarking on *Further Illustrations* (Matthew 1988) to cover all the species was an after-thought consequent upon the success of the previous illustrations.

TABLE 2. Summary of illustrations details in the various volumes

Volume & Year	Number of		Notes
	Plates	Species	
2 (1982)	960	882	
3 (1983)	111	14	(almost 100 plates are support for the keys).
4 (1988)	834	820	
	1,905	1,716	i.e. 84.25% of species illustrated

It is obvious that there is a real demand for illustrations, a sizeable proportion of which comes from persons not using the text of the *Flora*, but active in the field, (foresters, conservationists, etc.) who need to recognise plants without reading about them. Verdcourt's (1992) appeal to illustrate at least those species for Ceylon not covered in our *Flora* wisely underscores this point.

A.4 The Flora

A.4(a) Analysis of materials

It was axiomatic that research students make a *new* analysis of materials, after Dr W.B. Turrill's prescription that the best way to teach systematic botany was to assemble the collections into a room, remove all the books and lock the room with the student inside until (s)he put some order into the collections through a key, etc.

Training in studying any material (fresh, pickled or boiled) involving the writing of comparable descriptions (Leenhouts 1968), extracting common features into the higher categories, noting down diagnostic features for keys, guiding the draughtspersons into making sketches of significant structures to scale, etc. were various facets of this original exercise. And with good students, it was a voyage of discovery. Guiding the students from simpler to more complex groups, it was like symbiotic growth: that the guide was constantly working with the students was decisively important.

A.4(b) Taxonomy

As to the concept of *genus* and *species*: whereas onlookers were interested in how *many* new taxa, our philosophy was how *few* of them. A regional *Flora* should accept the commonly accepted limits in the light of good recent research, and look for novelty in the amount of *new information* to be put into the various taxa. Species descriptions were prepared *de novo* from our own materials from which common features were extracted out for genera and families.

It was obvious that assistance from experts was essential for achieving any qualitative improvement over existing *Floras*. I spent a year in overseas centres, especially at Kew, in four stints of three months each in tackling groups that could not be done satisfactorily in India, consulting experts and also scanning the literature. I gratefully acknowledge the assistance of some 50 experts (Matthew 1983: xxxv – xxxvii) who generously assisted.

A.4(c) Nomenclature and synonymy

Having arrived at a modern and balanced taxonomy of the groups with the assistance of experts, the nomenclature with the necessary synonymy (accounting for the synonyms in circulation in the region) is essential in a regional *Flora*. As for synonymy, we methodically scanned the following previous *Floras*: Roxburgh (1820, 1824 & 1832), Wight (1840 – 50, 1840 – 53), Wight & Arnott (1834), Beddome (1869 – 74, 1874), Hook.f. *et al.* (1872 – 97), Gamble & Fischer (1915 – 36) and Fyson (1932). The *local name* in the local script was given following the nomenclature (Matthew 1983: xxv-xxvii).

A.4(d) Keys

Two family keys (one based mainly on floral characters, the other on fruit characters) were prepared. In general, three sets of characters were looked for in each lead: vegetative, floral and fruiting. Qualitative characters were preferred to quantitative ones. We were often able to correlate breaks in continuity of characters with altitude of occurrence (in keys) because we had adequate field knowledge. The concept of *"replacement taxa"* revolved round this. Alternate keys were given when useful. (Matthew 1983: 599 for the key on *Memecylon*).

A.4(e) Field Notes

Table 3 (Matthew 1983: xxix) gives the format used to prepare this section, strictly comparable for all the included taxa. *Para 1:* Occurrence: gives the habitat, range of altitude of occurrence in the region, common, abundant or rare, concluding with a note on conservation status when called for. *Para 2:* Phenological information (i.e. centred around flowering time). This section is an important, original contribution.

TABLE 3. Format for compiling Field Notes

I: OCCURRENCE

(a)   Plains : Habitat .... Moisture availability
                Vegetational type .... Associations
                Primary ....
       Coastal : Mangrove, dune ....
(b)   Hills   : Alt (m) 200-400-600-800-1000-1200-1400
                        Abundance – Conservation status

II: NOTES

(a)   Habit + description
(b)   Size :   normal or below or above
(c)   Phenology :   leaf-fall-renewal
                Flower (diam); Month 1-2-3-4-5-6-7-8-9-10-11-12
                Fruit : measurements, colour changes.

A.4(f) (World) Distribution

Distributions were taken from recent monographs when available. Hara *et al.* (1979 – 82) was found useful. For species of restricted distribution, especially to the Peninsula, with little worthwhile recent research, Hooker *et al.* (1872 – 97) or Gamble & Fischer (1915 – 36) were the basis.

A.4(g) Phytogeographic conclusions

True, phytogeographic conclusions for small areas have only a limited validity, although findings from the Indian peninsula, terminating an entire subcontinent, will certainly have some interest. However, absence of good field work on the one hand, and the swarms of recently proposed new taxa not yet monographically tested on the other, have not made things easy. With these provisos, the following points are significant: (i) that *c.* 50% (486/990) of the indigenous genera were monotypic for the region (the figure is reduced to 33% when the Palni hills are included as was foreseen); (ii) Endemic pockets: *Hildegardia, Vanilla, Suregada angustifolia, Pamburus (Atalantia) missionis, Walsura trifolia, Derris ovalifolia,* as if the last sentinels of a bygone age, seem to occur in isolated pockets; (iii) Wild relatives of cultivated plants: *Mangifera indica* (Matthew 1983: 308-09) and *Artocarpus heterophyllus* (Matthew 1983: 1509) truly occur in the wild state in the area; (iv) our project can be considered a study of micro-evolutionary processes, and is a case-study of what is happening over larger tracts. As such, it has significance, although IUCN categories are better not applied to such small areas.

2. THE PALNI HILLS

B.1 Tract and Field Work

The Palni (Pulney) Hills, an eastward spur of the Western Ghats (Map 3), (area 2,068 sq. km.) were explored as the *montane* counterpart of the *plains* of Part I. As a hill station, the area had received early attention from the colonial days because not only was the vegetation rich but the location provided escape from the torrid plains (Wight (1837, 1846 – 51), Beddome (1858), Bourne (1897), Saulière (1914) and Fyson (1932)). My own association began in 1950 that resulted in The Exotic Flora (Matthew 1969). The fact that some of the early stalwarts were still alive then, or at least had reliable information from the early settlers, enhanced the historical value of the work.

The Nilgiris, though with a richer vegetation owing to the location, and better known, suffered excessive eco-degradation from mindless commercialisation and tourist invasion. The transfer of power in 1947 had hardly any natural history or environmental component, so that depletion of the primary vegetation (grasslands converted into wattle plantations and Sholas for all round exploitation) reached a point of no return within decades. The Palni Hills, on the other hand, happily escaped official attention (no rail road, and only one motor road until the 1970s), blessed with a tradition of natural history and an active environmental group like the Palni Hills Conservation Council, maintained better environmental sensitivity and implementation of eco-restoration programmes, so that today the Palni Hills

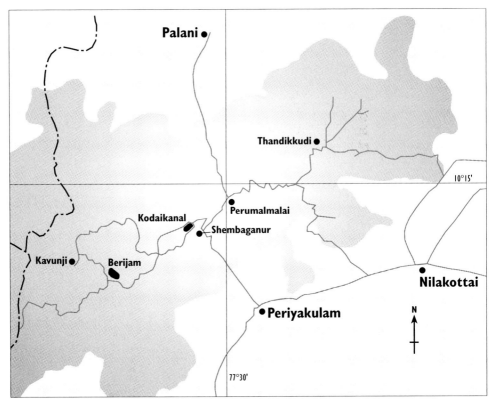

MAP 3. The Palni Hills

are not only better protected, but have a comprehensive database on natural history and environment (Matthew 1994).

Over and above the botanical exploration of the last century already referred to, the Sacred Heart College shaped up as a centre of natural history, so that on my arrival there in 1950, the foundations for work were already laid. It is obvious that a *Flora of the Palni Hills*, revising Fyson (1932) is a fitting montane counterpart of the *Flora of the Tamilnadu Carnatic*.

B.2 Documentation

The methodology (field work, naming of collections, illustrations, etc.) was identical with PART I. In addition to the five South Indian herbaria noted there, SHC, CAL and K were consulted especially in view of a more comprehensive documentation. In Kew itself, I spent three months screening the Palni Hills materials, especially of the Bournes. The comprehensive species citation will be incorporated into the text.

A total of 2,478 species will appear in the *Flora* of which 60% had appeared in the plains *Flora*. Hence there will be a number of devices to avoid repetition of details, without omitting any worthwhile information.

B.3 Illustrations

Another 1,150 species (none from the previous *Flora*) will be illustrated in identical format. 949 plates will appear in 1996 in the *Illustrations* volume and another 200 as supplement along with the text of the *Flora* in 1997.

B.4 The Flora

Nomenclature: The genus name is given as in Farr *et al. Index nominum Genericorum*, along with reference to any recent revision/monograph as well as four important *Floras* around the region: *Flora Malesiana, Revised Handbook for the Flora of Ceylon, Flora of Tropical East Africa* and *Flora of (West) Pakistan*; those of India and Bangladesh are yet in the early stages.

As for the species, those which had appeared in the Tamilnadu Carnatic, will have only the correct name, a reference to the *Flora* and any later reference(s); for species that did not appear in the *Flora*, the complete conventional coverage will be given.

The vernacular (Tamil) name, collected, or at least verified in the field, will be given for all the included taxa.

Keys will cover all the 2,478 taxa of the region including all those that appeared in the Flora.

Description: Family and genus descriptions will cover only the included taxa. When a family has only one genus, no family description will be given. When a genus has only one or two species, the genus description itself is avoided while giving descriptions for the included species. This practical exigency keeps the ordinary user in mind.

Notes will cover all the 2,478 species.

Distribution is given only for the species not mentioned in *Flora*. For the aliens, reference is given to *New Royal Horticultural Society Dictionary of Gardening* a device that ensures that this new publication is uniformly referred to.

TABLE 4. (Matthew 1994: 106 – 110) indicates the number of taxa of the forthcoming *Flora*

Families	Genera	Species				
	(monotypic in brackets)	Native	Naturalised	Cultivated	Garden	Total
202	1,137 (377)	1,758	161	344	215	2,478

CONCLUSION

I am happy to report on the successful completion of a 20-year, 2-stage, project for a regional illustrated *Flora*, rounded off with a Field *(Excursion) Flora*, initially in English and later in the vernacular to ensure the widest outreach possible. It can honestly be claimed that the region covered by the *Flora*, thanks to exhaustive field work of some 1,000 days, with an accumulated new collection of 50,000 field numbers widely distributed, is botanically better known than almost any other comparable region in India. And I do hope our methodology and standards will be a model for other third world countries (Radcliffe-Smith 1983 – 95).

(a) The fact that the *Carnatic Flora*, with the *Excursion Floras*, was produced by a private foundation without State support, at an astonishingly low outlay (Rs. 25,30,000 or £49,000 or $77,000; as for the Palni Hills *Flora*, an amount of Rs. 3,03,760 or £6,100 or $9,500 has been promised for preparing the manuscript for the press, while the printing costs of Rs. 9,05,000 or £18,200 or $ 29,000 have yet to be realised) has vital consequences for third world countries in demonstrating the wide chasm between what is possible with indigenous personnel and home funds (when leadership and motivation are present), and what generally obtains in the 'white elephant culture' (Matthew 1992: v-vi);

(b) the building up of an active research base with the requisite infrastructure (collections, library with international exchanges, relevant research programmes) has great possibilities for the future (Matthew 1992);

(c) our marketing, as an integral part of our research productions, ensures both that the titles will be available even in the distant future, and at affordable prices;

(d) the real worth of our efforts is being gradually realised even by the State Government that had not contributed a penny in all these 20 years. The new Gazetteer of the Tiruchirapalli District (appearing after a lapse of 90 years) will have an Appendix listing all the 1,300 species of plants of the district, each with the correct botanical name, local name, uses, phenology and conservation status, all based on our *Flora* and the collections;

(e) our recommendations have been sought in establishing the Palni Hills National Park and related conservation measures.

However, in spite of such tangible results, I am conscious that only a beginning has been made. On the one hand the leaders in the tropics, and educators and researchers in particular, should have sound ideas on their genuine biological and environmental research priorities. This is not easy as long as local personnel have not evolved a truly indigenous thinking but rather follow the initiatives of the advanced countries with which they are more familiar, and are advisers to their Governments in policy matters. And within these countries, the newly emerging middle class, by and large, would opt for the easy option at home, and even migrate into greener pastures overseas. On the other hand, the vast majority of ordinary people have never been involved in decision-making and need to be given a new vision of development in which they have a key role. And especially in today's context of the fast-depleting green cover of the planet, largely located in the tropics, the local people need to know their plants. Such has been our goal. We did the job often stoically in the sober realisation that *development* is a slow process, to assist in which is the best service an enlightened and trained citizen can do for the people, even though those in charge now may have little comprehension of what we had been trying to achieve.

REFERENCES

Beddome, R. H. (1858). Flora of the Pulney Hills as observed in September and October, 1857. Madras J. Litt. Sci. (N.S.) 3(5): 169 – 202.

—— (1869 – 74). The Flora Sylvatica of Southern India. Madras. 2 Vols.

—— (1874). Icones Plantarum Indiae Orientalis. Madras.

Bourne, A.G. (1897). List of species of plants which are supposed to be indigenous to, or denizens of, Southern India. Madras.

Fyson, P. F. (1932). The Flora of the South Indian Hill Stations. Madras.

Gamble, J. S. & Fischer, C. E. C. (1915 – 36). Flora of the Presidency of Madras. 3 Vols. London.

Hara, et al. (1978 – 82). An Enumeration of the Flowering Plants of Nepal. 3 Vols. London.

Hooker, J. D. et al. (1872 – 97). The Flora of British India. 7 Vols. London.

Leenhouts, P. W. (1968). A guide to the practice of herbarium taxonomy. Utrecht.

Matthew, K. M. (1969). The Exotic Flora of Kodaikanal. Rec. Bot. Surv. India 20: 1 – 241.

—— (1981). Materials for a Flora of the Tamilnadu Carnatic. Tiruchirapalli.

—— (1982). Illustrations on the Flora of the Tamilnadu Carnatic. Tiruchirapalli.

—— (1983). The Flora of the Tamilnadu Carnatic (3 parts). Tiruchirapalli.

—— (1988). Further Illustrations on the Flora of the Tamilnadu Carnatic. Tiruchirapalli.

—— (1991, 1995). An Excursion Flora of Central Tamilnadu, India. New Delhi.

—— (1992). A Manual for RHT and SHC Herbaria, India. Tiruchirapalli.

—— (1993). Tamil translation of the Excursion Flora by S. John Britto, Tiruchirapalli.

—— (1994). A Handbook of the Anglade Institute of Natural History, Shembaganur (Ed. 2). Tiruchirapalli.

Nair, N. C. et al. (1983 – 89). Flora of Tamil Nadu. 3 Vols. Coimbatore.

Radcliffe-Smith, A. (Book Reviews). in Kew Bull. 37: 699 – 701 (1983); 39: 848 (1984); 41: 474 – 476 (1986); 44: 752 (1989); 48: 627 (1993) & 50: 191 (1995).

Roxburgh, W. (1820, 1824, 1832). Flora Indica. 3 Vols. Calcutta.

Saulière, A. (1914). Plants from the Pulney Hills. Tiruchirapalli.

Verdcourt, B. (Book Review). in Kew Bull. 47: 783 – 784(1992).

Wight, R. (1837). Statistical Observations on the Vurragherries or Pulney Mountains. Madras J. Litt. Sci. (N.S.) 5(15): 280 – 289.

—— (1840 – 50). Illustrations of Indian Botany. 2 Vols. Madras.

—— (1840 – 53). Icones Plantarum Indiae Orientalis. 6 Vols. Madras.

—— (1846 – 51). Spicilegium Neilgherrense. 2 Vols. Madras.

—— & Walker Arnott, G. A. (1834). Prodromus Florae Peninsulae Indiae Orientalis. London

The Canopy Biology Programme in Sarawak

Hidetoshi Nagamasu[1], Tamiji Inoue[2] & Abang Abdul Hamid[3]

Summary. The Canopy Biology Programme in Sarawak (CBPS) was initiated as one of two programmes of an international co-operative project entitled the Long-term Forest Ecology Research Project at Lambir Hills National Park, Sarawak by the Forest Department Sarawak, Harvard University (U.S.A.), and several Japanese universities in 1992. The CBPS aims to clarify: 1) how unstable environmental changes at the global level influence phenology and reproductive systems, from flowering to seed dispersal, of component plants in mixed dipterocarp forests and 2) how life history strategies and population dynamics of animals that build mutualistic relationships with plants (pollinators, seed dispersers and ant-mutualists) and that feed on materials produced by plants (phytophagous animals) are affected by the environmental changes directly or indirectly through plant phenology.

In lowland mixed dipterocarp forest in Lambir Hills National Park, the Canopy Biology Plot of 8 ha was established. Two tree towers (33 m and 55 m in height) connected by nine aerial walkways (300 m in total length) were constructed to provide easy access to the forest canopy. To analyse changes of climate in tropical rain forest and its influence on plants and animals, the vertical structure of microclimate in the forest is being continuously monitored. The phenology of plants and seasonal dynamics of animals is also being observed. The phenological census of about 600 individually identified plants along walkways and towers, and the seasonal change of insect abundance was started in August 1992.

Introduction

The tropical rain forest is characterised by the richest biodiversity on earth. Although it covers only three to seven percent of whole land area, more than half the species of plants and animals live there. There are, however, few animals on the dark forest floor and dense foliage cover is far above the ground. It is not surprising, therefore, that a series of recent works have revealed that the canopy of tropical rain forest is the centre of most plant productivity and reproduction (Sutton *et al.* 1983; Whitmore 1984) and animal abundance (Erwin 1983, 1988; Rees 1983; Stork 1987, 1988) because of the rich distribution of food and shelter resources.

The canopy of tropical rain forest is, however, the least investigated ecosystem. Even such fundamental information on the ecosystem as leaf flushing, flowering and fruiting phenology of plants, seasonal population dynamics of animal species, and interactions or mutualistic relationships between plants and animals that would contribute to the biodiversity of tropical rain forests in the canopy, is little known due to the technical difficulties of canopy access and continued observations for a long period.

[1] Department of Natural Environment Sciences, Faculty of Integrated Human Studies, Kyoto University, Kyoto 606-01, JAPAN.
[2] Centre for Ecological Research, Kyoto University, Shimosakamoto 4, Otsu 520, JAPAN.
[3] Forest Research Branch, Forest Department Sarawak, 93660 Kuching, Sarawak, MALAYSIA.

In the last two decades, several methods have been developed to access the canopy 40 to 70 m above the ground, for example ropes and ascenders (Mitchell 1982; Perry 1978, 1984; Dial & Tobin 1994), a raft and airship (Hallé & Blanc 1990; Hallé & Pascal 1992), a crane (Joyce 1991; Illueca & Smith 1993), and towers and walkways (Mitchell 1982). Each has advantages and disadvantages; a system with towers and walkways is most suitable for long-term observation covering a wide area in West Malesian rain forests. Rafts have advantages in mobility, but have difficulties in landing on the irregular surface of the canopy created by emergent trees in West Malesia and in continuing observation over a long period. A crane cannot extend beyond its boom length. Both rafts and cranes need specialists for manipulating them through the observation period.

In 1991, we started a long-term ecological research project on the canopy biology of tropical rain forests in Sarawak (Canopy Biology Programme in Sarawak (CBPS)), using an observation system with towers connected by aerial walkways. This is one of two components of an international co-operative project entitled the Long-term forest Ecology Research Project at Lambir Hills National Park, Sarawak by the Forest Department Sarawak, Harvard University (U.S.A.), and several Japanese universities including Ehime University, Osaka City University and Kyoto University.

Mixed dipterocarp forest in West Malesia is one of the richest tropical rain forest areas in the world with a diversity of tree species (Richards 1952; Whitmore 1984). This area is believed to be under a rather stable, warm, humid climatic regime throughout a year. Recently, however, it has been found that global environmental changes such as the El Niño Southern Oscillation (ENSO) have a strong influence on tropical rain forest in Southeast Asia. It is also suggested that the drought caused by ENSO triggers general flowering which characterises the mixed dipterocarp forest of West Malesia (Ashton *et al.* 1988; Ashton 1993; Appanah 1993). The CBPS aims to clarify: 1) how unstable environmental changes at the global level influence phenology and reproductive systems, from flowering to seed dispersal, of component plants in the mixed dipterocarp forests, and 2) how life history strategies and population dynamics of animals, that build mutualistic relationships with plants (pollinators, seed dispersers and ant-mutualists) and that feed on materials produced by plants (phytophagous animals), are affected by the environmental changes directly or indirectly through plant phenology.

STUDY SITE

Lambir Hills National Park (4°20'N, 113°50'E, 60 m in altitude at the headquarters) is located about 30 km south of Miri, the capital of Miri Division, Sarawak (Figs. 1, 2). The park covers 6,949 ha (Figs 1,2). The highest peak, Bukit Lambir, is 465 m in altitude. The forest consists of two types of original vegetation common in Borneo, i. e. mixed dipterocarp forest and tropical heath forest. The former occurs on sites rich in clay at rather lower elevations, covering 85% of the total Park area. Tropical heath forest (kerangas) is restricted around Bukit Lambir covering the remaining 15% of the Park (Yamakura *et al.* 1995). Annual precipitation based on measurements from 1992 to 1993 near the Park Headquarters is 4940 mm, and is about twice of that at Miri Airport (Momose *et al.* 1994a).

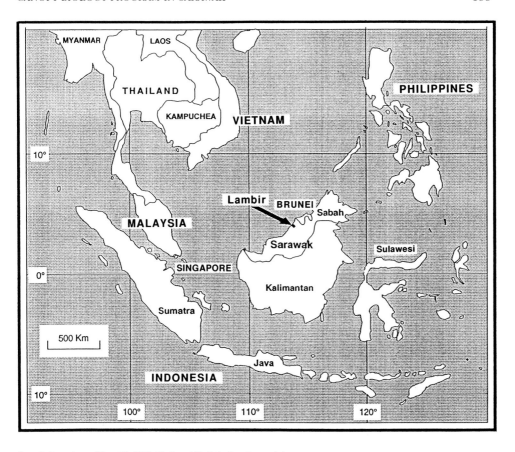

FIG. 1. Location of Lambir Hills National Park in Southeast Asia.

CANOPY OBSERVATION SYSTEM

The canopy biology plot (200 × 400 m, 8 ha) was established on a clay-rich soil near Park Headquarters. The plot is covered by intact mixed dipterocarp forest (Fig. 3). All trees >50 cm dbh in the plot and >10 cm dbh in a 2 ha inner plot, along walkways and around tree towers, were mapped and given an identification number.

We constructed two tree towers using the termite resistant timber, made with *belian* or Borneo Ironwood (*Eusideroxylon zwageri, Lauraceae*). Tree Tower 1 (T1) is built around an emergent tree of *Dryobalanops lanceolata* (*Dipterocarpaceae*) 72 m tall and 1.52 m diameter above the buttresses. The construction started in July 1992 and was finished in September. It has 11 wooden square platforms (4.88 m × 4.88 m) connected by wooden stairs, and observers can spirally step up around the trunk. The top platform is 33 m above the ground, above which we placed two wooden terraces on branches 45 m and 55 m above ground. These terraces are connected by aluminium ladders.

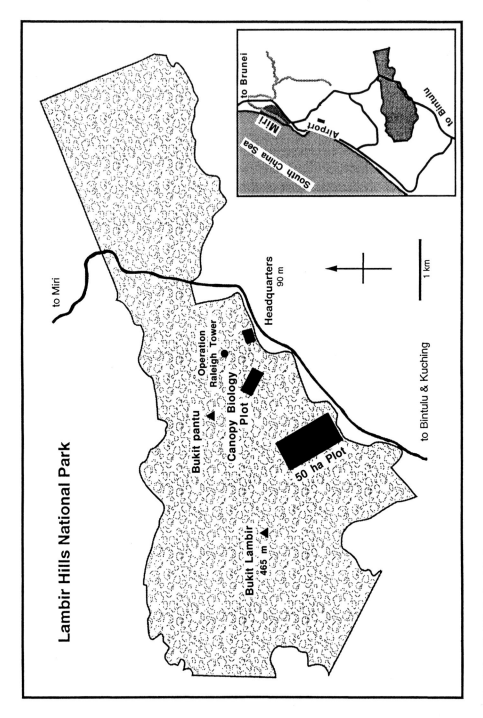

Fig. 2. Lambir Hills National Park.

Tree Tower 2 (T2) is constructed on one side of a canopy tree of *Dipterocarpus pachyphyllus* (*Dipterocarpaceae*) 48 m tall and 1.36 m dbh. The construction started in January 1993 and was finished three months later. T2 has 14 platforms connected by aluminium ladders, the top one being 48.6 m above the ground. The platform size (2.44 m × 2.44 m) is much smaller than that of T1, but the top platform is 16 m higher and exceeds the crown of the tree.

The two tree towers were connected by nine aerial walkways of suspending bridges with metal wires (16 mm diameter, 2000 kg test), which pass through the forest canopy from between 15 and 35 m above the ground. Eight emergent or canopy trees with large boles are used as piers. The length of each walkway span ranges from 25 to 54 m, depending on the distribution of pier trees. The total length of the nine walkways is about 300 m. Observers walk on wooden boards on aluminium ladders suspended by wire (diameter 3.5 mm). The construction of this walkway system started in September 1992 and was finished in April 1993.

MICROCLIMATE OF CANOPY

To obtain detailed microclimatic data in the canopy layer, we set automatic meteorological sensors (Yokogawa Weather Cooporation) for rainfall, temperature, humidity and solar radiation on the top platform of T1, 35 m above the ground. Climatic data are automatically recorded in a data logger every ten minutes. We started the census in May 1993. A similar microclimatic census at an open place at the Park Headquarters, about 500 m from T1, has been carried out from January 1992.

PLANT PHENOLOGY

The plant phenology census is an important component of the programme. The census is planned to reveal the following points:

a) *Influence of the unstable environmental changes on plant phenology.* The pattern of fluctuation of reproductive effort for each plant may directly reflect the influence of unstable environmental changes such as ENSO.

b) *Inter- and intra-specific relationships of plant in phenological behaviour.* Phenological behaviour is a part of plant life history strategy. We plan to categorise plants by their phenological behaviour. It is also important whether phenology is synchronised or not among conspecific plant individuals.

c) *Interactions with animals in mutualistic and antagonistic relationship.* The relationships between plant phenology and seasonal changes of animals such as predators, pollinators and dispersers are being examined.

We started phenological observations of about 600 individually identified plants in the canopy observation system in August 1992, just after a general flowering in March 1992. Both vegetative (leaves) and reproductive (flowers and fruits) phenology are being observed. We quantify the amount of leaves, flowers or fruits to estimate the condition of flushing, flowering and fruiting. If an individual has reproductive organs, a specimen is collected for identification and for improvement of the plant inventory (Momose *et al.* 1994b, 1994c; Yumoto *et al.* 1995).

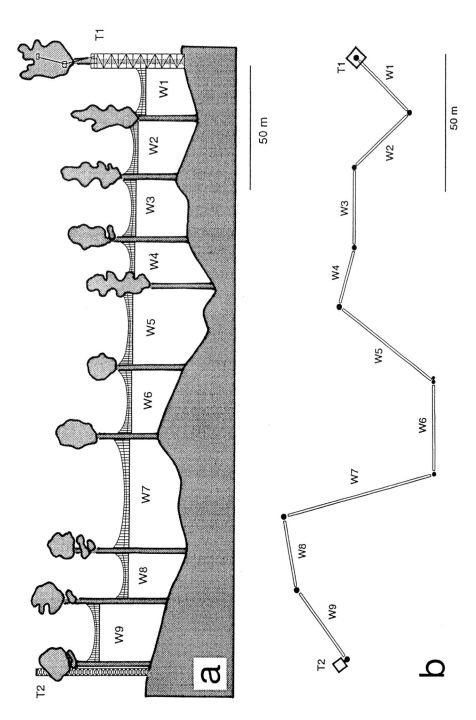

FIG. 3. a) Profile of the canopy observation system. b) Top view of nine spans of walkways. T: tree tower; W: walkway.

SEASONAL DYNAMICS OF INSECTS

The census of seasonal dynamics of animals is the counterpart of that for plant phenology. We started monthly monitoring of seasonal changes in insect populations in August 1992, using light traps, fruit-fly traps, Malaise traps, yellow-pan traps, Toda's banana traps, ant traps, fragrance traps, honey-water bait and mimosa sweeping. Only light traps and fruit-fly traps remain in operation. The others were only operated in the first census year.

UV light traps were operated monthly on T1 at various heights from 0 to 45 m above the ground. For 13 months, about a million insects were collected in total. On the basis of these collections, Kato *et al.* (1994, 1995) analysed the seasonality and vertical structure of light-attracted insect communities. The mean number of collected insects was highest at the canopy level and gradually decreased downwards. The most predominant nocturnal insects in the forest canopy were fig wasps, representing 84% of collections. The vertical distribution showed different patterns depending on insect groups.

FLORISTIC AND FAUNAL INVENTORY

Plant materials collected from the canopy biology plot and surrounding areas were identified and a preliminary inventory was presented (Nagamasu *et al.* 1994). General collections continue and more than 1000 specimens a year are collected. We established a herbarium in the laboratory near Park Headquarters in 1993. The identified specimens are sorted and arranged in alphabetical order. The herbarium is open to any researchers at the Lambir.

Most of the approximately one million insect collections were pinned and sorted to at least order level. A few groups such as ants (257 spp, Yamane & Nona 1994) and stingless bees (27 spp and forms, Inoue *et al.* 1994) were identified to species, but the bulk of the collection awaits specialists.

POLLINATION ECOLOGY

In tropical rain forests most plants are pollinated by animals, and wind-pollinated plants are rare (Bawa *et al.* 1985). Thus, tropical angiosperms need mutualistic interactions for their fertilisation process. In CBPS, pollination syndromes and the pollination ecology of various taxonomic groups in different tree layers were studied.

Momose & Inoue (1994) revealed that there are five pollination syndromes in tropical rain forests in Lambir, of which main visitors are beetles, medium-sized bees, small diverse insects, *Megachile* (*Megachilidae*) and *Amegila* (*Anthophoridae*). Kato (1994, 1996) studied the pollination syndrome at the forest floor in detail, and showed that pollination systems there are distinguished from those in the higher layers by the prevalence of specific interactions.

Several pollination systems have been studied in detail. These include the pollination of an emergent tree, *Dryobalanops lanceolata* by stingless bees (Momose *et al.* 1994d), epiphytic parasites of *Loranthaceae* by spider-hunters (Yumoto *et al.* 1997), several epiphytic and terrestrial orchid species by stingless bees (Inoue 1994), small sized trees in understory *Popowia* spp. (*Annonaceae*) by thrips (Momose, in prep.), an unusual gymnosperm *Gnetum gnemon* in the understory by nocturnal moths (Kato & Inoue 1994), and a geoanthic tree *Uvaria* sp. (*Annonaceae*) flowering on the ground by cockroaches (Nagamitsu & Inoue 1997).

REFERENCES

Appanah, H. (1993). Mass flowering of dipterocarp forests in the aseasonal tropics. J. Bioscience 18: 457 – 474.

Ashton P. S., Givnish, T. J., & Appanah, S. (1988). Staggered flowering in the *Dipterocarpaceae*: new insights in to floral induction and the evolution of mast fruiting in the aseasonal tropics. Am. Nat. 132: 44 – 46.

Ashton P. S. (1993). The community ecology of Asian rain forests, in relation to catastrophic events. J. Bioscience 18: 501 – 514.

Bawa, K. S, Bullock, S. H., Perry, D. R., Coville, R. E. & Grayum, M. H. (1985). Reproductive biology of tropical rain forest trees. II. Pollination systems. Am. J. Bot. 72: 346 – 356.

Dial, R. & Tobin, S. C. (1994). Description of arborist methods for forest canopy access and movement. Selbyana 15(2): 24 – 37.

Erwin, T. L. (1983). Beetles and other insects of tropical forest canopies at Manaus, Brazil, sampled by insecticidal fogging. In: S. L. Sutton, T. C. Whitmore & A. C. Chadwick (eds), Tropical rain forest: Ecology and management. pp. 59 – 75. Blackwell Scientific Publications, Oxford.

Erwin, T. L. (1988). In tropical forest canopy — The heart of biotic diversity. In: E. O. Wilson & F. M. Peter (eds), Biodiversity. pp. 123 – 129. National Academy Press, Washington, D.C.

Illueca, J. E. & Smith, A. P. (1993). Exploring the upper tropical forest canopy. Our Planet 5: 12 – 13.

Inoue, T., Nagamitsu, T., Momose, K., Sakagami, S. F. & Hamid, A. A. (1994). Stingless bees in Sarawak. In: T. Inoue, & A. A. Hamid (eds). Plant reproductive systems and animal seasonal dynamics — long term study of dipterocarp forests in Sarawak. pp. 231 – 236. Center for Ecological Research, Kyoto University, Otsu.

Joyce, C. (1991). A crane's eye view of tropical forests. New Scientist 21: 30 – 31.

Kato, M. (1994). Plant-pollinator interactions at the forest floor of a lowland mixed dipterocarp forest in Sarawak. In: T. Inoue & A. A. Hamid (eds), Plant reproductive systems and animal seasonal dynamics — Long-term study of dipterocarp forests in Sarawak. pp. 150 – 157. Center for Ecological Research, Kyoto University, Otsu.

—— (1996). Plant-pollinator interactions in the understorey of a lowland mixed dipterocarp forest in Sarawak. Am. J. Bot. 83: 732 – 743.

Kato, M. & Inoue, T. (1994). Origin of insect pollination. Nature 368: 195.

Kato, M., Inoue, T., Hamid, A. A., Itino, T., Merdek, M. B., Nona, A. R., Nagamitsu, T., Yamane, S. & Yumoto, T. (1994). In: T. Inoue & A. A. Hamid (eds), Plant reproductive systems and animal seasonal dynamics — Long-term study of dipterocarp forests in Sarawak. pp. 199 – 221. Center for Ecological Research, Kyoto University, Otsu.

Kato, M., Inoue, T., Nagamitsu, T., Hamid, A. A., Merdek, M. B., Nona, A. R., Itino, T., Yamane, S. & Yumoto, T. (1995). Seasonality and vertical structure of light-attracted insect communities in dipterocarp forest in Sarawak. In: L. H. Seng, P. S. Ashton & K. Ogino (eds), Long-term ecological research of tropical rain forest in Sarawak. pp. 132 – 141. Ehime University, Matsuyama.

Mitchell, A. W. (1982). Reaching the rain forest roof — A handbook on technique of access and study in the canopy. UNEP.

Momose, K. & Inoue, T. (1994). Pollination syndrome in the plant-pollinator community in the lowland mixed dipterocarp forests of Sarawak. In: T. Inoue & A. A. Hamid (eds), Plant reproductive systems and animal seasonal dynamics — Long-term study of dipterocarp forests in Sarawak. pp. 119 – 141. Center for Ecological Research, Kyoto University, Otsu.

Momose, K., Nagamitsu, T., Sakai, S., Inoue, T. & Hamid, A. A. (1994a). Climate data in Lambir Hills National Park and Miri Airport, Sarawak. In: T. Inoue & A. A. Hamid (eds), Plant reproductive systems and animal seasonal dynamics — Long-term study of dipterocarp forests in Sarawak. pp. 26 – 39. Center for Ecological Research, Kyoto University, Otsu.

Momose, K., Nagamitsu, T., Yumoto, T., Nagamasu, H., Inoue, T., Itino, T., Kato, M., Kohyama, T. & Hamid, A. A. (1994b). Manual for monitoring plant phenology in Lambir Hills National Park, Sarawak. In: T. Inoue &. A. A. Hamid (eds), Plant reproductive systems and animal seasonal dynamics — Long-term study of dipterocarp forests in Sarawak. pp. 57 – 105. Center for Ecological Research, Kyoto University, Otsu.

Momose, K., Nagamitsu, T., Yumoto, T., Nagamasu, H., Inoue, T., Itino, T., Kato, M., Kohyama, T. & Hamid, A. A. (1994c). Plant phenology in a non-general flowering period of 17 months from August 1992, in Lambir Hills National Park, Sarawak. In: T. Inoue & A. A. Hamid (eds), Plant reproductive systems and animal seasonal dynamics — Long-term study of dipterocarp forests in Sarawak. pp. 106 – 110. Center for Ecological Research, Kyoto University, Otsu.

Momose, K., Nagamitsu, T. & Inoue, T. (1994d). Reproductive ecology of an emergent tree, *Dryobalanops lanceolata, Dipterocarpaceae*, in a non-general flowering period in Sarawak. In: T. Inoue & A. A. Hamid (eds), Plant reproductive systems and animal seasonal dynamics — Long-term study of dipterocarp forests in Sarawak. pp. 158 – 162. Center for Ecological Research, Kyoto University, Otsu.

Nagamasu, H., Momose, K. & Nagamitsu, T. (1994). Flora of the Canopy Biology Plot and surrounding areas in Lambir Hills National Park, Sarawak. In: T. Inoue & A. A. Hamid (eds), Plant reproductive systems and animal seasonal dynamics — Long-term study of dipterocarp forests in Sarawak. pp. 47 – 56. Center for Ecological Research, Kyoto University, Otsu.

Nagamitsu, T. & Inoue, T. (1997). cockroach pollination and breeding system of *Uvaria elmeri* (*Annonaceae*) in a lowland mixed-dipterocarp forest in Sarawak. Am. J. Bot. 84: 208 – 213.

Perry, D. R. (1978). A method of access into the crowns of emergent and canopy trees. Biotropica 6: 155 – 157.

Perry, D. R. (1984). The canopy of the tropical rain forest. Sci. Am. 251: 114 – 122.

Rees, C. J. C. (1983). Microclimate and the flying Hemiptera fauna of a primary lowland rain forest in Sulawesi. In: S. L. Sutton, T. C. Whitmore & A. C. Chadwick (eds), Tropical rain forest: Ecology and management. pp. 121–136. Blackwell Scientific Publication, Oxford.

Richards, P. W. (1952). The tropical rain forest. Cambridge University Press, Cambridge.

Stork, N. E. (1987). Canopy fogging, a method of collecting living insects for investigations of life history strategies. J. Natur. Hist. 21: 563–566.

Stork, N. E. (1988). Insect diversity: facts, fiction and speculation. Biol. J. Linn. Soc. 35: 321–337.

Sutton, S. L., Whitmore, T. C. & Chadwick, A. C. (1983). Tropical rain forest: Ecology and management. pp. 121–136. Blackwell Scientific Publication, Oxford.

Whitmore, T. C. (1984). Tropical rain forests of the Far East (2nd ed.). Clarendon, Oxford.

Yamakura, T., Kanzaki, M., Itoh, A., Ohkubo, T., Ogino, K., Chai, E. O. K., Lee, H. S. & Ashton, P. S. (1995). Forest architecture of Lambir rain forest revealed by a large-scale research plot. I. Topography of the plot. In: L. H. Seng, P. S. Ashton & K. Ogino (eds), Long term ecological research of tropical rain forest in Sarawak. pp. 2–20. Ehime University, Matsuyama.

Yamane, S. & Nona, A. R. (1994). Ants from Lambir Hills National Park, Sarawak. In: T. Inoue & A. A. Hamid (eds), Plant reproductive systems and animal seasonal dynamics — Long-term study of dipterocarp forests in Sarawak. pp. 222–226. Center for Ecological Research, Kyoto University, Otsu.

Yumoto, T., Itino, T. & Nagamasu, H. (1997). Pollination on hemiparasites (*Loranthaceae*) by spider hunters (*Nectariniidae*) in a Southeast Asian tropical rainforest canopy. Selbyana 18 (in press).

Solving problems in the taxonomy of *Aglaia* (*Meliaceae*): functional syndromes and the biological monograph

C. M. Pannell[1]

Summary. This paper explores the value of ecologically-based field studies for improving our understanding of complex taxonomic problems. It also proposes that more use be made of the concept of functional syndromes which is derived from such a combination, because of its contribution to understanding evolution, especially coevolution between plants and animals. A successful programme of conservation in many tropical ecosystems depends upon an understanding of plant – animal interactions so that the conservation of viable populations of interdependent organisms can be achieved. All of this is illustrated by the detailed study of the functional syndromes relating to dispersal in *Aglaia*. The genus *Aglaia* is widespread and common in forests of the Malesian region and, with 106 species, is the largest and taxonomically most intractable genus in the Meliaceae. Field studies in west Malesia revealed suites of characters in the infructescences, fruits and seeds, which are associated with the morphology, behaviour and dietary requirements of the dispersing animals. The main dispersers in west Malesia are primates, especially apes, and arboreal birds. Recent field work has confirmed that cassowaries disperse the seeds of some New Guinea species. Recommendations are made for methods of collecting which could lead to the description of functional syndromes in other groups of plants. It is suggested that the study of functional syndromes provides a biological explanation for some of the differences detected by molecular techniques and put into a theoretical framework by cladistics.

INTRODUCTION

Many disciplines in addition to ecology are used to broaden the results of a classification based on herbarium specimens. These include, among others, molecular biology, phylogenetics, ethnobotany, economic botany, palynology and biogeography. This paper makes the case for adding to these the study in the field of the functional significance of the characters used in classification, using the results obtained from a study of the genus *Aglaia* as an illustration.

When I started to work on *Aglaia*, it was to investigate the reproductive biology of species in the genus, without any intention of getting involved in the taxonomy. I had a vague idea that someone else would do that!

After a year in the field in the Malay Peninsula, it became obvious that the preliminary results of the field study were difficult to interpret because of the chaotic state of the taxonomy of the genus. So, the biological study was continued within the context of a taxonomic revision of the genus and this proved to be a powerful combination. Having completed a revision of the entire genus throughout its range (Pannell 1992), it is now clear that a study of the function of taxonomic characters provides an important tool in the classification of difficult plant groups. This is because

---

[1] 2, Wolvercote Court, Wolvercote Green, Oxford OX2 8AB.

a knowledge of the functional significance of some of the taxonomic characters can either make possible the resolution of complex patterns of variation or provide some explanation for the variation observed (see Pannell & White 1988 for examples). Hence, an understanding of living processes from field work can be used to interpret dead herbarium specimens. The monographic revision also means that ecological findings can be interpreted in a wider context than just for those species which are found in the small part of the range of the genus or family covered by the field work.

DISPERSAL IN AGLAIA

A. Fruits and functional syndromes

In *Aglaia*, the aspect of the ecology which brought the taxonomic revision to life was dispersal, because of its clear association with characters which are used in the classification of the genus. Hence, field observations led to the understanding of how fruits worked and revealed the existence of syndromes of architectural, morphological and biochemical characters associated with different dispersers (Pannell & Kozioł 1987). This discovery emerged in three stages. First, during field work in Sumatra and Peninsular Malaysia, it became clear that the position, structure and development of *Aglaia* fruits fell into two distinct groupings (i.e. a structural syndrome). Second, the dispersal agents associated with these structural syndromes indicated that there was a close correlation between them and the fruits (i.e. a functional syndrome). Third, biochemical studies of the edible portions of the fruits matched the association already detected between the fruit morphology and the dispersal agents (Pannell & Kozioł 1987). All of this meant that herbarium study of unseen species could now be interpreted in the context of these discoveries and questions such as 'what happened to the evolution of *Aglaia* in New Guinea where no primates are found' could be addressed (Pannell & White 1988).

There are two main types of dispersers associated with two main types of fruit in west Malesia. These are birds for the dehiscent fruits of section *Amoora* and primates for the indehiscent fruits of section *Aglaia*. There are no indigenous primates other than man to the east of Lydekker's Line (see Pannell & White 1988 and sect. B below) and many of the birds involved in dispersal are different from those in the West, so the details of the syndromes must often be different.

The two main functional syndromes in west Malesia can be described in detail because of the combination of field observations, photographs, collections, including fruits in spirit, biochemical analyses of the edible aril and taxonomic work on the genus.

Bird-dispersed species of *Aglaia* have infructescences of a few, dehiscent fruits. The colours of different layers of the pericarp and that of the aril surrounding the seed are contrasting and form an attractive display. The peduncle is short and birds can perch on the subtending branch or other nearby branch and remove seeds from the fruits. The bird swallows the seed whole, the aril is rich in lipids and is easily removed from the rest of the seed by the gentle action of the gizzard of fruit-eating birds. The seed is then regurgitated or passes quickly through the gut.

The infructescences of primate-dispersed species of *Aglaia* may be large or small and they usually hang from slender flexible branches. Primates can hang from a more sturdy branch and reach the fruits with their long arms. The fruits are indehiscent, but the primates manipulate the fruit to remove the pericarp. They are encouraged to swallow the whole seed, with its sweet juicy aril, because the aril adheres firmly to the testa and cannot easily be sucked or bitten off the rest of the seed. The aril is rich in reducing sugars and sweet-tasting amino-acids and has a pleasant odour.

More details of the features of the plants and animals which make up the functional syndromes are given in Tables 1 and 2.

TABLE 1. The arboreal-bird-dispersal syndrome in *Aglaia*

Dehiscent *Aglaia* fruits	Arboreal birds
Fruits in infructescences of few fruits near a sturdy branch	Most of the dispersing birds need to reach seeds from a perch
Fruits up to 10 cm diameter. Seeds up to 5.5 × 3.5 × 2 cm	Seeds are swallowed by hornbills, pigeons and, for the species with smaller seeds, by smaller birds
Pericarp inedible, contains latex, dehiscent	Birds have direct access to the seeds once the fruit has dehisced
Contrast in colours of dehisced fruit; pericarp brown, reddish-brown or pink, inner pericarp white; aril red	Bird vision is probably sensitive to red and birds are attracted to contrasting colours
Thin aril rich in lipid; large seed	Aril provides high energy food to compensate for the heavy, undigested seed being carried by a flying animal
Aril easily removed from the seed	Gentle action of the gizzard removes the aril. The seed is regurgitated or defaecated
No smell	Poor sense of smell
No flavour	Poor sense of taste

B. The significance of faunal boundaries in Malesia

Only 12 of the 106 species of *Aglaia* occur on both sides of Lydekker's Line. Eight of these are widespread and complex or variable species (Pannell 1992). There are many factors which might have caused the distinct New Guinea species to evolve. One of these is the faunal differences between east and west Malesia. The primates and many of the bird families which disperse *Aglaia* seeds to the west of New Guinea are not present in New Guinea and the areas to the east, north and

TABLE 2. The primate-dispersal syndrome in *Aglaia*

Indehiscent *Aglaia* fruits in West Malesia	Primates, especially apes, in West Malesia
Infructescences hanging or borne on slender branches	Accessible to long-armed primates
Fruits rarely more than 5 cm diam. Maximum dimensions of seeds 2.9 × 1.4 × 1 c m	The maximum size of seed that primates can swallow is less than that for the large fruit-eating hornbills and pigeons of SE Asia
Indehiscent, inedible pericarp, contains latex	Primates can manipulate the fruit and remove the pericarp
Pericarp in some species orange or pinkish-orange throughout; aril translucent pink, orange or white	Old World apes and New World monkeys are probably attracted to orange.
Aril thin and firmly attached to the seed	Seed tends to be swallowed
Aril rich in reducing sugars and sweet-tasting amino acids	Primates have a 'sweet tooth'
Distinctive and pleasant odour	Good sense of smell
Distinctive and pleasant taste	Good sense of taste

south of New Guinea. In New Guinea there are, instead, birds such as birds of paradise, bower birds and cassowaries, as well as a great diversity of pigeons. All of these are fruit eaters, as are some of the indigenous marsupials. The New Guinea species of *Aglaia* may therefore have evolved different dispersal syndromes because of the different dispersers available in this part of its range (Pannell & White 1988).

This prediction has recently been confirmed by the work of Mack (1995a, 1995b). An enormous fruit (up to 18 cm in diameter) of *Aglaia*, which either falls to the ground and releases the seeds or sheds its seeds from the tree, is dispersed by the Dwarf Cassowary. A larger fruit and seed, a difference in colour distribution and a difference in timing of dehiscence and release of seeds turns the arboreal-bird-dispersal syndrome into a cassowary-dispersal syndrome (Table 3).

C. Coevolution between plants and animals

The importance of the functional approach to taxonomy is explored more thoroughly in a book initiated by Frank White and currently nearing completion (Fortune Hopkins *et al.*). Each of the five main authors has worked on a biological monograph of a group of tropical plants. In each case, field studies of the functional significance of the characters used in classification has formed an important part of the taxonomic treatment. It is suggested that some coevolution with dispersers, pollinators or symbionts has taken place in four of these groups.

TABLE 3. The cassowary-dispersal syndrome in *Aglaia*

Tardily dehiscent *Aglaia* fruits in New Guinea	Cassowaries
Fruits either dehisce on the tree or fall and crack open on hitting the ground	Flightless ground feeders
Fruits up to 18 cm in diameter Seeds are at least 10.5 cm long, 7 cm wide, 6 cm thick and weigh an average of 114 g	Able to swallow such enormous seeds
The seeds fall out of the fruit	Feed on the individual seeds on the ground
Contrast in colour between scarlet aril and large pale yellow attachment scar	Seeds are conspicuous on the ground

The corresponding features of the plants and the animals which are associated with them are described. The fifth group is the Ebenaceae, for which two species complexes, *Euclea natalensis* and *Diospyros natalensis,* and the functional significance of their variation in relation to the physical environment are analysed. A range of problems to which the study of functional syndromes has been applied is thereby illustrated. It is clear from these examples that the scope and usefulness of this approach is much wider than has hitherto been recognised.

Coevolution between plants and the animals which disperse their seeds is usually diffuse rather than one to one (which is sometimes found in pollination) and this is discussed by Wheelwright & Orians (1982), Howe (1985) and Fortune Hopkins *et al.* (in prep). The combination of field observations and monographic study in *Aglaia* has shown that the two types of fruit are associated with two broad groups of dispersers. Arising from these, it is a reasonable inference that some coevolution has taken place between two groups of species within the genus and two main groups of dispersal agents. These are birds for species with dehiscent fruits and primates for some species with indehiscent fruits. In Borneo, Leighton (1986 and in Terborgh 1990) believes that the largest of the bird-dispersed fruits belong to a class of heavily protected dehiscent fruits comprising some 75 species of predominantly meliaceous and burseraceous trees. He suggests that these fruits are specialised for dispersal exclusively by hornbills and that an important function of the bird's laterally-flattened bill is to prise open the splitting husks to extract the seeds, which are covered with lipid-rich flesh. These seeds are a particularly important source of food for cooperative breeding species during fruit-poor times (Leighton & Leighton 1983). Similarly, the modification of the dehiscent fruits of a few New Guinea species of *Aglaia* which is associated with dispersal by cassowaries, suggests that there has been some coevolution with this group of birds. Further field work is likely to reveal that different animals are associated with as yet undetected differences in the dispersal syndromes in different parts of the range of the genus, especially given the marked differences in fauna in east and west Malesia. The biological monograph provides an opportunity for recognising such coevolution, because suites of characters in the morphology and biochemistry of the fruits of related plant species can be identified which are associated with the nutritional requirements, morphology and behaviour of groups of dispersing animals.

D. Conservation

There are important practical implications for a general understanding of mutually beneficial plant-animal interactions. It is now clear that conservation of tropical forest communities should be based, to a large extent, on such knowledge (Pannell 1989; Terborgh 1990; Mack, 1997). This would ensure that the large plants and animals in conserved areas are not biologically dead through a failure to sustain sufficiently large breeding populations along with viable populations of the organisms upon which they depend. If this association is not understood, attempts at conservation are in danger of failing.

IMPLICATIONS FOR THE FUTURE

A. Collecting methods

Collecting methods are an important consideration when information is needed on the biology of a group in order to make some sense of its taxonomy. To my knowledge, the 18 cm diameter cassowary fruit has not been collected or described by any botanist. The non-fruiting material originally sent to me for identification differed slightly from A. flavida in indumentum and in the number of lateral veins on the leaflets. Aglaia flavida is common in the Solomon Islands, but its fruits are only up to 8 cm long and 5.5 cm in diameter.

The written information about the fruit (A. Mack, in litt.) and fruit and seed specimens sent subsequently confirmed that two different species are involved (A. flavida Merr. et Perry and A. mackiana Pannell. Since the minor differences in indumentum and other leaflet characters are correlated with the differences in fruit structure, it is possible to distinguish the two species, even when fruiting material is not available (Pannell, 1997). Without the difference in the fruits, giving such weight to the leaf characters would be difficult to justify. If such characters were routinely used for species delimitation in Aglaia, we would return to the situation generated in the past by the numerous new species described by taxonomists such as Merrill (e.g. 1918a & 1918b) and Elmer (1937). These new species were often based on characters such as length of petiole or size of leaflets or inflorescences, characters which are now known to vary considerably within many species. They can only be used for separating species if they are correlated with other characters such as the structure and density of the indumentum, the venation pattern, texture of the leaflets and structure of the flowers or fruits.

There is therefore an urgent need for a shift in priorities for collecting. In all genera with fleshy fruits or seeds eaten by animals, a ripe fruit should never be passed over, even if it is large and heavy. It should whenever possible be collected into spirit to preserve its size, shape and the characters of the edible aril. Notes on mode of presentation of the infructescence, number of fruits in the infructescence, colours of the different layers of the fruit and seeds, texture of the aril, smell and, where appropriate taste, should accompany these collections. In most species of *Aglaia*, collections of ripe fruits in spirit, accompanied by dried specimens of the leaves, are urgently needed so that full descriptions of the fruits for each species can be made and the characters associated with dispersal studied.

Also seriously lacking in herbarium collections of dioecious plants is the female flower. When a male flower is collected, if another tree is found with a smaller inflorescence of slightly larger flowers, this is likely to be the female. A collection of this to accompany the male material from the same population would be of great value — but not, of course, under the same collecting number.

B. Conclusion: the importance of the study of functional syndromes

Molecular biology gives us the hard scientific base for identifying differences in genotypes. Cladistics gives us a theoretical and intellectual base for estimating phylogenies and/or relationships within or between taxonomic groups and for producing classifications which reflect the proposedphylogenies and relationships. Functional syndromes gives us biological explanations for differences detected by molecular biology and put into a theoretical framework by cladistics. It is a discipline which can also be used for identifying the criteria by which conservation of biodiversity can be achieved. It gives us the methodology for investigating plant/animal interdependence, plant/plant interdependence and the effect of physical factors on plant morphology. The interplay of the many factors which influence every organism in an ecosystem is thereby better understood and can be placed into a wider taxonomic, ecological and evolutionary framework.

ACKNOWLEDGEMENTS

I am grateful to the late Frank White for his interest and encouragement during the period in which the ideas presented in this paper were formulated and to Christopher McCrudden and Iain Prance for their critical comments on the manuscript.

REFERENCES

Elmer, A. D. E. (1937). A century of Philippine *Meliaceae*. Leafl. of Philipp. Bot. 9: 3274 – 3325.

Fortune Hopkins, H. C., Huxley, C. R., Pannell, C. M., Prance, G. T. & White F. (in prep.). The biological monograph: the importance of field studies and functional syndromes for taxonomy and evolution.

Howe, H. F. (1985). Constraints on the evolution of mutualisms. Amer. Nat. 123: 764 – 777.

Leighton, M. (1986). Hornbill social dispersion: variations on a monogamous theme. In: Rebenstein, D. I. & Wrangham, R. W. (eds.). Ecological Aspects of Social Evolution. Birds and Mammals: 108 – 130. Princeton University Press, Princeton.

Leighton, M. & Leighton, D. R. (1983). Vertebrate responses to fruiting seasonality within a Bornean rainforest. In: Sutton, S. L., Whitmore, T. C. & Chadwick, A. C. (eds.). Tropical Rain Forest: Ecology and Management pp. 181 – 196. Blackwell, Oxford.

Mack, A. (1995). Seed dispersal by the Dwarf Cassowary, Casuarius bennetii, in Papua New Guinea. Ph. D. dissertation, University of Miami.

Merrill, E. D. (1918a). New species of Bornean plants. Philipp. J. Sci. 13: 76 – 80.

Merrill, E. D. (1918b). New or noteworthy Philippine plants. Philipp. J. Sci. 13: 289 – 296.

Pannell, C. M. (1989). The role of animals in natural regeneration and the management of equatorial rain forests for conservation and timber production. Commonwealth For. Rev. 68: 309 – 313.

Pannell, C. M. (1992). A taxonomic monograph of the genus *Aglaia* Lour. (Meliaceae). Kew Bull. Add. Ser. 16. London: HMSO.

Pannell, C. M. & Koziol, M.J. (1987). Ecological and phytochemical diversity of arillate seeds in *Aglaia* (*Meliaceae*): a study of vertebrate dispersal in tropical trees. Phil. Trans. R. Soc. Lond. B 316: 303 – 313.

Pannell, C. M. & White, F. (1988). Patterns of speciation in Africa, Madagascar and the tropical Far East: regional faunas and cryptic evolution in vertebrate-dispersed plants. Monogr. Syst. Bot. Missouri Bot. Gard. 25: 639 – 659.

Terborgh, J. (1990). Seed and fruit dispersal — commentary. In: Bawa K. & Hadley M., Reproductive ecology of tropical rain forest plants. Man & the Biosphere series, 7: 181 – 190. UNESCO, Paris.

Wheelwright, N. T. & Orians, G. H (1982). Seed dispersal by animals: contrast with pollen dispersal, problems of terminology, and constraints on coevolution. Amer. Nat. 119: 402 – 413.

Generic delimitation in *Grammitidaceae* (Filicales)

Barbara S. Parris[1]

Summary. The 700 or more species of *Grammitidaceae* are in general easily defined, but their grouping into higher taxa poses difficulties which have not been resolved by cladistic methods. Reticulate evolution does not appear to be responsible for these problems because, with the exception of a very few tetraploids, the species are diploid. Some genera are clear-cut, such as *Acrosorus, Adenophorus, Calymmodon* and *Scleroglossum* in the Southeast Asia-Malesia-Pacific Islands region, *Ceradenia, Melpomene* and *Zygophlebia* in Africa and the New World, *Grammitis* sensu strictissimo in the eastern Pacific, New World and Africa and *Lomaphlebia* in the New World, which have at least one unique generic character occurring in all taxa, together with other characters present in all taxa of a genus but not restricted to it. Other recently described or emended essentially New World genera with a few species in Africa (*Cochlidium, Enterosora* and *Terpsichore*) or the eastern Pacific Islands (*Lellingeria*) or Southeast Asia-Malesia (*Micropolypodium*), are defined by unique assemblages of several correlating characters, each of which may be widespread in the family. Character distribution in most Old World taxa is such that generic definition by character grouping would produce more than 25 genera, 10 of which are monotypic.

INTRODUCTION

The fern family *Grammitidaceae* is a characteristic and important component of tropical montane rainforest in both hemispheres. It also extends to both the north and the south temperate regions, but is much more diverse in the latter. In general the family is under-collected, partly because many species are small and partly because many of them closely resemble each other to the inexperienced eye. For its size, the family is the least studied in the Old World, although the situation is rather better in the New World. Currently it is estimated to contain more than 700 species, in 17 genera, but little is known of the relationships between the genera. At present I am working on the systematics of the 450 or more Old World species including re-definition of the genera and their inter-relationships.

The family is an important one in Malesia to the extent that about 300 species occur there and approximately one in eight of montane Malesian fern species belong to it. More than half of the Malesian species are found in New Guinea, where it is the largest fern family apart from *Thelypteridaceae*. Like most other large fern families in New Guinea, it has a significant proportion of endemic species. With 59% endemicity, it ranks only below *Cyatheaceae*, with 84% of its species endemic, and *Theypteridaceae*, with 67% endemic taxa.

In the Old World eight genera are usually recognised, *Acrosorus, Adenophorus* (Bishop 1974), *Calymmodon, Ctenopteris, Grammitis* (including *Oreogrammitis*),

[1] Fern Research Foundation, 21 James Kemp Place, Kerikeri, Bay of Islands, New Zealand.

Prosaptia, Scleroglossum (including *Nematopteris*) and *Xiphopteris*, but my current studies on the re-definition of generic limits have led to the recognition of several additional genera which will be described soon.

In recent New World Floras, authors have maintained all species in a single genus, *Grammitis*, or separated *Cochlidium* from it. Recent revisions have led to the description of five new genera, *Ceradenia* (Bishop 1988), *Lellingeria* (Smith, Moran & Bishop 1991), *Melpomene* (Smith & Moran 1992), *Terpsichore* (Smith 1993) and *Zygophlebia* (Bishop 1989), and another five genera, *Cochlidium* (Bishop 1978), *Enterosora* (Bishop & Smith 1992), *Grammitis* sensu strictissimo (Bishop 1977), *Lomaphlebia* (Bishop 1989, Smith 1993) and *Micropolypodium* Hayata (Smith 1992) are recognised there.

Some of the genera are obviously monophyletic and clearly circumscribed. *Acrosorus, Adenophorus, Calymmodon* and *Scleroglossum* in the Southeast Asia-Malesia-Pacific area, *Ceradenia* and *Zygophlebia* in the New World and Africa, *Grammitis* sensu strictissimo in the Pacific, New World and Africa, and *Lomaphlebia* in the New World, have at least one unique generic character separating them from the rest of the family, together with other characters present in all taxa of the genus, but not restricted to it. *Calymmodon* has sori protected by the folded pinnae, *Acrosorus* has sori deeply sunken in pouches in the lamina, *Scleroglossum* has sori deeply sunken in two grooves in the lamina, *Grammitis* sensu strictissimo has a black sclerotic frond border, *Lomaphlebia* has a marginal vein, while *Adenophorus, Ceradenia* and *Zygophlebia* each have a distinctive unique type of hair.

Other recently described or emended essentially New World genera with a few species in Africa (*Cochlidium, Enterosora, Melpomene* and *Terpsichore*), the eastern Pacific (*Lellingeria*) or eastern Asia (*Micropolypodium*), are deemed to be monophyletic by their American authors because they are defined by a unique assemblage of several correlating characters, each of which may be widespread in the family (Smith & Moran 1992). For example, *Melpomene* has rhizome scales which are clathrate throughout, entire except for the apex which has 1 – 10 minute glandular cells, cordate at base and attached at a point rather than across the width of the scale. Each one of these characters is found elsewhere in the family, but the association of all four is used to define *Melpomene* as a genus.

Pragmatically there is no problem in recognising the first group of genera with single unique characters, nor is there with the second group, because all of the genera can be readily identified.

The majority of Old World species are retained at present in the three genera *Ctenopteris, Grammitis* s.l. and *Xiphopteris* which are defined in an artificial manner by the degree of dissection of the lamina. *Grammitis* contains species with simple fronds which usually have one row of sori on each side of the mid-vein (e. g. *G. fasciata*), but sometimes have up to four rows on each side (e. g. *G. sumatrana*). In *Xiphopteris* the fronds are pinnately divided with each segment containing a single sorus (e. g. *X. hieronymusii*). In *Ctenopteris* the lamina is pinnately divided with each pinna or lobe bearing more than one sorus. The pinnae may be entire or divided. The degree of laminar division is not consistently correlated with other characters, however. The bipinnately divided species *C. yoderi* and *Adenophorus montanus* may look superficially very similar, but differ markedly in the details of frond indumentum. The former has the dark simple eglandular hairs which are widespread in *Grammitidaceae*, while the latter has the unique glandular hairs of *Adenophorus*, which is endemic to the Hawaiian Islands. Another bipinnately divided species, *C. heterophylla*, looks rather similar to *C. yoderi* and *Adenophorus montanus*, but in characters of indumentum and rhizome scales it is very close to the simple-fronded *Grammitis billardierei*, with which it hybridises, and its numerous relatives (Parris 1977). Likewise *G. sumatrana* has the same type of rhizome scales, frond indumentum and sporangial hairs as the pinnately divided *C. longiceps* and hybrids are known between the two (Parris 1984).

METHODS

When laminar dissection is accepted as not being of prime importance in generic definition, greater use can be made of other characters to assess relationships, as has been done in the recent New World classifications. Characters of the rhizome, rhizome scales, stipe and rachis, types of frond hair, venation patterns, soral arrangement and sporangial ornamentation have often been considered important in delimiting species and arranging them in species groups within the complex of *Ctenopteris/Grammitis* s.l./ *Xiphopteris* in the Old World (e. g. Parris 1983).

These characters have been recorded for more than 170 species, mainly from Malesia, Australasia and the Pacific Islands. They include whether the rhizome is dorsiventral, producing fronds on the upper surface and roots on the lower surface, or whether it is radial, producing fronds and roots in whorls of three or six (Hovenkamp 1990), and details of the types of hair present. Hairs for example can be 'simple eglandular', a widespread hair type restricted to the family, in which the cell walls are not evident at maturity, 'catenate eglandular', in which the cell walls are obvious, 'simple glandular', or be branched in various ways. These represent a small part of the variation in hairs which have been recorded.

In the hope that a cladistic approach might throw some light on the evolutionary history of the family I prepared data on 32 characters for 27 species from the Southeast Asia-Australasia-Pacific area, including some from the distinct genera *Acrosorus*, *Calymmodon* and *Scleroglossum*. The related family *Polypodiaceae* was used as an outgroup to determine the polarity of characters and character states. The data were run on PAUP (Swofford 1990) and HENNIG86 (Farris 1988).

RESULTS AND DISCUSSION

Unfortunately neither cladistics programme produced useful results. After five hours PAUP had tried 2,522,443 trees and saved 4642 of them. HENNIG86 organised the species in a straight line with two pairs of species grouped together. My reaction to this was that cladistics had nothing to offer the student of *Grammitidaceae*.

The reasons for the failure of the cladistics approach are not immediately evident, but reticulate evolution may be one possibility. In some fern families, for example *Aspleniaceae* and *Dryopteridaceae*, various levels of polyploidy exist and reticulate evolution has been well documented. Very few chromosome counts have been made on *Grammitidaceae*, in sharp contrast to the fern families better represented in the temperate regions of the northern hemisphere. Walker (1985) points out that more than a quarter of the chromosome numbers (of the some 30 species whose counts were known at the time he wrote) are quoted as approximations. Counts for only 35 species, of the 700+ species in the family, are known to me, 19 from the Old World and 16 from the New World. Of the 35, 29 are diploid, and only six tetraploids have been reported; *Prosaptia contigua* and *Xiphopteris cornigera* in Sri Lanka (Manton & Sledge 1954), and *Cochlidium serrulatum* (Walker 1985), *Lellingeria delitescens, Melpomene moniliformis* and *Terpsichore semihirsuta* (Walker 1966) in Jamaica. *Cochlidium serrulatum* is also known as a triploid in Jamaica and Trinidad (Walker 1985). With so few counts available it is impossible to rule out the possibility that reticulate evolution has contributed to the diversity of the family, but on the morphological evidence available there appears to be none of the characteristic species complexes which would lend credence to this theory.

A range of base numbers is known in *Grammitidaceae*, with $x = 32, 33, 35, 36$ and 37 reported definitely and substantiated by photographic evidence (Walker 1985). A base number of $x = 37$ is found in many genera of *Grammitidaceae* including *Adenophorus* (Wagner 1963), *Ctenopteris* (Brownlie 1958), *Grammitis* s.l. (Brownlie 1958, Brownlie 1961, Wagner 1963), *Ceradenia* (Smith 1992), *Enterosora* (Bishop & Smith 1992), *Lomaphlebia* (Walker 1966), *Melpomene* (Smith & Moran 1992), *Micropolypodium* (Smith 1992), *Prosaptia* (Manickam & Irudayaraj 1992; Manton & Sledge 1954), *Terpsichore* (Smith 1993) and *Xiphopteris* (Manton & Sledge 1954). Base numbers of $x = 32, 33$ and 37 are known in *Lellingeria* (Smith et al. 1991; Walker 1966), while $x = 33$ (Wagner 1980) and 35 (Walker 1985) are reported for *Cochlidium*, and $x = 36$ is known in *Grammitis* s. l. (Manickam & Irudayaraj 1992). Much of the cytological diversity in the family is concentrated in the New World, where base numbers of $x = 32, 33, 35$ and 37 have been reported. In contrast, only two base numbers, $x = 36$ and 37 are known from the Old World. The former is known only in *Grammitis attenuata,* a distinctive species of southern India and Sri Lanka, while the latter count is the most common in the family, being known for 23 species. Further cytological studies may help to resolve some of the taxonomic problems in the family, but with our present knowledge it is not very useful. Molecular studies on the family are non-existent, but certainly should be tried.

Although cladistic and cytological studies have not contributed to our knowledge of generic inter-relationships in the family, the useful array of morphological characters to hand for Old World taxa can be employed to provide a pragmatic solution to the generic re-allocation of the species presently treated in *Ctenopteris* and *Xiphopteris*, and many in *Grammitis*, which is needed for Flora Malesiana and other major regional Floras.

The segregation into a new genus of a few species which have two types of hair unique in *Grammitidaceae*, and not known outside the family, is straightforward.

There is a problem, however, with applying the American technique of defining genera by unique groups of characters to Old World species, which are very diverse, apparently far more so than those of the New World, and which have less well correlated characters. Useful characters in the New World are often less reliable in the Old World. For example, one of the characters which is important in defining New World genera is the presence or absence of hydathodes on the endings of the veins on the adaxial surface of the lamina. In Malesian-Pacific species the presence or absence of hydathodes is much less reliable in that some species of an otherwise natural group (e. g. *Calymmodon*) may have them, others may not. In some species some plants may have them, but others do not (e. g. *G. reinwardtii*), or they may be present on vein endings on one part of the lamina and absent on another (e. g. *Ctenopteris allocota*).

Another problem in the Old World species is the presence of a strikingly different character in a few species which appear to have little else in common apart from characters which are widespread in the family, for example spiral thickening in the rhizome scale cells of *Ctenopteris thwaitesii* from Sri Lanka and *Grammitis dolichosora* and its undescribed close relative in Malesia. *Ctenopteris thwaitesii* has deeply pinnatifid fronds with pinnately branched veins, glabrous sporangia and branched hairs, while *Grammitis dolichosora* and its relative have simple fronds with few-forked veins, setose sporangia and simple hairs. *Ctenopteris thwaitesii* does not seem closely related to any other taxon, but *Grammitis dolichosora* and its relative would fit very comfortably into a group of more than 60 species were it not for the spiral thickening in the rhizome scales. My solution would be to treat these three species as comprising two genera. There are other examples of similar distribution of unusual characters, most of which can only be detected easily with a binocular microscope. This kind of solution seems preferrable to retaining the present artificial genera which are based on the degree of dissection of the lamina, but would produce more than 25 genera, ten of which are monotypic. Whether these genera would be widely accepted is another matter.

REFERENCES

Bishop, L. E. (1974). Revision of the genus *Adenophorus* (*Grammitidaceae*). Brittonia 26: 217 – 240.

——— (1977). The American species of *Grammitis* sect. *Grammitis*. Amer. Fern J. 67: 101 – 106.

——— (1978). Revision of the genus *Cochlidium* (*Grammitidaceae*). Amer. Fern J. 68: 76 – 94

——— (1988). *Ceradenia*, a new genus of Grammitidaceae. Amer. Fern J. 78: 1-5.

——— (1989). *Zygophlebia*, a new genus of Grammitidaceae. Amer. Fern J. 79: 103 – 118.

——— & Smith, A. R. (1992). Revision of the fern genus *Enterosora* (*Grammitidaceae*) in the New World. Systematic Botany 17: 345 – 362.

Brownlie, G. (1958). Chromosome numbers in New Zealand ferns. Trans. Roy. Soc. New Zealand 85: 213 – 216.

——— (1961). Additional chromosome numbers — New Zealand ferns. Trans. Roy. Soc. New Zealand (Bot.) 1: 1 – 4.

Farris, J. F. (1988). HENNIG86. Published privately.

Hovenkamp, P. (1990). The significance of rhizome morphology in the systematics of the *Polypodiaceae* (sensu stricto). Amer. Fern J. 80: 33 – 43.

Manickam, V. S. & Irudayaraj, V. (1992). Pteridophyte Flora of the Western Ghats, South India. 653 pp. B. I. Publications Pvt. Ltd, New Delhi.

Manton, I. & Sledge, W. A. (1954). Observations on the cytology and taxonomy of the pteridophyte flora of Ceylon. Phil. Trans. Roy. Soc. B, 238: 127 – 185.

Parris, B. S. (1977). A naturally occurring intergeneric hybrid in *Grammitidaceae* (Filicales): *Ctenopteris heterophylla* × *Grammitis billardierei*. New Zealand J. Bot. 15: 597 – 599.

—— (1983). A taxonomic revision of the genus *Grammitis* Swartz (*Grammitidaceae*: Filicales) in New Guinea. Blumea 29: 13 – 222.

—— (1984). Another intergeneric hybrid in *Grammitidaceae*: *Ctenopteris longiceps* x *Grammitis sumatrana*. Fern Gazette 12: 337 – 340.

Smith, A. R. (1992). A review of the fern genus *Micropolypodium* (*Grammitidaceae*). Novon 2: 419 – 425.

—— (1993). *Terpsichore*, a new genus of *Grammitidaceae* (Pteridophyta). Novon 3: 478 – 489.

—— & Moran, R. C. (1992). *Melpomene*, a new genus of *Grammitidaceae* (Pteridophyta). Novon 2: 426 – 432.

——, —— & Bishop, L. E. (1991). *Lellingeria*, a new genus of *Grammitidaceae*. Amer. Fern J. 81: 76 – 88.

Swofford, D. (1990). Phylogenetic analysis using parsimony. Version 3.0. Illinois Natural History Survey, Champaign, Ill.

Wagner, F. S. (1980). New basic chromosome numbers for genera of neotropical ferns. Amer. J. Bot. 67: 733 – 738.

Wagner, W. H. (1963). A biosystematic survey of United States ferns — preliminary abstract. Amer. Fern J. 53: 1 – 16.

Walker, T. G. (1966). A cytotaxonomic survey of the pteridophytes of Jamaica. Trans. Roy. Soc. Edinburgh 66: 169 – 237.

—— (1985). Cytotaxonomic studies of the ferns of Trinidad 2. The cytology and taxonomic implications. Bull. B. M. (Nat. Hist.) Bot. 13:149 – 249.

The vegetation of a submontane moist forest on Mt Kinasalapi, Kitanglad Range, Mindanao, Philippines

JOHN J. PIPOLY III[1] & DOMINGO A. MADULID[2]

Summary. The Philippine Plant Inventory (PPI) conducted an inventory of submontane moist forest in Mt Kinasalapi, Kitanglad Mountain Range, located in the north-central portion of Bukidnon Province, island of Mindanao. A permanent plot, 500 × 20 m, further subdivided into twenty-five, 20 × 20 m subplots, was established. All trees ≥ 10 cm DBH were measured, tagged and collected, as well as lianas with an averaged DBH ≥ 10 cm over the first two 2 m above ground. This plot design was used to facilitate subsequent use of the plot by field biologists in other subdisciplines, and to maximize documentation of diversity. Forty-three species were found, of which 47% of them are endemic to the Philippines. In addition, there was extremely high tree density, totalling 1028 individuals, with a combined basal area of 55.41 m². The density figure is one of the highest known, from actual rather than extrapolated data sets, and the basal area figure is superseded in the Philippines only by data from Bicol National Park in Luzon Province. Further studies are needed in Philippine submontane and montane forests to determine if this sample is at all representative of other forests found in these life zones.

INTRODUCTION

One hectare, semi-permanent or permanent plots are rapidly becoming the tool of choice for field biologists to document biological diversity, particularly in the tropics. In tropical moist, wet and rain forests, a variety of methodologies have been used, most notably, square plots (Brown 1919; Korning *et al.* 1990; Erwin *et al.* 1991), belt transects (Prance *et al.* 1976; Boom 1986; Campbell *et al.* 1986), and the Point-Centered Quarter Method (Cottam & Curtis, 1956) used by Mori *et al.* 1983; Balslev *et al.* 1987; Mori & Boom 1987). An interesting study conducted by Korning *et al.* (1990) showed that the Point-Centered Quarter methodology used along line transects reflects maximum diversity and provides average values of density and tree size, whereas the quadrat plot reflects the local structure and composition of the forest within the plot.

Studies of lowland rain, moist and wet forests have steadily increased in the past 20 years, and with few exceptions (Meijer 1959; Whitmore 1975), most claim that Amazonian forests are the most species diverse (Gentry 1988; Valencia *et al.* 1994). However, there have been relatively few permanent or subpermanent plots in the premontane (or submontane) life zone in either the Paleo- or Neotropics. Meijer (1959) carried out an inventory near Tjibodas, in west Java, but did not have Importance Value Index (IVI) and related data calculated. Perhaps the best comparisons can be drawn from the work of Yamada (1975, 1976a, 1976a, 1976b, 1977) from Mt Pangrango, West Java, with that of Madsen & Øllgaard (1994) from

[1]Botanical Research Institute of Texas, 509 Pecan Street, Fort Worth, TX 76102–4060, U.S.A.
[2]Botany Division, Philippine National Museum, P.O. Box 1126, Manila 1051, Philippines.

Podocarpus Park in Ecuador. In the Philippines, Long Term Ecological Research (LTER) sites have been established at Irawan (Palawan), Bicol (Luzon), Manila Bay (Luzon) and this study. Among these, only Kitanglad is in the submontane environment. The only relatively comparable work done in the Philippine submontane life zone was that of Brown (1919) which was performed on Mount Makiling. The present study reports the first results from an attempt to launch a strategic sampling programme to remedy the dearth of information regarding this critical life zone, one that is not the most species rich in terms of tree species, but one that does harbour many endemics. We eventually seek to document whether the overall plant species richness (i.e. trees, shrubs, lianas and herbs) is large, as has been seen in the Neotropics by Henderson *et al.* (1991).

The Philippine flora is unequivocally one of the most endangered on the planet (Madulid 1987, 1991, 1994a). While it is not the most species-rich in terms of numbers of species, it exhibits extremely high endemism, with estimates ranging from approximately 76.5% (Merrill 1923 – 26) to 40 – 44% (Madulid 1994a, 1994b). The high endemism, concomitant with the paucity of intact primary forests, currently estimated at less than 188,000 ha (Madulid 1994a), obviates the necessity to quantitatively sample each vegetation type in order to understand the life cycles of the remaining forests and their highly endemic taxa. Furthermore, of the 18 most critical plant sites extant in the Philippines (Madulid, 1995), 11 contain sizeable forests belonging to the sub-, lower and upper montane forest categories (1000 – 2400 m elevation) of Madulid (1994a). The physiognomy and abiotic components of these forests appear to be best matched with the premontane moist, wet and rainforests of the Holdridge (1947) classification used in analyses of neotropical vegetation, the life zone most frequent in the areas proposed by Henderson *et al.* (1991) as the most rich. For these reasons, it seems imperative that we prioritize quantitative sampling of these forests, using our present Philippine Plant Inventory Project as a start.

METHODOLOGY

Study Area

The one ha study plot is located in the Kitanglad National Protected Area, on Mt Kinasalapi, one of several ranges of mountains in the Kitanglad Mountain Range. The 31,360 ha reserve ranges in altitude from slightly over 100 m to 2379 m in elevation, containing lowland residual evergreen moist forest, lower and upper montane moist forest, grassland, and shrubland. Mean annual rainfall data from the nearby town of Malaybalay indicates that the area receives from 2500 – 3500 mm, with a rainy season (230 – 330 mm per month) from May to October, and a dryer season (110 – 200 mm per month) from November to April. Average montly temperatures range from 22.8 – 24.6°C, with relative humidities of 71 – 86%.

The plot is located along a slope from 2065 – 2360 m, on the southwest flank of Mt Kinasalapi, Municipality of Lantapan, Sitio Sungco, 08°00'05"N, 125°30'26"E, in primary forest on volcanic soils. The area is of the highest conservation priority, because it is the site of the greatest diversity of mammals and birds in the Philippines, with 134 and 58 species, respectively (Heaney 1993).

Methods

The study plot (500 × 20 m) was divided into 20 × 20 m squares, marked with orange flags and nylon rope. All trees ≥ 10 cm DBH at 135 cm above the ground were tagged with aluminium numbered plates, their DBH, height and position in the plot were recorded. Six duplicate vouchers were made for all trees in sterile condition, and 12 made for fertile ones. Special effort was made to collect both juvenile sucker shoots as well as mature ones from the upper canopy, because several taxa, most notably *Gordonia luzonica* and *Litsea diversifolia* had strikingly different leaf morphologies between the two phases in their life cycle. The two complete sets for the study are deposited in the herbaria of the Philippine National Museum (PNH) and the Botanical Research Institute of Texas (BRIT). Identifications were performed by the authors and we are now awaiting confirmation by taxonomic specialists associated with the Flora of the Philippines programme.

Lianas were included only if they were rooted in the plot and if their average stem diameter (thickest vs. thinnest point) through the first 2 m above ground was ≥ 10 cm. This eliminated over-census of lianas due to large basal thickenings; a wood callus frequently results from tension wood production, broken bark and wound response consequent to the weight of the liana on the curved portion of the stem just above the ground. We maintain that these growths are developmental anomalies not biologically significant in terms of the biomass the rest of the liana contributes to the forest, and therefore, lianas with those growths (one other individual of *Alyxia* in this study) are routinely excluded from our census.

RESULTS AND DISCUSSION

Diversity and endemism

Among the trees ≥ 10 cm DBH, 43 species were encountered, of which 40 are dicotyledonous, one is a liana and two species are tree ferns. This is extremely low when compared to lowland tropical forests, which often range from well over 140 to over 300 species per hectare (Valencia *et al.* 1994). Among montane forests, figures range from 59 to 67 (Madsen & Øllgaard 1994) down to ten (Yamada 1976a, 1976b) for the same size class over one hectare. However, as had been mentioned in the introduction, the Philippine flora is highly endemic, and this inventory confirms the estimates of Madulid (1994a) with 20 species, or 47% of the tree taxa surveyed endemic to the Philippines. The low species richness, coupled with high endemism, presents a unique problem in conservation biology which will require novel solutions in biodiversity management.

Density

Tree density was high, consisting of 1028 individuals (Table 1), a figure nearly 30% higher than that normally reported for lowland forests in the Neotropics (Balslev *et al.* 1987; Campbell *et al.* 1986; Korning *et al.* 1991; Valencia *et al.* 1994), and from 20 – 25% greater than average from neotropical montane forests (Grubb *et al.* 1963; Pipoly & Cogollo, unpubl.), with the exception of the results of

TABLE 1. Species Composition of a Submontane Forest, Mt Kinasalapi,
Kitanglad, Mindanao

Species	Dens.	Basal Area (cm²)	Rel. Dens.	Rel. Freq.	Rel. Dom.	IVI
ACERACEAE						
Acer niveum Blume	36	17326.36	3.5	4.03	3.13	10.66
ACTINIDIACEAE						
Saurauia sp.	1	36.72	0.1	0.34	0.01	0.45
ANNONACEAE						
Annonaceae indet.	6	4300.04	0.58	1	0.78	2.36
APOCYNACEAE						
Alyxia sp.	1	86.2	0.1	0.34	0.02	0.46
AQUIFOLIACEAE						
Ilex apoensis Elmer	8	1994.34	0.79	1	0.36	2.15
ARALIACEAE						
Aralia bipinnata Blanco	5	975.09	0.49	1	0.18	1.67
CHRYSOBALANACEAE						
Licania sp.	1	78.5	0.1	0.34	0.01	0.45
CORNACEAE						
Mastixia tetrapetala Merrill	31	33489.45	3.02	4.03	6.04	13.09
CYATHEACEAE						
Cyathea cf. *brevipes* Copeland	116	22499.56	11.28	7.98	4.06	23.32
DAPHNIPHYLLACEAE						
Daphniphyllum buchananiifolium Hall. f.	3	351.51	0.39	1.00	0.06	1.45
DICKSONIACEAE						
Dicksonia blumei (Kunze) Moore	34	8735.97	3.31	2.68	1.58	7.57
ELAEOCARPACEAE						
Elaeocarpus calomala (Blanco) Merrill	18	17059.03	1.76	1.68	3.08	6.52
ESCALLONIACEAE						
Polyosma sorsogonensis Elmer	93	27858.37	9.05	6.38	5.03	20.46
Polyosma verticillata Merrill	34	5916.08	3.31	3.02	1.07	7.40
Quintinia apoense (Elmer) Schlechter	40	12873.72	3.89	3.36	2.32	9.57
EUPHORBIACEAE						
Homalanthus populneus Pax	8	1604.33	0.78	0.67	0.29	1.74
Mallotus sp.	1	422.52	0.1	0.34	0.08	0.52
Euphorbiaceae indet.	1	78.5	0.1	0.34	0.01	0.45
FAGACEAE						
Lithocarpus woodii (Lance) A. Camus	25	19308.02	2.43	2.68	3.48	8.59
LAURACEAE						
Litsea diversifolia Blume	62	73545.11	6.03	4.7	13.27	24
LEEACEAE						
Leea sp.	4	772.36	0.39	1	0.14	1.53

TABLE 1 continued

Species	Dens.	Basal Area (cm²)	Rel. Dens.	Rel. Freq.	Rel. Dom.	IVI
MELASTOMATACEAE						
Astronia sp.	1	174.28	0.1	0.34	0.03	0.47
MORACEAE						
Ficus sp.	18	772.36	1.75	2.68	0.14	4.57
Ficus sp.	1	183.76	0.1	0.34	0.03	0.47
MYRICACEAE						
Myrica javanica Blume	8	2400.65	0.78	1.68	0.43	2.89
MYRSINACEAE						
Myrsine amorosoana Pipoly	49	9700.74	4.77	5.37	1.75	11.89
MYRTACEAE						
Eugenia sp.	2	1555.22	0.19	0.67	0.28	1.14
Xanthomyrtus diplycosiifolius (C. B. Rob.) Merr.	20	9189.56	1.95	2	1.66	5.61
PITTOSPORACEAE						
Pittosporum pentandrum (Blanco) Merrill	12	3908.91	1.17	2.35	0.71	4.23
PODOCARPACEAE						
Dacrydium pectinatum de Laubenfels	11	17300.6	1.07	2	1.9	4.97
Phyllocladus hypophyllus Hook. f.	83	153016.1	8.07	4.7	27.62	40.39
Podocarpus pilgeri Foxworthy	23	7029.31	2.24	4.02	1.27	7.53
ROSACEAE						
Prunus preslii var. *vulgaris* Koehne	30	8636.32	2.92	3.7	1.56	8.18
RUBIACEAE						
Psychotria sp.	1	151.67	0.1	0.34	0.03	0.47
STAPHYLEACEAE						
Turpinia ovalifolia Elmer	89	21263.2	8.66	5.03	3.83	17.52
Turpinia sp.	4	1772.2	0.39	0.67	0.32	1.38
SYMPLOCACEAE						
Symplocos confusa Brand	9	3728.24	0.88	2	0.67	3.55
Symplocos lancifolia Siebold & Zuccarini	8	2303.71	0.78	2	0.42	3.2
Symplocos montana Vidal	5	1390.58	0.49	1.34	0.25	2.08
Symplocos sp.	34	5916.08	3.31	3.02	1.07	7.4
THEACEAE						
Adinandra dumosa Jack	36	17355.31	3.5	3.69	3.84	11.03
Adinandra elliptica C.B. Robinson	31	7458.64	3.02	3.69	1.35	8.06
Eurya acuminata DC.	39	9635.8	3.79	4.03	1.74	9.56
Gordonia luzonica Vidal	20	25823.05	1.95	2.68	4.66	9.29
TOTALS	1028	554061.99	100.17	103.23	99.49	302.89

Madsen & Øllgaard (1994), who found 1273 individuals per hectare in their study. In Malesia, Yamada found 1516 individuals per hectare at 2400 m elevation but the figure was derived from extrapolation. In our plot, the species with the highest density was *Cyathea* cf. *brevipes*, with 116 individuals, or 11.3% of all stems. Only seven species, *C. brevipes, Polyosma sorsogonensis, Turpinia ovalifolia, Phyllocladus hypophyllus, Litsea diversifolia, Myrsine amorosoana* and *Quintinia apoense*, comprise 52% of all individuals with 532 stems. Eight taxa, *Saurauia* sp., *Licania* sp., *Astronia* sp., *Ficus* sp., *Psychotria* sp., *Mallotus* sp., *Euphorbiaceae* indet., and *Alyxia* sp., were represented by only one individual. *Alyxia* sp. was the only liana censused in the study.

Frequency

Relative frequency figures, when compared to relative density, indicate that there is not an appreciable amount of clumping among the seven most abundant species. The only slight difference in ranking of species in relative density vs. relative frequency is that of *Myrsine apoensis*, which has 40 individuals less than *Turpinia ovalifolia*, but has a slightly higher relative frequency (5.37 for the former vs. 5.03 for the latter). This indicates that the distribution of *Turpinia ovalifolia* is more clumped than that of *Myrsine apoensis*. Likewise, *Phyllocladus hypophyllus* has 19 more individuals than *Litsea diversifolia*, but they share the same relative frequency, 4.7. The latter figure indicates that *Phyllocladus* is more clumped than *Litsea*.

Dominance

The combined basal area of 55.4 m² (Table 1), is a relatively high figure for this forest type. Madsen & Øllgaard (1994) found 44 m², while Yamada (1976a, 1976b) found 56 m². The average basal area per species is 1.3 m², and the average diameter per individual was 21.4 cm. Two species, *Phyllocladus* and *Litsea*, comprise more than 50% of the basal area of the plot. However, *Phyllocladus*' basal area is 2.7 times greater than that of *Litsea*, and approximately four times greater than that of the species who have the next greatest combined basal areas, Rubiaceae indet. and *Cyathea*. A recent study of the lowland forests at Bicol (*Madulid*, 1994b), revealed only 635 trees, with a combined basal area of 59.45 m².

Importance Value Index (IVI)

The species with the highest IVI was clearly *Phyllocadus hypophyllus* (*Podocarpaceae*), with a value of 40.39 (Table 1). This figure compares favourably with other tropical tree inventories for both lowland and upland forests, which range from 12.5 to 52.4 (Cain *et al.* 1956; Grubb *et al.* 1963; Mori *et al.* 1983a; Boom 1986; Campbell *et al.* 1986; Mori & Boom 1987; Balsev *et al.* 1987; Korning *et al.* 1991; Madulid 1994b; Madsen & Øllgaard 1994; Pipoly & Cogollo, in prep.; Yamada 1976a, 1976b). When compared to the Bicol study (Madulid 1994b), there is no taxon which even approaches the figure we have for *Phyllocladus*. The second most important species is *Cyathea* cf. *brevipes*, which was unexpected.

Structure

The submontane moist forest on Mt Kinasalapi is similar in general physiogonomy to the midmontane forest studied on Mt Makiling by Brown (1919), the montane forests studied by Yamada (1976a, 1976b), Madsen & Øllgaard (1994), and the premontane pluvial forests studied by Pipoly & Cogollo (in prep.) in the Colombian Chocó , where relatively few species occupy the canopy, even fewer are emergent, and the rest belong to two layers in the understorey. However, in the Kinasalapi forest, some emergents of *Phyllocladus* exceed 35 m in height, although the trunks are often variously contorted. Canopy species include *Acer niveum*, *Elaeocarpus calomala*, *Lithocarpus woodii*, *Litsea diversifolia*, *Symplocos lancifolia*, *Phyllocladus hypophyllus*, *Adinandra dumosa*, *Adinandra elliptica* and *Mastixia tetrapetala*. Of the canopy species, *Elaeocarpus*, *Lithocarpus*, *Litsea*, *Phyllocladus* and *Mastixia tetrapetala* can be emergent. In the uppermost layer of the understorey, *Annonaceae* indet., *Licania* sp., *Dacrydium pectinatum*, *Xanthomyrtus diplycosiifolius*, *Podocarpus pilgeri*, *Turpina ovalifolia*, *Symplocos confusa*, *Eurya acuminata*, *Cyathea brevipes* and *Gordonia luzonica* are generally found. Finally, the next layer of understorey includes *Ilex apoensis*, *Aralia bipinnata*, *Dicksonia blumei*, *Euphorbiaceae* indet., *Myrica javanica*, *Myrsine amorosoana*, *Eugenia* sp., *Pittosporum pentandrum* and *Prunus preslii* var. *vulgaris*. The rest of the species tagged belong to the layer just above the shrub layer, and rarely exceed 4 m in height. Overall, the average tree height for all tagged trees is 11.4 m, but extremes range from 4 to well over 35 m. Only one liana, *Alyxia* sp., qualified for tagging, but one unidentified *Cucurbitaceae*, and one individual of *Embelia floribunda* (*Myrsinaceae*) were found. This inventory differs from that of most tropical montane or lowland inventories principally by the total lack of palms and *Fabaceae*. It seems to be most like the *Quercus-Neolitsea* Association described by Brown (1919) wherein the canopy averaged 18 m, with emergents up to and sometimes exceeding 22 m, concomitant with DBH in excess of 50 cm. While the Brown (1919) study showed the prominence of *Cyathea* on Makiling, Kitanglad houses many more individuals of greater stature. Some individuals of *Cyathea* attained a height of nearly 15 m, and a DBH of up to 34.1 cm, another striking difference from other inventories.

Conclusions

The submontane forest on Mt Kitanglad is primary, and clearly dominated by *Phyllocladus hypophyllus*. Using trees ≥ 10 cm DBH as an inventory criterion, the forest has the highest density (1028 stems) and largest basal area (55.4 m²) for any submontane or premontane forest published from the Philippines and certainly ranks high among those forest types generally. Despite the fact that only 43 species were found among stems ≥ 10 cm DBH, 47% of them are endemic and another eight are partially identified and may turn out to be other endemics. Further detailed studies of Philippine submontane forests will be necessary to determine if this inventory is at all representative among the mountains in the adjacent area (such as Mt Apo). Brown's (1919) study of Mt Makiling, using 0.25 ha quadrats, yielded very different results when compared to our study, so it will be necessary to repeat the inventory using the same methodology.

ACKNOWLEDGEMENTS

This study is carried out under the auspices of the Philippine Plant Inventory Project, a joint endeavour of the Botany Division, Philippine National Museum, and the Botanical Research Institute of Texas, supported by NSF Grant DEB 9300910. We would like to thank the DENR of Malaybalay, Bukidnon, Central Mindanao University (in particular, Dr Victor Amoroso) and the Talaandig tribe from Sungco for their cooperation and logistical support.

LITERATURE CITED

Balslev, H., Luteyn, J., Øllgaard, B. & Holm-Nielsen, L. (1987). Composition and structure of adjacent unflooded and floodplain forest in Amazonian Ecuador. Opera Bot. 92: 37 – 57.

Boom, B. (1986). A forest inventory in Amazonian Bolivia. Biotropica 18: 287 – 294.

Brown, W. (1919). Vegetation of Philippine Mountains. The relation between the environment and physical types at different altitudes. Manila Bureau of Printing.

Cain, S., De Oliveira Castro, M., Murça Pires, J. & da Silva, M. T. (1956). Application of some phytosociological techniques to Brazilian rain forest. Amer. J. Bot. 43: 911 – 941.

Campbell, D., Daly, D., Prance, G. T. & Maciel, U. (1986). Quantitative ecological inventory of terra firme and várzea tropical forest on the Rio Xingú, Brazilian Amazon. Brittonia 38: 369 – 393.

Cottam, G. & Curtis, J. (1956). The use of distance measures in phytosociological sampling. Ecology 37: 451 – 460.

Erwin, T. L. (1991). Establishing a tropical species co-occurence database. Parts 1 – 3. Mem. Museo Hist. Nat., U.N.M.S.M. (Lima) 20: 1054.

Gentry, A. H. (1988). Changes in plant community diversity and floristic composition on environmental and geographicl gradients. Ann. Missouri Bot. Gard. 75: 1 – 34.

Grubb, P., Lloyd, J., Penningon, T. & Whitmore, T. (1963). A comparison of montane and lowland rain forest in Ecuador. I: The forest structure, physiognomy and floristics. J. Ecol. 51: 567 – 601.

Heaney, L. (1993). Survey of vertebrate diversity in Mt Kitanglad Nature Park, Mindanao. Unpublished manuscript, Philippine National Museum. Manila. 26 pp.

Henderson, A., Churchill, S. & Luteyn, J. (1991). Neotropical plant diversity. Nature 351: 21 – 22.

Holdridge, L. R. (1947). Determination of world plant formations from simple climatic data. Science 105: 367 – 368.

Korning, J., Thomsen, K. & Øllgaard, B. (1990). Composition and structure of a species rich Amazonian rain forest obtained by two different sample methods. Nord. J. Bot. 11: 103 – 110.

Madsen, J. & Øllgaard, B. (1994). Floristic composition, structure and dynamics of an upper montane rain forest in Southern Ecuador. Nord. J. Bot. 14: 403 – 423.

Madulid, D. (1987). Guide to botanical collecting density in the Philippines. Acta Manilana 36: 59 – 65.

—— (1991). The endemic genera of flowering plants in the Philippines. Acta Manilana 39: 47 – 58.

—— (1994a). Plant diversity in the Philippines. In: C.-I. Peng & C. H. Chou (eds), Biodiversity and Terrestrial Ecosystems. Institute of Botany, Academica Sinica, Monograph Series No. 14. Taipei, Taiwan. Pp. 105 – 109.

—— (1994b). A floristic and vegetation study of the Bicol National Park. Manuscript, Philippine National Museum. Manila. 109 pp.

—— (1995). Philippines. In: S. Davis, V. Heywood and A. Hamilton (eds), Centres of Plant Diversity- A guide and strategy for their conservation. Volume II. World Wildlife Fund & IUCN. London.

Meijer, W. (1959). Plant sociological analysis of montane rainforest near Tjibodas, West Java. Acta Bot. Neerl. 8: 277 – 291.

Merrill, E. D. (1923 – 26). An Enumeration of Philippine Flowering Plants. Bureau of Printing. Manila.

Mori, S., Boom, B., de Carvalho, A. M. & dos Santos, T. (1983). Ecological importance of *Myrtaceae* in an eastern Brazilian moist forest. Biotropica 15: 68 – 69.

—— & Boom, B. (1987). The Forest. In: S. A. Mori & collaborators. The *Lecythidaceae* of a lowland Neotropical forest: La Fumeé Mountain, French Guiana. Mem. New York Bot. Gard. 44: 9 – 29.

Pipoly, J. & Cogollo, A. (In prep.) Composition, diversity and structure of a promontane pluvial forest, Parque Nacional Natural "Las Orquídeas", Antiquia/Chocrd. 44: 9 – 29.

Prance, G. T., Rodrigues, W. & da Silva M. F., (1976). Inventário florestal de um hectare de mata de terra firme, km 30 Estrada Manaus-Itacoatiára. Acta Amazônica 6: 9 – 35.

Valencia, R., Balslev, H., Paz, G. y NiSilva (1976). Inventário florestal de um hectare de mata de terra firme, km 30 Estrada Manaus-Itacoa. In: T. Whitmore (ed.), Tropical Rain Forests of the Far East. Clarendon Press Oxford, 282 pp.

Yamada, I. 1975. Forest ecological studies of the montane forest of Mt Pangrango, West Java. I: Stratification and floristic composition of the montane rain forest near Cibodas. Tonan Ajia Kenkyu (Southeast Asian Studies) 13: 402 – 426.

—— 1976a. Forest ecological studies of the montane forest of Mt Pangrango, West Java. II. Stratification and floristic composition of the forest vegetation of the higher part of Mt Pangrango. Tonan Ajia Kenkyu (Southeast Asian Studies) 13: 513 – 534.

—— 1976b. Forest ecological studies of the montane forest of Mt Pangrango, West Java. III. Litter fall of the tropical montane forest near Cibodas. Tonan Ajia Kenkyu (Southeast Asian Studies) 14: 194 – 229.

—— 1977. Forest ecological studies of the montane forest of Mt Pangrango, West Java. Floristic composition along the altititude. Tonan Ajia Kenkyu (Southeast Asian Studies) 15: 226 – 254.

Character states and taxonomic position of the monotypic Sri Lankan *Schizostigma* (*Rubiaceae–Isertieae*)

C. Puff[1], A. Igersheim[2] & R. Buchner[1]

Summary. The following character states of the herbaceous, monotypic genus *Schizostigma*, endemic to the forests of Southwest Sri Lanka, were thoroughly investigated: vegetative characteristics, including the genus' trend towards anisophylly; inflorescence morphology; floral morphology and anatomy, notably features of the gynoecium; fruit morphology and anatomy, including fruit development; seed morphology and anatomy, with a detailed study of the exotesta structure; palynology, and chromosomes. Virtually all of *Schizostigma*'s character states fall within the range of variation found within the large genus *Sabicea*, a genus hitherto only known from tropical America, Africa and Madagascar. *Schizostigma* is, therefore, transferred to *Sabicea* and thought of as an "Asiatic extension" of that genus [*Schizostigma hirsutum* Arn. ex Meisn. *Sabicea ceylanica* Puff, nom. nov.]. A brief overview of the genera traditionally placed in the *Isertieae* and occurring in Asia shows that the tribal position of most of these is doubtful, insufficiently documented or possibly incorrect. Only the position of *Mussaenda* and *Schizomussaenda* in the tribe *Isertieae* is undisputed; the genus pair is, however, not likely to be closely allied to the Sri Lankan taxon.

Introduction

In both old and recent standard works dealing with rubiaceous genera, the status of *Schizostigma* Arn. ex Meisn. as a monotypic genus endemic to Sri Lanka has never been questioned. Also its placement in the tribe *Isertieae* as presently circumscribed (cf. Robbrecht 1988) has never been disputed. As regards its generic relationships, various older authors have expressed rather deviating views (see Discussion for details).

In an attempt to clarify the status and relationships of *Schizostigma*, a thorough investigation of its character states was carried out. In the course of this study it became obvious to us that *Schizostigma* shows numerous, conspicuous character state agreements with *Sabicea*, which we had investigated earlier (data unpublished).

Therefore in the present paper we not only present detailed information on *Schizostigma* but also compare its character states with those of *Sabicea*.

Materials and Methods

Detailed investigations were carried out on the following samples preserved in 70% ethanol (figure numbers in brackets; vouchers deposited at WU): *Schizostigma hirsutum* Arn., Sri Lanka, Sabaragamuwa Prov., Ratnapura Distr., Sinharaja forest, *Puff* 910803-1/1 (Fig. 1, 3A,C – E,G, 4A – B, D – E, 5A – B, 7A – C), 910806-1/3 (Fig. 5C – D, 6A – E). In addition, the collections of *Schizostigma* in the herbaria PDA, W and WU were investigated.

[1]Institute of Botany, University of Vienna, Rennweg 14, A-1030 Wien, Austria.
[2]Present address: Institute of Systematic Botany, University of Zurich, Zollikerstrasse 107,CH-8008 Zurich, Switzerland.

For comparison, herbarium specimens of *Sabicea* (mostly from WU) were studied, and the following preserved samples were investigated microscopically: *Sabicea discolor* Stapf (Guinea, *Puff & Bangoura* 900413-1/1, Fig. 3B); *Sabicea vogelii* Benth., (Sierra Leone, *Puff* 810728-2/3; Fig. 6H); *Sabicea* sp. (Guinea, *Puff & Bangoura* 900413-1/11, Fig. 3F and 900414-1/1, Fig. 4C, 5E, 7D).

Herbarium samples of *Pseudosabicea mildbraedii* (Wernh.) N. Hallé, Zaire, *Breyne* 2363 (BR) (Fig. 6F – G), were reconstituted for subsequent microtome sectioning using the method described in Puff *et al.* (1993a). Otherwise, methods follow Igersheim (1993).

Although *Schizostigma hirsutum* will be formally transferred to *Sabicea* (see Discussion), the generic name will be retained when comparing its character states with *Sabicea*. As *Schizostigma* is monotypic, the specific epithet is omitted throughout.

RESULTS

Growth form, habit and leaves

Schizostigma is a creeping perennial herb with adventitious roots at the lower nodes. The branched, prostrate stems produce numerous short, ± erect flower- and fruit-bearing branches (c.10 – 20 cm long). The plants are normally present in large colonies, comprised of entangled shoots of different individuals.

All vegetative parts are hirsute. Hairs are ± straight, uniseriate and multicellular, and c.1 – 2 mm long (all other hairy parts, except for the inside of the corolla, have this kind of hairs).

The decussately arranged leaves of *Schizostigma* are ± indistinctly to clearly anisophyllous (see Fig. 1 for a documentation of the variability in the expression of anisophylly both within an individual and between individuals).

The leaf blades are ovate-lanceolate, from c. 25 × 15 mm to c. 80 × 30 mm, and (densely) hairy on upper and lower surface, but especially dense along the veins and midrib. The hairy petioles are 5 – 25 mm long.

The stipules are entire, (broadly) ovate, acute at apex, membranaceous, ± persistent, often ± recurved and variable in size (sometimes only 2 × 4 mm, or bigger, 4.5 – 7.5 × 7 – 9 mm). They are glabrous or ± sparsely hairy on the outside, always glabrous inside, and often (very) hairy on the margins.

With regard to growth form, the large genus *Sabicea* shows a rather broad spectrum. According to Wernham (1914: 5) the majority of species are [climbing] shrubs, two are trees, and some "small, prostrate, and more or less herbaceous". However, it should be noted that some of the species Wernham (1914: 5) gave as examples for a ± herbaceous habit, have since been transferred to the allied genus *Pseudosabicea* (Hallé 1963).

Several *Sabicea* species (and also *Pseudosabicea* species originally described as *Sabicea*) show very conspicuous anisophylly (with one leaf of a pair being several times as large as the other; e. g. *Sabicea diversifolia*), but most are isophyllous.

According to Kirkbride (1979), pubescent *Sabicea* species have "incompletely septate" hairs, but our investigations of *Sabicea* species showed that the structure of the external hairs is the same as in *Schizostigma*.

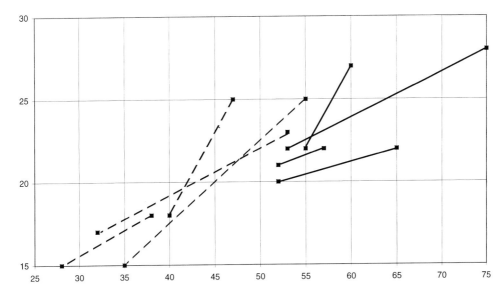

FIG. 1. Diagram documenting anisophylly in two *Schizostigma* plants from a population in Sinharaja forest (*Puff 910803-1/1*). Leaf blade lengths (x-axis) and widths (y-axis) in mm of small ("minus") and large ("plus") leaves at individual nodes. The "plus" and "minus" leaf from each node is connected by a straight line to show size relationships (solid and interrupted lines, respectively, refer to data from two different plants). The longer the connecting line between "plus" and "minus" leaf, the bigger the leaf blade length difference at a given node; the steeper the line, the bigger the width difference.

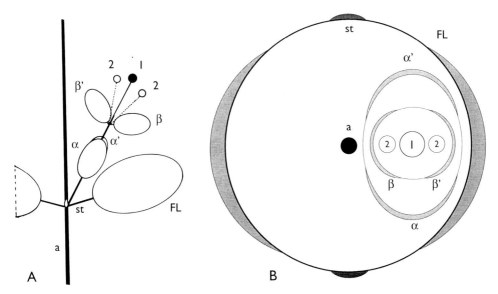

FIG. 2. Schematic presentation of the axillary inflorescence of *Schizostigma* (**A**) and corresponding ground plan (**B**). Abbreviations and explanations: a, axis; st, stipule; FL, foliage leaf; α, α', prophylls of lateral axis; β, β', pair of bracts (may be missing; bracts may not be strictly opposite); 1, terminal flower; 2, 2', lateral flowers (one or both are often absent). In the living plant, internodes in the inflorescence region are much congested and pedicels are subobsolete (cf. Fig. 3G); for greater clarity, they are shown elongated.

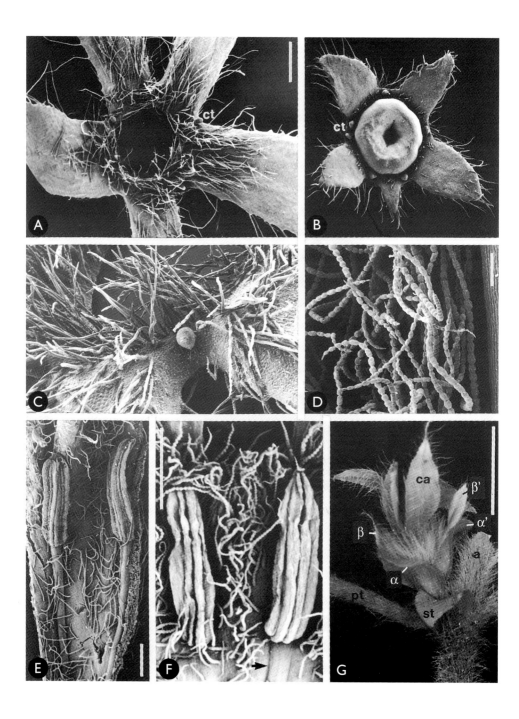

Sabicea, too, has entire stipules. They are sometimes similar to those of *Schizostigma* (e. g. *S. aristeguietae.* Steyermark 1967: Fig. 34C – D).

Neither vegetative nor fertile parts of *Schizostigma* and the studied *Sabicea* species have raphides.

Inflorescence

The inflorescences of *Schizostigma* are axillary and usually 1-flowered (Fig. 2, 3G); 2- or 3-flowered inflorescences are less common. Near the base of the peduncle there is a pair of prophylls (α, α' in Fig. 2), followed by another pair of bracts (β, β' in Fig. 2) which may either be no longer strictly opposite or be missing altogether. Prophylls and bracts are similar in shape and size to stipules in the vegetative region. Internodes in the inflorescence region are very short in bud and at anthesis. After fertilization, the pedicel of the developing fruit usually starts elongating (cf. figure of a presumably very young fruit with persistent calyx lobes in Beddome 1874: plate 95); in fully mature fruits, pedicels may reach a length of c. 6 mm. Peduncles and pedicels are densely hairy.

Sabicea also has axillary inflorescences. They are pedunculate or sessile, several- to many-flowered, panicle-like or capitate. The inflorescences of *Schizostigma* come closest to those of *Sabicea* sect. *Laxae* (sensu Wernham 1914).

Flower

Calyx

The 5-merous calyx of *Schizostigma* is differentiated into a very short basal tubular part and free calyx lobes. The latter are broadly to narrowly lanceolate, c. 6 – 9 × 1.5 – 4 mm, acute at the apex and (slightly) unequal in size and shape (within a solitary flower, calyx lobe length : width ratio can range from c. 2 : 1 to 4 : 1). The lobes are hairy, primarily on the margins, and a conspicuous, dense band of long hairs is present at the transition between the tubular part of the calyx and the free lobes (Fig. 3A, C). In the sinuses between the lobes there are ± sessile colleters (often only 1 per sinus; Fig. 3C).

The calyx, erect at first (bud, anthesis), starts spreading as the fruit develops. In fruit, the persistent calyx retains its distinctly leaf-like nature and does not become fleshy as in certain other *Rubiaceae–Isertieae* with berries (e.g. *Mycetia* spp., or *Leucocodon reticulatum*, where both persistent calyx and fruit wall become equally fleshy).

FIG. 3. Floral characteristics of *Schizostigma* (A, C – E, G) and, for comparison, of *Sabicea* (B, F). **A – B** calyx of *Schizostigma* (A) and *Sabicea discolor* (B); in A, the calyx has been removed from the ovary (thus the disk, visible in B, cannot be seen), and only part of its basal tubular part is left; note, in both A and B, the small colleters (ct) in the sinuses of the calyx lobes; **C** detail of calyx showing "external indumentum" and colleter; **D** moniliform hairs ("internal indumentum") from the inside of the corolla (transition mouth - base of corolla lobes); **E** artificially opened corolla tube showing anthers and strands of filament-like tissue running down the interior of the tube towards the base; **F** as E, but of *Sabicea vogelii* ; note strands of filament-like tissue (arrow) and moniliform hairs; **G** node with one-flowered, axillary inflorescence of *Schizostigma* [pt, petiole; ca, calyx lobe of flower shortly after anthesis (ovary hidden between bracts); other abbreviations as in Fig. 2]. Scale bars: 10 mm (G), 1 mm (A = B; E; F); 0.1 mm (C; D).

Sabicea, too, has a calyx with an indistinct to very short basal tubular part and variously shaped and sized free lobes. As already noted by Wernham (1914: 10), the lobes are frequently somewhat unequal (also cf. Fig. 3B). There are also colleters in the sinuses between the calyx lobes (and on adjacent parts of the calyx; Fig. 3B).

Corolla

The 5-merous corolla of *Schizostigma* is valvate in bud. The corolla tube, c.10 – 12 mm long, is (narrowly) funnel-shaped, c. 1 – 1.5 mm diam. at the base and c. 3 – 3.5 mm at the throat. The corolla lobes, c. 8 – 10 mm long, are narrowly triangular and spreading to recurved. Earlier descriptions, such as that of Schumann (1891) indicating the presence of "hooded" corolla lobe apices ("... mit 5 an der Spitze kappenförmig zusammengezogenen einwärts gekrümmten Lappen"), could not be confirmed.

The entire outside of the corolla is covered with hairs of the same type as on all other hairy parts; the inside (especially the lobes), however, has moniliform ("beaded") unicellular hairs (Fig. 3D), similar to those in e.g. *Pseudopyxis* (Puff 1990: Fig. 2F and 5E).

Sabicea has 4 – 5-merous corollas, valvate in bud, with ± cylindrical to narrowly funnel-shaped tubes and lobes which, in most species, are shorter than the tube. Corolla sizes vary greatly, the total length of the corollas ranging from < 10 mm to 10 cm. As in *Schizostigma*, moniliform hairs were noted in the interior of the corolla (Fig. 3F).

Androecium

The 5 stamens of *Schizostigma* are almost entirely included in the corolla tube, only the tips of the anthers are slightly exserted. The ± linear, dorsifixed anthers, c. 2.7 – 3.2 mm long, have a small, indistinct apical connective appendage. The actual free filaments, inserted below the middle of the anthers, are short, but filament-like strands (the vascular supply of the filaments), fused to the interior of the corolla tube, can be traced down to its base (Fig. 3E).

In *Sabicea*, the 4 – 5 stamens are included and inserted at or above the middle of the corolla tube. In short-styled morphs of distylous species, stamens are inserted near the throat, and the tips of the anthers may or may not be exserted (cf. Bridson & Verdcourt 1988: *Sabicea orientalis*, Fig. 71, 5 – 6). As in *Schizostigma*, filament-like strands, attached to the corolla, extend to the base of the tube but the actual, free filaments are short (Fig. 3F).

Gynoecium

In contrast to calyx, corolla, and androecium, the gynoecium of *Schizostigma* is no longer strictly 5-merous but 4 – 7-carpellate; that of *Sabicea* is 4- or 5-carpellate.

Style and Stigma

Schizostigma has a common style and filiform, spreading stigma lobes only c. 3 mm long and much shorter than the style. The uppermost part of the style and the stigmas are always exserted (the flowers are definitely not distylous!). The stigma lobes have their receptive surface, comprised of papillae/short hairs, only on the "upper" (adaxial) side.

The base of the style is surrounded by an annular disk, situated on the roof of the ovary (Fig. 4B).

Sabicea resembles *Schizostigma* in having a common style which is much longer than the filiform stigma lobes. The latter may or may not be distinctly exserted. Some species of *Sabicea* are distylous; in short-styled morphs, style and stigmas are always included (cf. Bridson & Verdcourt 1988: *Sabicea orientalis*, Fig. 71,6).

Ovary

The ovary of *Schizostigma* most commonly is 5-locular, occasionally only 4-locular (cf. Fig. 4D), or sometimes the number of locules is increased to 7. Baillon's (1880) characterization of *Schizostigma*'s ovary as being "5-locular (rarely 2 – 7-locular)" is misleading as it refers to *Schizostigma* s.l. (i.e., *Schizostigma*, including the 2-carpellate genus *Pentaloncha*).

The subglobose ovary is densely covered with ± long, straight, whitish hairs (the latter of the same structure as those in the vegetative region).

The placentas, each with numerous ovules, are T-shaped in ± median cross-section (Fig. 4A, D). In tangential section, the placentas appear oblong or very indistinctly cordiform. The latter is the case if the apical part of a placenta becomes 2-partite. Longitudinal ovary sections reveal that the area of insertion of a placenta to the septum extends from the upper end of the septum downwards to below the middle (Fig. 4B).

Vascularisation of the placentas is by means of bundles ascending from the base of the ovary through the septum (Fig. 4B) and then branching out into the placentas. Moreover, conspicuous, intensively stained pollen transmitting tissue is always visible in the upper part of the placenta (Fig. 4A). Longitudinal sections show that it eventually runs from the placenta upward (through the uppermost part of the septum) and then enters the base of the style (Fig. 4B).

The ovules, basically ± ellipsoid but often somewhat compressed laterally (Fig. 4E), are anatropous, with the micropylar canal and the micropyle itself pointing towards the placenta (Fig. 4B, 5A – B); no distinct obturator is present.

The ovary cross-section in Beddome (1874: plate 95) is erroneous and gives a misleading picture of the placentation and ovules.

Sabicea, having 4- or 5-locular ovaries with numerous ovules on the oblong (Hallé 1963) to cordiform (Kirkbride 1979) placentas, does not significantly differ from *Schizostigma*. Ovary cross-sections reveal a very similar situation (Fig. 4C). Also see ovary sections and preparation of various *Sabicea* species in Hallé (1963: pl. 2).

Fruit

Schizostigma has subglobose, white, many-seeded berries, c. 8 – 12 mm in diam. The fruits are densely covered with long hairs, and crowned with persistent calyx lobes.

During fruit development, the mesocarp becomes a very extensive and thick, loose, parenchymatic tissue which is distinctly watery or "spongy" (compare Fig.

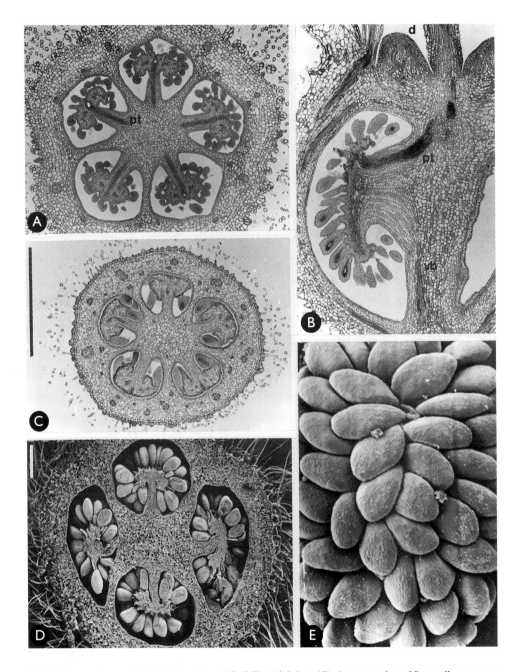

FIG. 4. Ovary and young fruit of *Schizostigma* (A-B, D-E) and *Sabicea* (C). **A** cross-section of 5-carpellate ovary at anthesis (pt, pollen transmitting tissue); **B** longitudinal section of ovary at anthesis showing insertion of placenta on septum, pollen transmitting tissue (pt), also note disk (d), base of style, corolla and calyx; the vascular supply of placentas and ovules (vb, vascular bundles) comes from the base of ovary; **C** as A, but of *Sabicea vogelii*; **D** SEM of cross-section of very immature fruit (4-carpellate); **E** SEM of very immature seeds attached to placenta (the latter hidden by the immature seeds). Scale bars: 1 mm (A = B = C; D), 0.1 mm (E).

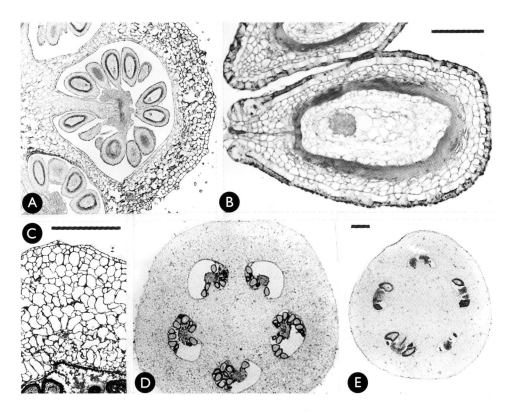

FIG. 5. Cross-sections of immature and mature fruits of *Schizostigma* (A-D) and *Sabicea* (E). **A** portion of immature fruit; note increase in size as compared to flowering stage (cf. Fig. 4A); **B** immature seed (same developmental stage as A), integument still several layered, integument epidermis (later exotesta) dark-stained (tannin-filled), also note endosperm differentiation into intensively stained (dark), dense outer and unstained, loose inner layers and developing embryo (dark; in endosperm cavity); **C** portion of fruit wall from mature fruit; note marked increase in fruit wall thickness as compared to A (same magnification); **D-E** mature fruits of *Schizostigma* (D) and *Sabicea vogelii* (E); note that in both not only the actual fruit wall but also all other parts (septa, central part) show a loose, large-celled parenchymatic tissue (soft and juicy in nature); as compared to earlier developmental stages (cf. A), these tissues have markedly increased in size so that the locules now appear relatively small. Scale bars: 1 mm (A = C; D = E), 0.1 mm (B).

4A,D, 5A and 5C – D). The tissue forming the septa, as well as the tissue in the centre of the fruit, undergo a similar change. Consequently, the diameter of the fruit increases manifold between the onset of fruit development and the mature fruiting stage. Since the size of the locules remains unchanged, they appear relatively small in the mature fruit (Fig. 5D). Due to the size increase of all other tissues, the placentas — also having become somewhat fleshy but not markedly enlarged — become rather indistinct (Fig. 5D).

Mature fruits do not show a trace of a distinct, sclerenchymatic endocarp and thus are "true" berries (as opposed to rubiaceous, many-seeded berry-like fruits with a thin, yet distinct sclerenchymatic endocarp layer).

Sabicea has virtually identical fruits (also ± globose, "true" many-seeded berries crowned by persistent calyx lobes), ranging in colour from white to red, purplish, dark violet and even blackish.

Anatomically, its fruits are indistinguishable from those of *Schizostigma* (cf. Fig. 5E, and also relevant figures in Hallé 1963: pl. 2). When dealing with the *Sabicea–Pseudosabicea–Ecpoma* alliance Hallé (1963), working on African material, noted that *Sabicea* is the only genus of the three in which a red juice is exuded when mature fruits are squashed (although it is unknown whether this is true for the entire genus).

Seed

Schizostigma has light brown seeds which are somewhat variable and irregular in shape (already noticeable in the young seeds depicted in Fig. 4E) and size (c. 0.4 – 0.7 mm long). They are often ovoid or ellipsoid, ± angled at the micropylar end and frequently laterally compressed (Fig. 4E, compare with Fig. 6A).

The centrally located embryo (ratio radicle : cotyledons c. 1:1) is relatively large, and surrounded by only 2 – 3 layers of large, starch-filled endosperm cells (Fig. 6A,E; also cf. Rodriguez 1976: photo 4).

The exotesta cells are elongated and ± irregular in outline (Fig. 6B). Cross-sections (Fig. 6D – E) reveal that the radial walls have extensive secondary thickenings restricted to the lower two thirds (or lower half) of the walls with knob- or wart-like outgrowths (also see Fig. 6B – C); the uppermost parts of the walls lack any secondary thickenings. The tangential walls are unthickened, except for occasional, thickened transverse "bridges" on the inner tangential walls (cf. surface views, Fig. 6B – C).

In microtome sections stained with toluidine blue, the lumen of the exotesta cells appears very dark blue (except for some empty, "white" areas), indicating that they are filled with (presumably) tannins. Since the lower parts of the radial walls show massive thickenings but their uppermost parts and the outer tangential wall are unthickened, the dark cell contents of each cell appear T-shaped in section (Fig. 6E). After artificial removal of the outer tangential walls and the unthickened upper parts of the radial walls followed by critical-point drying, the cell contents have largely fallen out, although coagulated remnants are sometimes still visible (Fig. 6B). On careful investigation, it is seen that there are some crushed endotesta layers below the exotesta, which — like the lumina of the exotesta cells — are dark stained (Fig. 6E – F). Their presence often makes it difficult to correctly interpret the testa structure.

FIG. 6. Seed and exotesta of *Schizostigma* (A-E) and, for comparison, of *Sabicea* (H) and *Pseudosabicea* (F-G). A cross-section of mature seed showing, from out- to inside, exo- and endotesta, endosperm and, in the endosperm cavity, the sectioned cotyledons of the embryo; **B** SEM of exotesta (outer tangential walls removed). Note remnants of coagulated cell contents (arrow); **C** as B, detail of exotesta cells showing "bridges" on inner tangential walls (arrow); **D** SEM-graph of exotesta cells in cross-section, note unthickened upper halves of the radial walls (arrow) and unthickened outer tangential walls; **E** corresponding microtome section; because of the dark-stained contents of the exotesta cells, the unthickened upper halves of the radial walls are hardly discernible (arrow), the thickened parts of the radial wall appear light coloured; crushed endotesta layers (appearing dark) are visible below the exotesta and, at the bottom, the outermost cell layer of endosperm; **F** as E, but of *Pseudosabicea mildbraedii*; **G** tangential section of exotesta cells of *Pseudosabicea mildbraedii*; **H** as C, but of *Sabicea vogelii*. Scale bars: 0.1 mm (A; B; G), 10 µm (C = D = H; E = F).

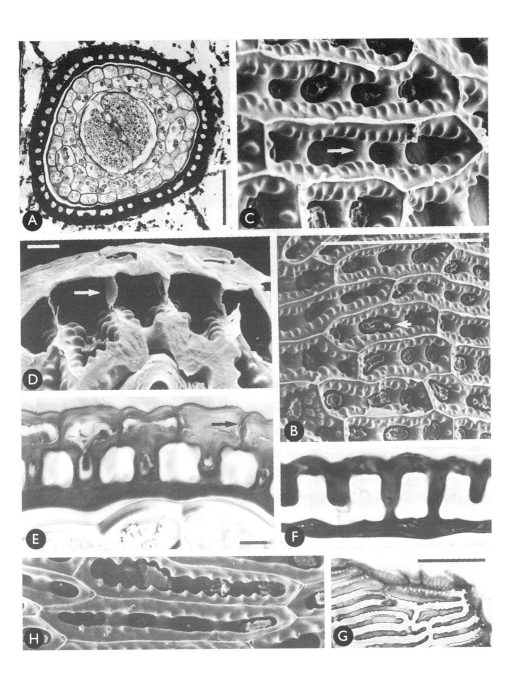

The seeds of *Sabicea* are virtually indistinguishable from those of *Schizostigma*. They are of equal size (0.4 – 0.6 mm in neotropical taxa: Kirkbride 1979) and shape. Their exotesta, too, is similar although there is some variation with regard to the shape of the cells (either longer, more elongated: e. g. in the African *Sabicea vogelii*, Fig. 6H, or shorter, irregularly rectangular: e.g. in the neotropical *Sabicea umbellata*: Kirkbride 1979: Fig. 16). Both in *Sabicea* and in the African segregate genus *Pseudosabicea* (Fig. 6F – G), the exotesta cells are anatomically almost identical to *Schizostigma* in that they also possess they same kind of local secondary thickenings (thickenings not encompassing the entire length of the radial walls; presence of knob-like outgrowths on the thickened radial wall portions; locally thickened "bridges" on the inner tangential walls: cf. Kirkbride 1979: Fig. 16).

Pollen

The pollen of *Schizostigma* (Fig. 7A – C) is tetra- or pentacolpate (grains with 4 and 5 apertures are found within an individual flower) and suboblate (P = 19 – 30 µm; E = 15 – 23 µm). The colpi are short (length : width ratio c. 2:1). The exine is finely reticulate.

Sabicea appears to vary to some extent. While pollen of *Sabicea vogelii* (Fig. 7D) agrees with *Schizostigma* in size, shape, number and nature of the apertures and exine structure, Graham (1987), having studied both fossil and modern pollen of American *Sabicea* species, reported some deviating features (notably oblate pollen shape, and the presence of three apertures). His description of the pollen being "tricolpate/porate" actually refers to the presence of short colpi as he gives an aperture length : width ratio of c. 2: 1. With regard to pollen shape, it is not possible to confirm his data as he only included P-values (32 – 36 µm).

Chromosomes

For *Schizostigma*, Kiehn (1986) reported a chromosome number of 2n = 4x = 40 – 44 ("tetraploidy on a basic number of x = 10 or 11").

All karyologically known taxa of *Sabicea* are also tetraploid (Kiehn 1985, 1986, and literature cited therein). However, *Sabicea* has two base numbers, either x = 9 or x = 11. X = 9 is known from Africa and Madagascar and x = 11 from both America and Africa. As chromosome numbers are only available for two American, three African and one Malagasy species (i.e. for less than 10% of the species), it remains to be seen whether the geographic distribution of these two base numbers really "holds", and whether they can be linked to any morphological or other characters.

Distribution

Schizostigma is endemic to evergreen rain forest areas of southwest Sri Lanka. *Sabicea* is widely distributed. In the Neotropics, it occurs from Mexico south to Peru, Bolivia and Brazil (Kirkbride 1979). In the Paleotropics, it is known from West to East tropical Africa and also from Madagascar (four species only). The genus has

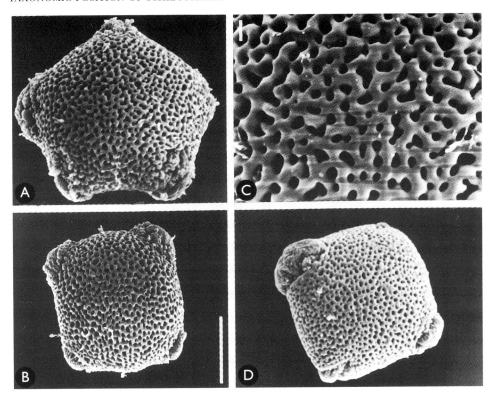

FIG. 7. Pollen of *Schizostigma* (A-C) and *Sabicea* (D). **A-B** 5- and 4-colpate grain from one and the same flower, equatorial view; **C** apocolpium; **D** *Sabicea vogelii*, 4-colpate grain, equatorial view. Scale bars: 10 μm (A = B = D), 1 μm (C).

c.120 species which are "about equally divided between tropical America and tropical Africa with 4 in Madagascar" (Bridson & Verdcourt 1988). Wernham (1914: 45) had doubtfully included Mauritius in the distribution range but, according to Verdcourt (1989), *Sabicea* is not recorded from the Mascarene Islands.

DISCUSSION

In older literature, various — in part very diverse — ideas have been expressed with regard to the relationships of *Schizostigma*.

Arnott (1839) presented a very good description of the genus which he stated "may be placed near *Sabicea*". This important statement was apparently overlooked later on.

Hiern (1877) suggested a relationship between the African genus *Pentaloncha* and *Schizostigma* (however, there is little evidence to support this view; see below).

Baillon's (1880) concept of *Schizostigma* was very broad. He included both *Pentaloncha* and *Temnopteryx* in the genus, stating they present "sections africaines" of *Schizostigma* with multiflorous inflorescences (Baillon l.c.: 320). However, he also noted on the same page that *Schizostigma* [s.l.] is very closely allied to the African genus *Stipularia* (differing in lacking the large involucral bracts of the latter) and, in turn, pointed out the close relationship between *Stipularia* and *Sabicea*. Consequently, in the systematic part of his work the relevant genera appear in the following sequence (pp. 451 – 453): [... 110. *Adenosacme* (= now *Mycetia*) –] 111. *Sabicea* – 112? *Stipularia* – 113. *Schizostigma* (including *Pentaloncha* and *Temnopteryx*) [– 114. *Urophyllum* (s. l.; URO., PAU.)[1]].

Hooker (1873) had apparently different ideas about the circumscription of *Schizostigma* and about relationships of the genera in question (all of which are grouped under tribe *Mussaendeae* by him). The relevant sequence (pp. 71 – 74) is: ... 120. *Urophyllum* (s. l.; URO., PAU.) – 121. *Sabicea* – 122. *Schizostigma* – 123. *Temnopteryx* – 124. *Pentaloncha* – 125. ? *Patima* (= now *Sabicea*) ... – 127. *Lecananthus* – 128. *Stipularia*.

Schumann (1891), again, rearranged the genera. They are all placed in "*Cinchonoideae - Gardeniinae - Mussaendeae*" and appear in the following sequence (pp. 65 – 69): ... 119. *Schizostigma* – 120. *Pentaloncha* – 121. *Coccocypselum* (COC.) – 122. *Temnopteryx* ... – 126. *Urophyllum* (s. l.; URO., PAU.) – 127. *Sabicea* – ... – 130. *Patima* [= now *Sabicea*] ... – 137. *Stipularia*.

Wernham, in his *Sabicea* monograph (1914: 14 & Fig. 2), suggested that section *Laxae* of *Sabicea* shows affinities to *Urophyllum*, and section *Capitatae* affinities to both *Coccocypselum* (COC.) and *Stipularia* (with *Sabicea dewevrei* being a transitional species).

Hallé (1966) dealt with several African "*Mussaendeae*" (*Isertieae*) genera mentioned above. He divided the tribe into the two subtribes *Mussaendinae* and *Urophyllinae*, and amongst each, he informally distinguished between 2- and 3 – 5-carpellate genera. As 3 – 5-carpellate *Mussaendinae* genera he listed the following taxa that concern us here: *Temnopteryx*, *Stipularia* and *Sabicea*. In the *Urophyllinae* he included genera some of which were previously included in *Urophyllum* (*Pauridiantha* and allies) and also (among the 2-carpellate taxa) the genus *Pentaloncha*. When dealing with the latter (p. 229) he also commented on Hiern's (1877) suggested relationship between *Pentaloncha* and *Schizostigma* noting that, in his opinion, it merely refers to resemblance in growth form and habit ("... fondé sur une ressemblance de port et non sur des structures véritablement parentes"). We agree with the view of Hallé (1966). In our opinion there is no solid evidence for a direct alliance between the African *Pentaloncha* and the Sri Lankan *Schizostigma*.

The above historical survey shows that *Schizostigma* has, in general, often been (loosely) associated with *Sabicea* or, more widely, the *Sabicea* genus complex (*Sabicea* and African allies or "satellites" *Stipularia*, *Temnopteryx*, and *Pseudosabicea* and *Ecpoma*; for the delimitation of the latter see Hallé 1963). Our comparative investigations suggest that in *Schizostigma* and *Sabicea* there are really no character states by which the two genera can be reliably separated. The only reason that could be brought forward in favour of keeping them as separate entities would be geographical distribution. On the other hand, we are inclined to believe that this is not a very convincing argument. Since *Sabicea* (as previously circumscribed) is

---

[1]Current tribal position (three letter acronyms after Robbrecht 1988, 1993); only given for those genera which at present are placed in tribes OTHER THAN *Isertieae* (syn. *Mussaendeae*).

widely distributed (America – Africa – Madagascar), inclusion of Sri Lanka in its range would only mean a further eastward extension, possibly an old "Gondwanaland extension". The derived character states of the Sri Lankan taxon (perennial herbacous habit, reduced inflorescences, etc.) could indicate that its ancestral stock arrived early and underwent drastic changes in the course of time. Our general conclusion is that *Schizostigma* needs to be merged with *Sabicea*. The genus and species is formally transferred below.

If the Sri Lankan *Schizostigma* is included in *Sabicea*, what then are its allies in Southeast Asia? This question is not easy to answer because the status and position of several Asiatic "*Isertieae*" genera is problematic.

Acranthera has previously been associated and included in the *Isertieae* (e.g. Robbrecht 1988), but the presence of a complicated anther–style–stigma complex (cf. Puff *et al.* 1995) makes it unlikely that this can be upheld. *Aphaenandra* is so poorly documented that, at present, it is not possible to draw any conclusions on its tribal position. *Indopolysolenia* (syn. *Polysolenia*), a genus with lid-capsules, is probably misplaced in the *Isertieae*. Lo (1993) transferred the type species (*I. (P.) wallichii*) to *Leptomischus*, a genus of the tribe *Hedyotideae*. *Keenania* (including *Campanocalyx*) and *Myrioneuron* (the latter possibly congeneric with the former) need reinvestigation and detailed documentation of their character states. We would not be surprised if such a study yields evidence for their exclusion from the *Isertieae* and transfer to the *Hedyotideae*. *Mycetia* (syn. *Adenosacme*), widely distributed from India east and south-eastward to South China, Indochina and West Malesia, does show some character state agreements with *Sabicea* (e.g. anatomically very similar "true" berries; often a clear trend towards anisophyllous leaves; Puff *et al.*, unpubl. and Puff pers. obs.). However, the genus differs markedly in other characteristics (e.g. the presence of raphides) so that we presently exclude the possibility of a close phylogenetic relationship between the two genera. *Mycetia* may even be misplaced in the *Isertieae*; detailed investigations are required for confirmation.

The position of the remaining two genera in the *Isertieae* occurring in Asia, *Mussaenda* (Afro-Asiatic; c.100 spp.) and its close ally, the monospecific Asiatic *Schizomussaenda*, on the other hand, is undisputed. However, the genus pair is not likely to be closely allied to *Sabicea* (incl. *Schizostigma*) as it considerably deviates in numerous character states (e.g. the frequent presence of enlarged, petaloid showy calyx leaves or fruit structure; cf. Puff *et al.* 1993b).

Sabicea *Aubl.*, Pl. Guian. 1: 192, tab. 75, 76 (1775).

Schizostigma Arnott ex Meisner, Pl. Vasc.Gen. 1: 164, 2: 115 (1838); Arnott in Ann. Nat. Hist. 3: 20 (1839). **Synon. nov.**

Sabicea ceylanica *Puff* nom. nov. Type: "Ceylon", Finlayson [Wall. Cat. 8463] (holotype K-WALL!).

Schizostigma hirsutum Arnott ex Meisner, Pl. Vasc.Gen. 1: 164, 2: 115 (1838); Arnott in Ann. Nat. Hist. 3: 20 (1839); Trimen, Handb. Fl. Ceylon 2: 327 (1894).

In all of the old literature, Arnott is given as the author of *Schizostigma*. As explained by Bremekamp (1947: 265) for *Acranthera*, another genus attributed to Arnott by older authors, this is incorrect. Arnott sent his description to Carl Friedrich Meisner (not Meissner; hence "*Schizostigma* Arn. ex Meissner", as in Farr et al. (1979) and Mabberley (1987), needs to be corrected).

A new name is necessary because there is a Central American species *Sabicea hirsuta* H.B. & K., Nov. Gen. & Sp. 3: 417 (1818).

ACKNOWLEDGEMENTS

Support of this study by the "Fonds zur Förderung der Wissenschaftlichen Forschung" (Project P10499-BIO) is gratefully acknowledged. Dr. Karin Gruber is thanked for the microtome work and technical assistance.

REFERENCES

Arnott, G. A. W. (1839). Descriptions of some new or rare Indian plants. Ann. Nat. Hist. 3: 20 – 23.

Baillon, H. (1880). Histoire des plantes 7. Hachette, Paris.

Beddome, R. H. (1874 [1868-1874]). Icones plantarum Indiae orientalis; or plates and descriptions of new and rare plants, from southern India and Ceylon. ... Vol. 1. Gantz Brothers, Madras.

Bremekamp, C. E. B. (1947). A monograph of the genus *Acranthera* Arn. ex Meisn. J. Arnold Arb. 28: 261 – 308.

Bridson, D. M. & Verdcourt, B. (1988). *Rubiaceae* (part 2): 415 – 747. In: R. M. Polhill (ed.), Flora of Tropical East Africa. Rotterdam, Brookfield, A.A. Balkema.

Farr, E. R., Leussink, J. A. & Stafleu, F. (1979). Index Nominum Genericorum (Plantarum). Regn. Veget. 100 – 102: 1-1896.

Graham, A. (1987). Fossil pollen of *Sabicea* (*Rubiaceae*) from the lower Miocene Culebra Formation of Panama. Ann. Missouri Bot. Gard. 74: 868 – 870.

Hallé, N. (1963). Délimitation des genres *Sabicea* Aubl. et *Ecpoma* K. Schum. en regard d'un genre nouveau: *Pseudosabicea* (*Mussaendeae* – *Rubiaceae*). Adansonia, sér. 2, 3: 168 – 177.

——— (1966). Flore du Gabon 12. Famille des Rubiacées (1re partie). Mus. Nation. Hist. Nat, Paris.

Hiern, W. P. (1877). Order LXX. *Rubiaceae*: 33 – 247.D. In: D. Oliver (ed.), Flora of tropical Africa 3. Reeve, London.

Hooker, J. D. (1873). Ordo LXXXIV. *Rubiaceae*: 7 – 151. In: G. Bentham & J. D. Hooker (eds), Genera plantarum ad exemplaria imprimis in herbariis kewensibus servata defirmata, 2. Reeve & Co, London.

Igersheim, A. (1993). The character states of the Caribbean monotypic endemic *Strumpfia* (*Rubiaceae*). Nordic J. Bot. 13: 545 – 559.

Kiehn, M. (1985). Karyosystematische Untersuchungen an *Rubiaceae*: Chromosomenzählungen aus Afrika, Madagaskar und Mauritius. Pl. Syst. Evol. 149: 89 – 118.

Kiehn, M. (1986). Karyosystematic studies on *Rubiaceae*: Chromosome counts from Sri Lanka. Pl. Syst. Evol. 154: 213 – 223.

Kirkbride, M. C. G. (1979). Review of the neotropical *Isertieae*. Brittonia 31: 313 – 332.

Lo, H.-S. (1993). A revision of the genus *Leptomischus* Drake. Acta Phytotax. Sinica 31(3): 273 – 276. [in Chinese]

Mabberley, D. J. (1987). The plant-book. A portable dictionary of the higher plants [reprinted with corrections 1989]. Cambridge Univ. Press, Cambridge.

Puff, C.(1990 ["1989"]). Observations on the Japanese endemic *Pseudopyxis* (*Rubiaceae-Paederieae*). Pl. Sp. Biol. 4(2): 131 – 144.

——, Andersson, L., Rohrhofer, U. & Igersheim, A. (1993a). The tribe *Schradereae* (*Rubiaceae*) reexamined. Bot. Jahrb. Syst. 114: 449 – 479.

——, Igersheim, A. & Rohrhofer, U. (1993b). *Pseudomussaenda* and *Schizomussaenda* (*Rubiaceae*): close allies of *Mussaenda*. Bull. Jard. Bot. Nat. Belg. 62: 35 – 68.

——, ——, Buchner, R. & Rohrhofer, U. (1995). The united stamens of Rubiaceae. Morphology and anatomy; their role in pollination ecology. Ann. Missouri Bot. Gard. 82: 357 – 382.

Robbrecht, E. (1988). Tropical woody *Rubiaceae*. Opera Bot. Belg. 1: 1 – 271.

—— (1993). Supplement to the 1988 outline of the classification. Index to genera. In: E. Robbrecht (ed.) Advances in *Rubiaceae* Macrosystematics. Opera Bot. Belg. 6: 173 – 196.

Rodriguez, P. (1976). Estudios sobre frutos carnosos y sus de las *Rubiaceae* de Venezuela. Acta Bot. Venez. 11: 283 – 383.

Schumann, K. (1891). *Rubiaceae*: 1 – 156. In: A. Engler & K. Prantl (eds), Die natürlichen Pflanzenfamilien 4(4). Engelmann, Leipzig.

Steyermark, J. A. (1967). *Rubiaceae*. In: Maguire, B. and collaborators: Botany of the Guayana Highlands – Part VII. Mem. New York Bot. Gard. 17: 230 – 436.

Verdcourt, B. (1989). 108. Rubiacées: 1 – 135. In: J. Bosser, T. Cadet, J. Guého & W. Marais (eds), Flore des Mascareignes. Sugar Ind. Res. Inst. Mauritius.

Wernham, H. F. (1914). A monograph of the genus *Sabicea*. Brit. Mus. (Nat. Hist), London.

The continuing story of *Spatholobus* (*Leguminosae-Papilionoideae*) and its allies

Jeannette Ridder-Numan[1]

Summary. A phylogenetic reconstruction of *Spatholobus* and the allied genera *Butea* and *Meizotropis* is given. This cladogram, together with cladograms of the genera *Fordia*, *Genianthus*, and *Xanthophytum*, is used for a biogeographical analysis of Southeast Asia. The geological background is summarized, as well as other historical information such as fossil evidence. The distribution areas (based upon the common distributions of the genera mentioned) coincide with several geological terranes in the west Malesian Archipelago, but it is not yet clear if this is the result of historical distributions or other factors. *Leguminosae* in the area originate from the Eocene, and according to fossil evidence they arrived there relatively late. The cladogram indicates an initial speciation event on mainland Southeast Asia, followed later by events in the Malesian archipelago and Peninsular Malaysia.

INTRODUCTION

During the Second Flora Malesiana Symposium in Yogyakarta I presented the initial results of my project on the phylogeny and biogeography of *Spatholobus* and its allies (*Leguminosae-Papilionoideae*). This paper outlines the continuation of this research.

The genus *Spatholobus* consists of 29 species of woody climbers in Southeast Asia and the west Malesian Archipelago. Two closely allied small genera, *Butea* and *Meizotropis*, occur on the mainland of Southeast Asia. (*Kunstleria* is chosen as outgroup for the phylogenetic analysis, because of the resemblance in flower and in the shape of the leaves; furthermore it has some less advanced characters, e.g. usually more than three leaflets per leaf and a flat, strap-like pod.) Revisions of these groups are provided respectively by Ridder-Numan & Wiriadinata (1985), Ridder-Numan (1992), Sanjappa (1987) and Ridder-Numan & Kornet (1994).

The phylogeny is used for the cladistic biogeographical analysis in combination with the distributions of taxa. Apart from *Spatholobus* and allies, I have used other groups occurring in the region for this analysis: *Fordia* (*Leguminosae-Papilionoideae*), of which a phylogenetic reconstruction is provided by Schot (1991), the genus *Genianthus* (*Asclepiadaceae*) revised by Klackenberg (1995), and the genus *Xanthophytum* (*Rubiaceae*) by Axelius (1990).

[1] Rijksherbarium/Hortus Botanicus, Postbus 9514, 2300 RA Leiden, The Netherlands.

MATERIAL & METHODS

For the phylogenetic analysis, 97 characters including ten leaf anatomical and seven pollen morphological characters were used. This provided a data matrix of 33 taxa (all species of *Spatholobus* except one for which no flower characters were available, *Butea*, *Meizotropis*, and *Kunstleria* as a whole and scored as a single taxon) and 97 characters, several of which are polymorphic multistate characters.

The analysis was carried out using PAUP (Swofford 1991; Mac version). Because of the size of the data matrix it was only possible to run a heuristic search (TBR, Tree-Bisection-Reconnection, which is the most extensive heuristic search option). All character states were used unordered. The option 'addition sequence random' was used instead of 'addition sequence simple', because in this way more most parsimonious trees were found. With 'addition sequence simple' it turned out that the analysis led to only one 'island' of parsimonious trees, which were not even necessarily the most parsimonious ones. Using 'addition sequence simple' and different reference taxa led to different sets of most parsimonious trees.

For the biogeographical analysis PAUP was used again. In order to obtain a more general area cladogram, phylogenetic reconstructions of other taxa, mentioned in the introduction, were added to the analysis. The matrix for this analysis contained 170 taxa (including the ancestors-nodes) and 30 areas (including the all-zero outgroup).

A map with the areas (Fig. 4) was constructed on the basis of common distributions of taxa in the analysis. Some of the areas contain endemics, others comprise only parts of more widespread taxa. The areas were chosen in such a way that they do not overlap. Part of the cladogram of *Xanthophytum* was not used in the analysis. This was a monophyletic group of taxa inhabiting areas occurring east of Sulawesi. As the introduction of areas inhabited by only one taxon causes them to be placed in a basal position in the area cladogram, this group was left aside in the analysis. In addition, all areas were defined specifically to fit the taxa that are used in the analysis. Using other groups can change the area definition, e.g. Sulawesi could be split into two parts. However, for the taxa used in this analysis this would be uninformative, and could cause the same problems with an artificial low place in the cladogram.

See Ridder-Numan (1996) for more details.

PHYLOGENETIC ANALYSIS

The results of the phylogenetic analysis were two sets of three nearly similar trees. The length of the trees was 589 steps and the consistency index was 0.46.

The trees were different in only a few clades. The first tree, Fig. 1a, has *Spatholobus pottingeri* placed basal to the rest of *Spatholobus*, and *S. parviflorus* included in the *S. ferrugineus* clade. In the other two trees (Figs 1b&c), *S. pottingeri* has a position between the *ferrugineus*-clade near *S. suberectus* and *S. pulcher*. *S. parviflorus* is situated in these two trees (Figs 1b&c) basal to all the other species of *Spatholobus*. The position of *S. pottingeri* and *S. parviflorus* in cladograms b and c is preferable on morphological grounds to that in a. There are no good arguments to choose between cladograms b and c.

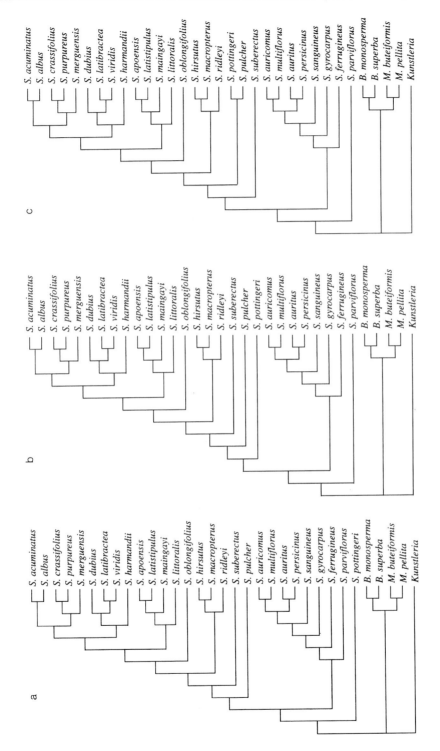

FIG. 1. Cladograms for *Butea*, *Meizotropis*, *Spatholobus* with *Kunstleria* as outgroup. Length = 389 steps, ci = 0.46.

The cladogram chosen for the biogeographical analysis is shown in Fig. 2. For a more extensive exposé on the phylogeny see Ridder-Numan (1996).

The genus *Spatholobus* is distributed in two slightly overlapping areas, which contain both widespread and endemic taxa. These areas are India and continental Southeast Asia up to the border of southern Thailand, and the West Malesian Archipelago including the Malay Peninsula up to the Isthmus of Kra. When this distribution is plotted on the cladogram (Fig. 2) we see that the lower branches contain species occurring on the mainland of Southeast Asia, and that from a certain point in the cladogram speciation occurs mainly in the area south of the Isthmus of Kra. There are a few exceptions. One clade is entirely continental (*S. crassifolius* clade), and a few species occur on both sides (*S. acuminatus, S. harmandii* and *S. gyrocarpus*). From this individual area cladogram one could conclude that speciation started on the continent, that an invasion of the Archipelago took place (or a vicariance event leading to continued separation of the Archipelago), and that a few times there must have been contact again with the continent (probably in one of the Ice Ages during low sea levels).

BIOGEOGRAPHICAL ANALYSIS

For the biogeographical analysis the above mentioned cladogram of *Spatholobus* and allies and their distribution was used, as well as those for *Fordia* (Schot 1991), *Genianthus* (Klackenberg 1995), and *Xanthophytum* (Axelius 1990).

When the areas were coded zero in the case of missing taxa, the analysis resulted in 19 trees of 366 steps and a consistency index of 0.45 (Fig. 3). The 19 cladograms are different in only a few clades.

On the area cladogram a widespread species of *Spatholobus*, *S. ferrugineus*, is plotted. This species occurs in the archipelago (except the Philippines), and the cladogram shows, in a very coarse way, the areas with a more monsoon type of climate versus those with a mainly wet climate.

Some areas may be more basally placed in the cladogram than they should be. This is most likely to be artificial and caused by the fact that only one or two taxa occur in that area. Thus the Nicobar Islands is a 'one-taxon' area.

In this cladogram the continental areas are found low in the cladogram, while the areas in the Malesian Archipelago and on the Malay Peninsula are placed higher up. This implies that these are sister areas and that they had their own history after an event which either split up the areas, or was a period of possible exchange of taxa. Furthermore, there is a division higher in the tree between the Bornean terranes and the Malay Peninsula implying that these two sets of areas also had their own history after a certain point in time.

It may seem rather strange to have Southeast Sumatra as a sister area to East Malaya, while the rest of Sumatra is more basal, but there is some geological support for this, e.g. the presence of a similar fossil flora.

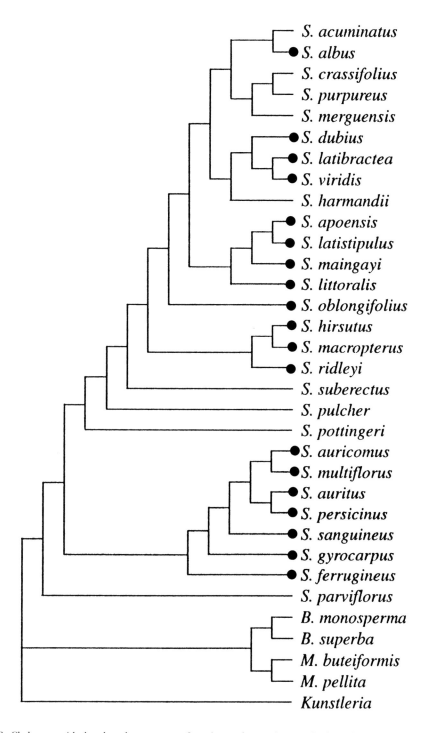

FIG. 2. Cladogram with the plotted occurrence of species on the continent or in the archipelago (●)

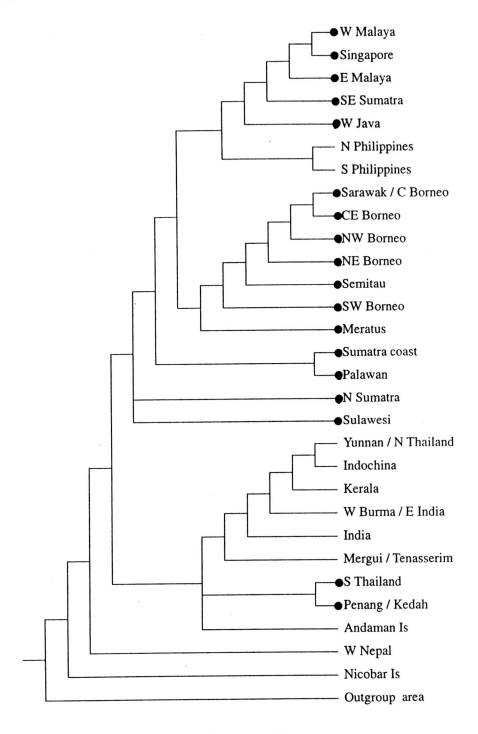

Fig. 3. Generalised area cladogram. Length = 366 steps, ci = 0.45.

HISTORICAL INFORMATION

In order to get an idea of the possible historical explanations for the area cladograms the geological background will be given, as well as some information on the fossil evidence and past climates. Some of the main tectonic structures are shown in Figure 5.

Geological background

There have been three major rifting events from the Gondwanaland margin, bringing slivers of this margin towards what is now Southeast Asia (Metcalfe 1994).

The first rifting event (Fig. 6), which took place in the Palaeozoic, was that of North China, South China and Indochina together with the East Malay Peninsula and south-east Sumatra. On their way northwards Indochina sutured against South China along the Song Ma suture (Fig. 5).

The second event (Fig. 7) was that of Sibumasu (east Burma, west and south Thailand, the West Malay Peninsula, and west Sumatra, Fig. 5), Lhasa, Qiantang and Iran and Turkey. They were united in the Cimmerian continent in their first rifting phase in the Permian, but later this continent broke up and Iran and Turkey rifted more to the west. In the Early Triassic, Sibumasu sutured to Indochina along the Uttaradit-Nan and the Bentong-Raub sutures (Fig. 5).

The third rifting phase (Fig. 8), in the late Jurassic, was that of west Burma, parts of the coast of Sumatra, Mangkalihat in Borneo and possibly West Sulawesi. West Burma sutured to Southeast Asia along the Sagaing Fault (Fig. 5).

At the same time India, a much larger part, separated in the late Jurassic and moved rapidly northwards. India started to collide with Eurasia in the Eocene, about 55 Ma (Fig. 9). The collision did not immediately cause an uplift of the Himalayas; this happened later (late Eocene; 45 Ma). The indentation of India in Eurasia resulted in rotation of the Southeast Asiatic region. This rotation was followed by the opening of basins such as the Thai basin and the South China Sea. The last opening caused rifting of the Chinese margin of North Palawan/ Mindoro, parts of Borneo, the Reed Bank and Dangerous grounds. The elements rifted southwards towards Borneo.

Looking at Borneo, the West Borneo Basement Block may have come from Indochina due to rifting caused by the opening of the Proto-South China Sea (Hutchison 1989 & 1992). The Meratus range, which accreted to it in the south-east, formed once part of the South China magmatic belt, under which the Pacific and Indian Plates were subducting (Fig. 7). To the west, the Semitau terrane collided with the West Borneo Basement Block in the Cretaceous, and accretion to its north took place after that. Later, the Mangkalihat terrane joined Borneo after the third rifting event from the Gondwanic margin (Metcalfe 1994). Other parts came rifting from the Chinese margin after the opening of the South China Sea. In later times, parts of Borneo, like the Meratus, were uplifted.

Comparing the tectonically different terranes (Fig. 5) with the map of areas based on the distributions of the genera used in this research (Fig. 4) a striking resemblance is apparent. In the distribution map, the line separating West and East Malaysia is easily recognisable and can even be followed into Sumatra, thus the same as the Bentong-Raub Fault along which the Malaysian part of Sibumasu is sutured against East

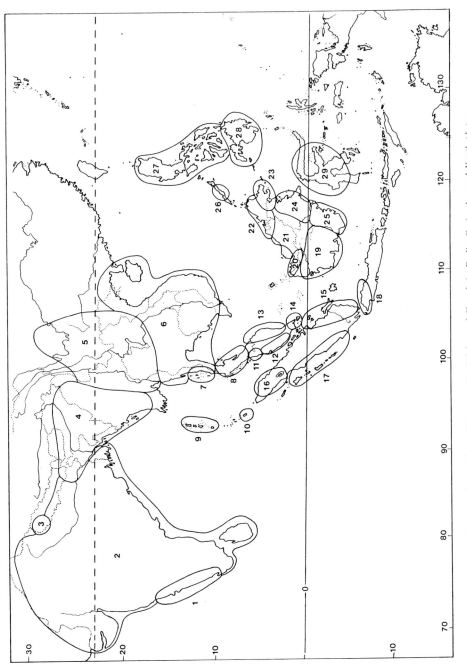

Fig. 4. Map with the areas of distribution of *Butea, Meizotropis, Spatholobus* and *Kunstleria, Fordia, Genianthus* and *Xanthophytum*.

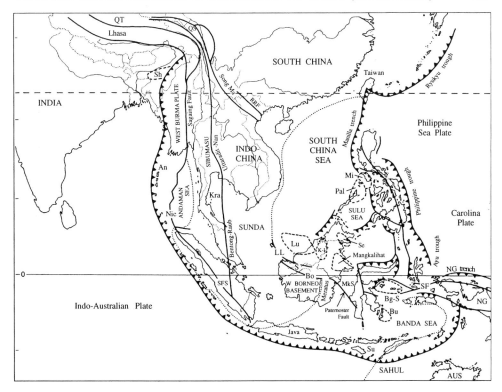

FIG. 5. Map showing the main geological features in Southeast Asia.
Abbreviations: An: Andaman Islands; AUS: Australia; Bg-S: Banggai-Sula; K-L: Kelabit-Longbowan; Kra: Isthmus of Kra; LL: Lupar Line; La: Luconia; Mg: Mangkalihat; Mi: Mindoro; MkS: Makassar Strait; NG: New Guinea/Irian Jaya; Nic: Nicobar Islands; Pal: Palawan; QS: Quando-Simao terrane; QT: Quingtang terrane; RRF: Red River Fault; S: Semitan; Se: Segama; SF: Sorong Fault; SFS: Sumatra Fault System; Sh: Shilong Plateau; Su: Sumba.

Malaysia. Also, different parts in Borneo are recognisable: the Meratus and Semitau ranges, the West Borneo Basement Block, and the Mangkalihat terrane as well as the area near Tawau (although this is less clear). On the mainland, coincidence of areas of distribution and geological terranes is less evident.

Fossil evidence

The first reliable records of legume fossils from China and India (Guo & Zhou 1992; Awashti 1992) date back to the Eocene, although a few, but unreliable, fossils of wood and pollen from the Late Cretaceous are known. In contrast to Africa/ Madagascar and Tropical America there is a limited fossil record in China and India, suggesting that these areas were reached relatively late (Raven & Polhill 1981). Diversification took place during the Eocene and from the middle Eocene onwards all three subfamilies of the *Leguminosae* are present in the fossil flora. In the fossil flora of the Miocene legumes are abundant, and very well recognisable. Later floras contain fewer legume fossils, possibly due to a drier climate which led to diminishing of the tropical evergreen forest (Awashti 1992) and the loss of many legume taxa that had become adapted to it.

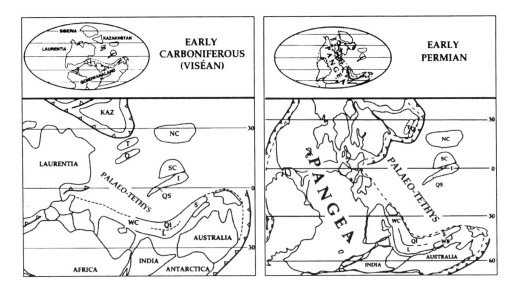

FIG. 6. First rifting event, Palaeozoic times, of North China, South China, Indochina and East Malaysia and South Sumatra (Metcalfe 1994).

Abbreviations: Ba: Banda allochton; Ba-Su: Banggai-Sula; B-S: Buru-Seram; BU: Buton; C: Cimmerian continent; ES: E Sulawesi; I: Indochina; L: Lhasa; M: Mangkalikhat; N: Natal; NC: N China; O: Obi-Bacan; QS: Qamdo-Simao; QT: Qiangtang; S: Sibumasu; SC: S China; SG: Songpan-Ganzi; Si: Sikuleh; Sm: Sumba; SWB: SW Borneo; T: Tarim; WB: W Burma; WC: W Cimmerian; WIJ: W Irian Jaya; WS: W Sulawesi

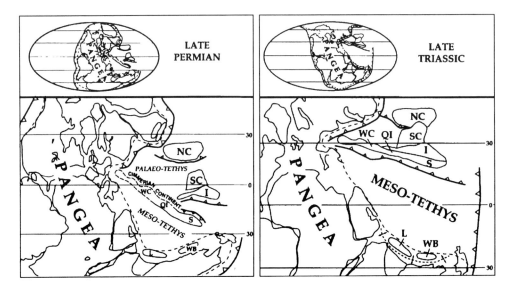

FIG. 7. Second rifting event, Permian, of Sibumasu, Lhasa, Qiangtang and Iran and Turkey. South China magmatic belt. See Fig. 6 for abbreviations.

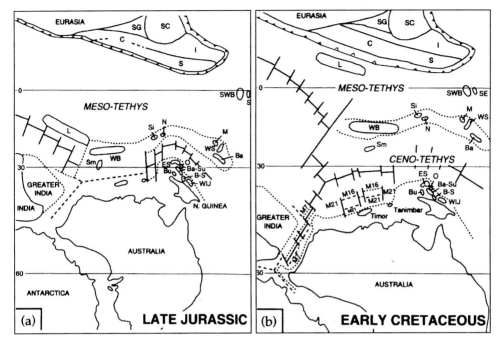

FIG. 8. Third rifting event, Jurassic, rifting of West Burma, Parts of Sumatra, Borneo and Sulawesi. Separation of India started as well. See Fig. 6 for abbreviations.

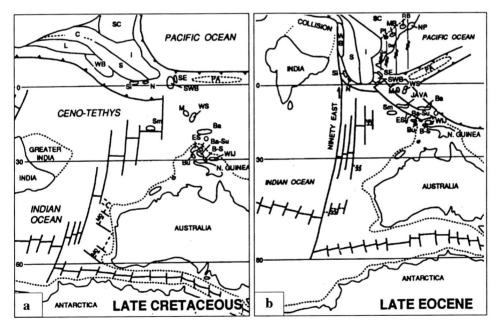

FIG. 9. Collision of India to Eurasia, and rotation of Indochina. Opening of Thai basin and South China Sea basin. Rifting of parts of Chinese continental margin towards Borneo. See Fig. 6 for abbreviations.

Climate and sea-level changes

Past climates and sea-level changes must have played an important role in the presence of plants in certain areas and their speciation. During the Ice Ages, sea-levels lowered and parts usually submerged (e.g. the Sundaland plateau) became available for colonisation. At the same time, a drier and cooler climate caused a lowering of the altitude limits of the montane flora.

Warm humid periods with very high sea-levels (up to 350 m above present sea-level) occurred during the Cretaceous and later again during the Mid-Miocene. During the Mid-Oligocene transgression the sea-level was as low as 250 m below the present day sea-level and the climate became cooler and drier. After the Mid-Miocene regression, with high temperatures and humidity, the climate stayed generally drier and more seasonal than before. The everwet forests were restricted to certain parts, and there was an expansion of deciduous forest and grasslands. During the Pleistocene Ice Ages, the sea-level again became low and the climate cooler and drier, causing a lowering of the forest limits and exposure of inundated areas.

CONCLUSIONS

The general area cladogram indicates that the areas on the continent are the oldest and that here the first speciation events took place. According to the fossil evidence, this could not have been earlier than the Eocene. At this time India was in collision with Eurasia, and the Himalayas had not yet been uplifted.

In addition, the area cladogram indicates later events which led to speciation in the Malay Peninsula and the west Malesian Archipelago. It is possible that this is the result of dispersal into the area from the continent during a relatively dry period (dry Sundaland Plateau). Another possibility is that ancestors occurred in the whole area already and that (by, for example, isolation due to higher sea-levels) speciation in both the Malay Peninsula and the Malesian Archipelago occurred.

There is coincidence between at least part of the geological terranes and the distribution areas. It is not certain if this coincidence has a historical background, and this has to be tested with other methods and different groups of taxa. The fossil evidence of the *Leguminosae* indicates a later origin than the age of most of the terranes and in addition suggests a relatively late presence of legumes in the region. It is very possible that most legumes reached south-east Asia from other parts in the world via a northern route (boreotropics hypothesis: Lavin & Lucknow 1993; Lavin 1995). The 'general' area cladogram does not contradict this.

ACKNOWLEDGEMENTS

Thanks are due to Prof. Kalkman for his stimulating support and critically reading of the manuscript. For technical assistance I would like to thank Ir. Jan Ridder and Jos Rietstap (TNO Institute of Applied Geophysics).

This research was supported by the Dutch Science Organisation (Life Sciences, project 805-40-041).

REFERENCES

Awashti, N. (1992). Indian Fossil Legumes. In: P. S. Herendeen & D. L. Dilcher (eds), Advances in Legume Systematics: Part 4. The fossil record. pp. 225 – 250. Royal Botanic Gardens, Kew.

Axelius, B. (1990). The genus *Xanthophytum* (*Rubiaceae*). Taxonomy, phylogeny and biogeography. Blumea 34: 425 – 497.

Guo Shuang-Xing & Zhou Zhe-Kun (1992). The megafossil legumes from China. In: P. S. Herendeen & D. L. Dilcher (eds), Advances in Legume Systematics: Part 4. The fossil record. pp. 207 – 223. Royal Botanic Gardens, Kew.

Herendeen P.S., Crepet, W.L. & Dilcher, D.L. (1992). The fossil history of the Leguminosae: Phylogenetic and biogeographic implications. In: P. S. Herendeen, & D. L. Dilcher (eds), Advances in Legume Systematics: Part 4. The fossil record. pp. 303 – 316. Royal Botanic Gardens, Kew.

Hutchison, C.S. (1989). Geological evolution of Southeast Asia. Oxford monographs on geology and geophysics, no.13. Oxford.

— (1992). The Eocene unconformity on Southeast and East Sundaland. Geol. Soc. Malaysia Bull. 32: 69 – 88.

Klackenberg, J. (1995). Taxonomy and phylogeny of the SE Asiatic genus *Genianthus* (*Asclepiadaceae*). Bot. Jahrb. 117: 401 – 467.

Lavin, M. (1995). Tribe *Robinieae* and allies; model groups for assessing early Tertiary northern latitude diversification of tropical legumes. In: M. Crisp & J. J. Doyle (eds), Advances in Legumes Systematics, part 7. Phylogeny. pp.141 – 160.

— & Lucknow, M. (1993). Origins and relationships of tropical North America in the context of the boreotropics hypothesis. Amer. J. Bot. 80: 1 – 14.

Metcalfe, I. (1994). Gondwanaland origin, dispersion, and accretion of East and Southeast Asian continental terranes. J. South Am. Earth Sciences 7 (3/4): 333–347.

Raven, P. H. & Polhill, R. M. (1981). Biogeography of the *Leguminosae*. In: R. M. Polhill & P. H. Raven (eds), Advances in Legume Systematics, part 1, pp. 27 – 34. Royal Botanic Gardens, Kew.

Ridder-Numan, J.W.A. (1992). *Spatholobus* (*Leguminosae-Papilionoideae*): a new species and some notes. Blumea 37: 63 – 71.

—— (1996). Historical biogeography of the Southeast Asian genus *Spatholobus* (*Leguminosae-Papilionoideae*) and it allies. Blumea supplement 10: 1 – 144.

—— & Kornet, D.J. (1994). A revision of the genus *Kunstleria* (*Leguminosae-Papilionoideae*). Blumea 38: 465 – 485.

—— & Wiriadinata, H. (1985). A revision of the genus *Spatholobus* (*Leguminosae-Papilionoideae*). Reinwardtia 10: 139 – 205.

Sanjappa, M. (1987). Revision of the genera *Butea* Roxb. ex Willd. and *Meizotropis* Voigt (*Fabaceae*). Bull. Bot. Surv. India 29: 199 – 225.

Schot, A.M. (1991). Phylogenetic relations and historical biogeography of *Fordia* and *Imbralyx* (*Papilionaceae: Millettieae*). Blumea 36: 205 – 234.

Swofford, D.L. (1991). PAUP: Phylogenetic analysis using parsimony, version 3.0s. Computer program formerly distributed by the Illinois Natural History Survey, Champaign, Illinois.

Flora Malesiana: a user's experience and view

Mien A. Rifai[1]

Summary. Being a research Flora, the scientific quality of the published Flora Malesiana is indeed very convincing, as is evident from a long personal experience in using these volumes for many kinds of practical purposes, especially in finding answers for various queries from the information-seeking public at large to the Herbarium Bogoriense. At the same time, however, it is sad to state that the Flora is not widely known among Indonesian biologists. The basic education of the present day Indonesian biologists is hardly adequate to enable them to grasp the meaning of the scientific terms used in the Flora. Moreover since the mastery of English by the majority of Indonesian university graduates is poor, this important source of knowledge represents a closed book to most of them. It has been noted that despite more rigorous publicity, the *PROSEA* volumes of economic plant treatments which by right should be much more sought after by the public as yet have the same fate as that of the Flora Malesiana. In view of the significance of Flora Malesiana as a source of information to manage one of the most important components of Indonesian biodiversity, to increase the usage of the Flora in the near future, attempts are now being made to improve the system of biology education in Indonesian universities, especially at the postgraduate level.

By training, by choice, and by profession it was as a mycologist that I joined the Bogor Herbarium in the mid-1960s. Among my responsibilities was the curation of a mere 21000 fungal specimens as compared to approximately 1,500,000 sheets of fern and flowering plant herbarium specimens kept in that institute. Shortly afterwards, however, it was these 1,500,000 specimens, which absorbed my whole attention and energy because due to unforeseen circumstances I was appointed as the Keeper of Bogor Herbarium. I soon found out that being in charge of a national herbarium belonging to a developing country I had to satisfy various kinds of inquiries from the information-seeking public, be they biological or agricultural researchers, policy or decision makers and planners, undergraduate and postgraduate students as well as school children, pharmacists and other people interested in herb medicines, exporters of plant products, garden or landscape architects, orchid hobbyists and other plant lovers, army officers undergoing training in jungle survival, boy scouts and girl guides who love roaming the countryside, foresters and so on. Naturally it was on the skeleton staff available at the Herbarium that I had to rely in finding the answers to these many questions, and by a lucky coincidence we had a technician (Mr. Nedi) specially trained by the late Prof. C. G. G. J. van Steenis to become capable of recognizing and identifying small scraps of plants by sight. Securing the identity of plants and getting their right scientific names were one of my main occupations to satisfy the clients of the Herbarium. The fact that I was a mycologist did not help very much in getting more related information about plants

---

[1] Herbarium Bogoriense, Puslitbang Biologi-LIPI, Jl. Ir. H. Juanda 22, Bogor 16122, Indonesia.

requested by the public. Understandably I had to turn to Floras and other treatises at my disposal, and it was then that I became familiar, later on very thoroughly, with the volumes of Flora Malesiana (van Steenis 1950). Since this is not a complete floristic treatment of the area under my jurisdiction, I had to find supplements to satisfy my needs. Therefore, the Flora of Java (Backer & Bakhuizen 1963 – 68), accounts of weed floras of tea (Backer & van Slooten 1924) and sugarcane plantations (Backer 1928 – 1934), the fern flora of Java (Backer 1939), books on fruits (Ochse 1927) and vegetables (Ochse & Bakhuizen 1931), and the well known treatment by Heyne of Indonesian useful plants (Heyne 1927), were also becoming my constant companions. To these were added Holttum's account of orchids (Holttum 1953), ferns (Holttum 1954), gingers (Holttum 1950) and bamboos (Holttum 1958) of the Malay Peninsula as well as Burkill's dictionary of economic products of the same area (Burkill 1935).

Being trained as a taxonomist I soon found that Flora Malesiana was the most reliable source of information to start the building up of answers to various inquiries from the public. Being a research Flora, the scientific quality of this series surpassed any other similar taxonomic work both in its depth and coverage as well as in the comprehensive presentation of data. Later I also found that these volumes were — and still are — very useful in finding small taxonomic problems to be investigated by the final year biology students in partial fulfilment of their first degree requirements.

Looking back to those bygone years I realized what a poor performance I would cut in doing my job if the Flora Malesiana project had never been launched. I would be also severely handicapped in dealing with questions coming from many people in Java if Backer had not completed his Flora of Java. These treatises have been the most used publications in my working room as well as in the library of the Herbarium Bogoriense.

From the very beginning I was so preoccupied in making the Herbarium function properly, that contacts with students and the educational sector in general were only sporadic. As the years went by I realized that it was only in the Herbarium Bogoriense that Flora Malesiana and Flora of Java have been so much used that new copies again and again have to be provided. In other parts of Indonesia, in universities or in the many government research establishments, the Flora is not widely known among Indonesian biologists especially in recent years. The advance of ecology, conservation and biotechnology in school and university curricula obviously have taken the toll by relegating taxonomy — and hence the need to know the flora — to a low priority and consequently has been ignored or avoided by students.

It is sad to say that Indonesian planners have had incredible views on how to develop the scientific capability of the country in that they could easily provide funds to build beautiful library buildings but have forgotten to provide annual budgets to buy books or scientific journals actually needed by a library. The absence of a budget required to fill the library with bibliographic materials makes it impossible for the managers of those libraries to procure books, and similarly they could not afford to subscribe to Flora Malesiana regularly. Although the Foundation Flora Malesiana originally made available about 300 copies of each published parts to various Indonesian institutions, after careful evaluation only about 70 of those were really sent out to certain selected libraries serving the higher learning and other scientific ventures. Even then, when scrutinized carefully it turned out that the copies sent out and received were not properly managed, a large number of them were missing from the bookshelves, some were kept under lock and key, and in one instance it was found that they had never been opened from the package they were sent in.

Jarvie and van Welzen (1994) wrote an admirable article analyzing the shortcomings of Flora writing in tropical countries, and Southeast Asia in particular. I concur wholeheartedly with all their conclusions and would like to reiterate the need to address the problems in order to make Floras more user friendly. Even if all the possible solutions put forward by Jarvie and van Welzen were taken care of, we in Indonesia will still have another problem, perhaps unique to us alone in that the mastery of English by the majority of Indonesian university graduates in the last ten years has become so poor that Flora Malesiana is, in effect, a closed book to them. Towards this end, attempts have been made to develop the Indonesian terminology needed and used in taxonomic works, to help Indonesians in understanding, using and later on in preparing floristic accounts themselves. This programme has been undertaken cooperatively with biologists from Brunei Darussalam and Malaysia through cooperation in the linguistic engineering of the Malay language.

What we have been aiming at through this programme is actually to improve biology teaching in primary and secondary education by giving due emphasis on biodiversity which has now become a "sacred cow". We are also trying to put biodiversity in the mainstream of our tertiary education, although not by stressing classical taxonomy as an important discipline, but rather by making use of the information provided by Floras and similar handbooks for solving the many daily problems faced by planners and decision makers in dealing with biodiversity questions which cover the aspects of research, uses and conservation. This action was initiated to encourage the use of Flora Malesiana as a major source of information by the majority of our graduates. This action is also stimulated by the fact that in spite of a more rigorous publicity the PROSEA volumes of economic plant treatments (Westphal & Jansen 1986) which by rights should be much more sought after by the public, as yet share the same fate as Flora Malesiana. Our copies of PROSEA volumes in the library of Herbarium Bogoriense are still in pristine condition signifying that they are rarely consulted. It is believed, however, that very soon PROSEA volumes will become much used by certain sectors of the public, whereas ecologists and other general biologists will continue to look for other sources of information. However, in view of the importance and the significance of Flora Malesiana, PROSEA volumes and similar handbooks as sources of information to manage the utilization and conservation of the most important components of Indonesian biodiversity, attempts are now being made to improve the system of biology education in Indonesian universities, especially at the postgraduate level. A workshop to address this problem and to develop action plans will be convened in conjunction with the XII National Congress of Biology in Jakarta.

During the past five years the Herbarium Bogoriense, in cooperation with Bogor Agricultural University, has offered a two year M.Sc. course in plant taxonomy. A total of 17 students have been enrolled, who mostly represent lecturers from various state universities responsible for teaching taxonomy and biodiversity. Six out the 17 were already admitted to the degree of M.Sc. At the beginning the students were permitted to write their theses in Indonesian. However, they are now required to prepare their dissertation in English because they are mostly investigating problems in conjunction with the Flora Malesiana Project. In this respect one of them has been working on *Araceae*, one on *Orchidaceae* and four on *Euphorbiaceae*, co-ordinated from Leiden. The idea is to encourage and increase local or regional participation in preparing the accounts for the Flora Malesiana volumes. The successful participants of the M.Sc. course who reach a certain level of academic achievement are invited to go further to do Ph.D. degree, planned to be conducted partly in the local institute and partly in a foreign herbaria, or incorporating a foreign professor as one of their supervisors. In the long run it is expected that these university lecturers will go back to their respective universities to instruct their students properly to enable them to use the Floras, both the user-friendly type or even the traditional and difficult ones already criticized by Jarvie & van Welzen (1994).

Realizing that problems due to the shortage of manpower in enhancing the use as well as the preparation of floristic works were also faced by the region, the Herbarium Bogoriense has run a short regional taxonomic training course on herbarium research methodology every 2 – 3 years since 1985, in cooperation with Rijksherbarium, Leiden and supported financially by the UNESCO/MAB Programme. The support from UNESCO has been dwindling in recent years to about half of the amount needed so that in the past three years the course has had to be postponed. A number of former participants of this course continued their study and have managed to obtain M.Sc. or Ph.D. degrees (Dr. L. G. Saw, Ms. Rugayah), so that the basic training given seems to be useful in stimulating and introducing the skills required for taxonomic work.

I hope that Flora Malesiana can be completed as soon as possible. We are striving very hard through our M.Sc. course to equip local botanists with the know-how to enable them to use and later on to contribute to the Flora, so that this ambitious project can be completed in the near future and that the by-products can be produced in large numbers and at higher frequency. For example, we need handy excursion Floras to every major island, especially in areas where the demand for information and knowledge are felt. If in the past local people have been passive users of the Flora, we hope that soon we will have more Soepadmos and more Latiffs, who are active contributors to the project. Outside assistance will always be needed so that close cooperation, as envisaged by the internationalization of the Foundation Flora Malesiana, should be our motto in trying to complete the Flora Malesiana.

References

Backer, C. A. (1928 – 1934). Onkruidflora der Javasche Suikerrietgronden. Surabaya.

—— (1939). Varenflora voor Java. Buitenzorg, 's Lands Plantentuin.

—— & R. C. Bakhuizen van den Brink (1963 – 68). Flora of Java. 3 Vols., Noordhof, Groningen.

Backer, C. A. & D. F. van Slooten (1924). Geillustreerd handboek der Javaansche theeonkruiden. Batavia.

Burkill, I. H. (1935). A dictionary of the economic products of the Malay Peninsula. 2 Vols., London, Crown Agents.

Heyne, K. (1927). De nuttige planten van Nederlandsch Indië. 2nd ed. 3 Vols. Departement van Landbouw, Nijverheid en Handel in Nederlandsch Indië.

Holttum, R. E. (1950). The *Zingiberaceae* of the Malay Peninsula. Gardens' Bull., Singapore 13: 1 – 249.

—— (1953). Orchids of Malaya. Govt. Printing Office, Singapore.

—— (1954). Ferns of Malaya. Govt. Printing Office, Singapore.

—— (1958). The bamboos of the Malay Peninsula. Gardens' Bull., Singapore 16: 1 – 35

Jarvie, J. K. & P. van Welzen (1994). What are tropical floras for in SE Asia? Taxon 43: 444 – 448.

Ochse, J. J. (1927). Indische vruchten. Volkslectuur, Weltevreden.

—— & R. C. Bakhuizen van den Brink (1931). Indische groenten. Batavia.

Steenis, C. G. G. J. van (ed.) (1950 –). Flora Malesiana (continuing series). Noordhoff-Kolff, Jakarta.

Westphal, E. & P. C. M. Jansen (eds) (1986 –). *PROSEA* (continuing series). Pudoc, Wageningen.

Kostermans: the man, his work, his legacy

Mien A. Rifai[1]

Summary. The role played by Prof. Dr. A. J. G. H. Kostermans in shaping some parts of Indonesian botany is elucidated based on the personal experiences of those who worked closely with him during his life time in Indonesia. His attitude to his chosen field of specialisation, his methods of addressing problems, botanical or others, and his love for his work are presented. Since Kostermans was also well known as a philanthropist, it is interesting to analyse his largesse and his underlying motives, his achievements and his failures, as well as his outlook on life in general. The picture reconstructed from this study is a very human Kostermans, a unique botanist whose legacy is most impressive to those who knew him personally. Only time will tell whether he will be ranked with Backer, Boedijn, Holttum, van Steenis and other makers of Malesian botany and remembered as such by future generations.

It is a well-known aspect of human nature to become curious and interested in knowing more about the lives of great, successful or otherwise exceptional fellow human beings whom we consider *primi inter pares*. History, biographies and obituaries have been written to elucidate the lives of these unusual personages, and we read them avidly, either because we believe that much can be learned from their experiences, or because of the appreciation and recognition of their meritorious services rendered to us. Occasionally exaggerated facts creep into these accounts so that tall stories often colour the pictures, making the whole thing appear larger than life.

Malesian botany is no exception. However incredible it may seem, we have been brought up with tales of how Blume published taxonomic treatises but kept the resulting printed materials to himself, how Teijsmann defied the order of the all powerful Governor General to make the garden around the Bogor Palace a scientifically sound botanic garden, how Koorders failed to recognise a living plant brought to him although he had been studying the herbarium specimen of the same species for months, or how Boedijn instructed his students by drawing details with his left hand on the black board while his right hand wrote the captions and simultaneously his mouth kept on pouring out more information very rapidly! Kostermans has become the source and object of similar stories, because his unique personality, his attitude, his position and his actions in the rapidly changing political and scientific developmental history of Indonesia when the Dutch left the country in the late 1950s, made him a living legend.

On the day of his funeral, the procession of his coffin bearers and the attending well-wishers who flocked the foyer of the Herbarium Bogoriense where he was laid in state, caused a temporary traffic jam around the Bogor Botanic Garden where he was buried. It so happened that on that eventful day of 11 July 1994 two other notable

---

[1]Herbarium Bogoriense, Puslitbang Biologi-LIPI, Jl. Ir. H. Juanda 22, Bogor 16122, Indonesia.

persons were buried in Bogor, namely the Vice Chancellor of Bogor Agricultural University, and Miss Kurniasih, the one time long associate of Kostermans who was formerly Head of the Administrative Office of Herbarium Bogoriense. But both in Bogor as well as in the national news coverage it was undoubtedly Kostermans who stole the day. Many days afterwards a number of local and national newspapers continued publishing articles about him.

The very act of burying him in Bogor Botanic Garden attracted comments from the mass media because that burial ground had been closed down about 100 years earlier, and only two botanists — the young explorers van Hasselt and Kuhl — were ever buried there. Even the great gardener Teijsmann who was responsible for making the Bogor Botanic Garden as we see it today was buried in a public cemetery far away from the Garden. Kostermans himself actually asked to be cremated and his ashes scattered in Bogor Botanic Garden. However, a number of persons, including his former students thought otherwise, and in this respect we were supported by the authority of the Indonesian Institute of Sciences as well as by the Bogor Municipality who issued a special permit for the occasion.

Although Kostermans never held a high official position which commanded authority and respect in the Indonesian bureaucracy, he was such a man who, by his own forceful personality, strong will, persuasive charm and proven achievement in scientific output, did make things happen around him. That was why old pensioners, life long friends, newly made acquaintances, professors, school children, policemen, housewives, researchers, and, of course, the whole families of his adopted sons from Bali, Yogyakarta, Bandung, Jakarta and Bogor all came, and most of them stayed until the funeral was over. Those who attended his funeral that day came to pay their final tribute to the man they loved, the antagonist they never managed to conquer, or even the enemy they hated but somehow respected. It is well known that Kostermans was an enigmatic person, likeable, but at the same time he could be very unpleasant to be associated with, simply because he dared to say what other people chose not to expound openly. This was the only non-Indonesian character of the man who otherwise tried to become at one with Indonesia, though perhaps not always wholeheartedly.

There were many Indonesians like myself, and other nationalities as well, who from the very beginning until the last day of his life had a kind of love-hate relationship with this paradoxical personality. There was not the slightest doubt in my mind that he did like me, because he took some pains to polish my education by finding for me a first class supervisor here in England through the help of Mr. L. L. Forman of Kew Herbarium. Moreover he loved me because out of the approximately ten students whom he handpicked himself to become plant taxonomists I have been the only one who chose to stay behind in Bogor and continue the tradition he started to do taxonomic work. Hence he advocated my appointment as the Keeper of Bogor Herbarium over my seniors. Nevertheless I knew very well that he disliked, and at times, hated me, because I often had my own opinion and my own way of doing things which were often in head-on collision with his often unconventional ideas. Yet at the same time he respected me for standing against him although he openly criticised my stand until the results of our differences showed that they were not always on his side.

Personal experiences of others closely working with him were comparable. With their permission I would like to mention some of these. For example, originally he disliked Dr. Soejatmi Dransfield (although not always openly) simply because she was a lady student who preferred to choose for herself lowly plants (it happened to be the difficult grass genera *Rottboellia* and *Cymbopogon*) rather than following his footsteps in working on taxonomic problems of Malesian tree genera. But he respected her because of her perseverance and tenacity in pursuing this study, especially after she obtained her Ph.D. through her own efforts. On the other hand, he loved Dr. I. G. M. Tantra because he saw in him someone to follow his steps in linking the Forestry Research Institute and the Bogor Herbarium, especially after Tantra completed his Ph.D. study on *Sterculia*, a genus close to the ones on which he had been working. You can imagine his disappointment when Tantra accepted the appointment as the Chief of Forestry Service in the Province of Bali. According to Kostermans, any work which did not promote taxonomy if performed by those especially trained for that purpose was a waste of talent and everything else. He prepared to change his attitude in this respect only if the overall results of that activity were overwhelmingly successful. Full success was the only yardstick he would use because he hated mediocre performance. Dr. E. Soepadmo was rather delayed in being sent abroad to obtain his Ph.D. because he and Kostermans were at odds almost all the time; among other things he disliked Soepadmo for originally intending to use the specific epithet *nunukanensis* as the name for Soepadmo's first new species of *Neesia* because according to him it was not appropriate. Soepadmo cleverly placated him by calling that species *Neesia kostermansiana*. In spite of this hard feeling, it was Soepadmo whom Kostermans expected to come home to head the Bogor Herbarium in the late 1960s.

For most of the time he was not wholly appreciated by his Indonesian associates, because of his often uncompromising attitude. Nevertheless everybody agreed that the little achievements of Indonesian taxonomic botany as contributed by Indonesian researchers would be less meaningful than they are today and completely different if he had not been there in the late 1960s. Almost single-handedly he educated the then fledgling Indonesian plant taxonomists, kept the Bogor Herbarium functioning so that in spite of serious handicaps its collections significantly increased through his own field activities, and the facilities for storage improved because he literally begged for support from anybody who would listen to him. His role was such that in those days a number of liberally minded Indonesian scientists wished that each Indonesian scientific institution had its own Kostermans to make the institute run as it should. Thanks to him plant taxonomy in Indonesia developed faster compared to other branches of biology such as genetics, microbiology or physiology, largely because he drilled his students vigorously, beginning by making them sweep the floor of their working room, processing their own collected specimens, and giving them a real example of how to do taxonomic research and to publish the results in English under the most adverse conditions. He was a truly exacting teacher, an inspiring although unofficial leader, a hard working researcher as well as a slave driver, so that he was feared and yet respected by his students and those who stood in his way to make taxonomic botany flourish further. The late Marius Jacobs did an admirable job in elucidating this aspect of his life when commemorating the 75th birthday of Kostermans in *Reinwardtia*.

Those who did not know him closely often felt embarrassed at his outbursts even in public meetings, or at his impatience when things were done slovenly, because, as suggested earlier, he would not desist from saying openly what he thought to be right at that moment. For example, from the first time I knew him, to my astonishment he said that he hated Indonesia and what it stood for, and yet there was no question whatsoever that he loved the country where he was born and where he spent the best years of his life. I was with him in Holland when he was supposed to stay there for one year but after two months he felt homesick and absconded from his contract, cut short his stay and went back to Indonesia to face all the shortcomings which he criticised all the time but where botanical problems were so bountiful that he said he could reign like a king in doing his beloved scientific work. This paradoxical thing was consistent with him, and already visible even when he was a child in Purworejo (a small town in Central Java) where he was born. Otherwise how can we explain a man who successfully completed his university training in animal physiology but chose plant taxonomy as his Ph.D. dissertation? How else can we understand the action of a dying man, who was painfully suffering from serious heart trouble and literally fighting for his life and yet four days before he died he took the trouble to write to Dr. Elizabeth Widjaja instructing her what to do with certain herbarium specimens he was working on for his last manuscript on Burmese *Cinnamomum*?

Because of this attitude it is easy to understand why Kostermans was exceptionally good when there was a challenge facing him. His favourite advice to youngsters around him was not to fear difficulty, especially if one wanted to be a researcher because doing research meant looking for difficulty to overcome. His success in obtaining government funds to complete the Bogor Herbarium building was much publicised because of the presence of strong competition for the funds which were scarce in the late 1960s, and because he dared to address the President of Indonesia directly. Even in doing his scientific work, Kostermans was at his best when he faced a challenge, especially when he found it necessary to criticise and correct poor taxonomic treatments by fellow botanists. His contributions in paving the way for future more solid work on *Durio, Mangifera, Spondias* and so on are such examples, which seem to outshine his other works when he was left to his own devices, because then he often produced only a routine type of taxonomic work.

However, let there be no mistake — this remark is not meant to belittle him and his contributions to the advancement of Indonesian botany. He was a great man in his own right and with his death the country lost one of her most talented citizens and colourful subjects, and Indonesian taxonomic botany lost one of its most prolific collectors and contributors. On the other hand one may wonder why such an acknowledged world expert on *Lauraceae* was not invited from the very beginning to prepare a treatise of this family for the account of *Flora Malesiana*? Perhaps van Steenis had foreknowledge about this, because although Kostermans was given a chance to finish an account of Malesian *Cinnamomum*, it ended in dismal failure.

In spite of the high opinion of himself, Kostermans never purposely imposed upon others to imitate him. He showed by real example what he expected others to follow in him rather than by verbal preaching alone, especially in dealing with his adopted children. But one thing he did stress to them: never expect thanks for whatever gifts or help you render to others. If you are helping people you do it simply because it pleases you to do it. That is your reward, and not the thanks you expect to be expressed by others to you. Actually there is an old Javanese teaching ("tanpa pamrih") as well as a Moslem expression to the same effect. However, it was from Kostermans that I learned the true meaning and significance of this teaching, even though I count myself as a good Moslem and have been preached several times about this by my Moslem teachers.

Kostermans was very unselfish in many respects and his household was run very democratically (one American even remarked that it was run following communist practice) because Kostermans shared everything he owned with the family of children he took under his wings. He received the same amount for himself of almost anything as the rest of the occupants of his house, be it food or clothing, and he insisted that the division must be equal to the last detail between the children and himself as well as the servants. He never instructed his adopted children to pursue certain studies, or to go to certain schools, but he would finance them as far as their academic ability could go, with the sky as the limit. Characteristically he never used his influence, his stature, and his unusual position in society to secure a place for his children in any select school, although he could easily have obtained special treatment if only he asked!

Although he did not profess to adhere to any particular religion he allowed his big family to practise the religion they chose, but then encouraged them to do it properly. He respected each one of them for practising daily religious rituals and asked the other members of his household to behave similarly. In so doing he expected that ultimately every one of the children he took under his wing would be able to fulfil a meaningful function in society according to their calling based on their own ability. Because of this attitude the professions chosen by his children vary greatly, from an army officer, technician, day labourer, factory manager, university professor, school teacher, accountant, medical doctor and many others.

It came to my notice that Kostermans did not really know, or never cared to know, the exact number of the children he adopted, the figure ranging between 80 to 120 according to his mood. Similarly he was inconsistent when asked about the number of good species he accepted in *Lauraceae*, his chosen family of specialisation, which varied from 2000 to 5000. Be that as it may, the number of the children he brought up was staggering by any standard, especially if one remembers that for most of his life Kostermans did not hold a position with fixed salary. He was silently discharged from his post in the Forestry Service without pension because of his conviction for sexual perversion in the Indonesian court of justice. But homosexuality was not the motive of his adopting so many children because I knew very well that he did not take advantage of all those boys. Like normal fathers he did care for those children as if they were his own; he shared their joys and their sorrows, he ate with these children and played or worked with them, everything conducted true to his preaching, that is without expecting any thanks or gratitude from them.

One cannot escape from the conclusion that to work closely with him, to see him run his household, and to understand his outlook about life in general, produces a picture of a very human person who was loveable or dislikeable depending on one's mood and outlook. Indonesians, like other Malay people, have a saying that "dead elephants leave behind their valuable ivory" as an inheritance to be remembered by future generations. As can be expected there are already many stories originating from their time relating to the acts of famous makers of Malesian botany such as C. A. Backer, K. B. Boedijn, R. E. Holttum and C. G. G. J. van Steenis. Their fame still lingers with us today largely because of the "ivory" they left behind in the form of *Flora of Java* of Backer and Bakhuizen v.d. Brink, the biology school textbook of Boedijn, treatises on orchids, ferns, bamboos and gingers by Holttum and the *Flora Malesiana* in the case of van Steenis.

It is unfortunate that the present day mastery of English by Indonesian students of biology leaves much to be desired so that the voluminous contributions by Kostermans, his "ivory", are known to very few people. His extensive treatise of Indonesian ricefield weeds, his important work on mango, and his admittedly difficult to use bibliography of *Lauraceae* are closed books to many Indonesian people. As it is, only time will tell whether or not Kostermans will be ranked with other famous makers of Malesian botany and remembered as one of them by future generations.

In his will he left everything he owned to the Rijksherbarium Leiden to be used in promoting Malesian plant taxonomy, characteristically stressing that it should exclude the utilisation of modern techniques such as cladistic analyses and the like. In view of the present trend in taxonomic research, once again only time will tell whether his wish will accelerate the development of plant sciences in the area. But one thing is certain: his legacy, as already enumerated by Jacobs, in shaping the development of Indonesian botany will become a burden, and as such a challenge to us, the Indonesian botanists of today.

Flora Malesiana: progress, needs and prospects

MARCO ROOS[1]

Summary. Progress in Flora Malesiana in recent years is summarised and an overview of the current network presented, including some recent achievements, new collaborators and projects, and fund raising activities. It is shown that a substantial acceleration in the production of family treatments has indeed been achieved, but that still much more is needed for a timely completion of Flora Malesiana. To keep the present momentum a number of factors hampering further acceleration have to be overcome. The results of the annual questionnaires are used to develop realistic expectations of family treatments to be completed in the next decade. This and information on the existing systematic resources and collections serve to illustrate the threats to plant systematics in general and future progress in Flora Malesiana in particular, and how we may deal with these. Possible strategies and actions to improve the present situation are discussed. Special attention is paid to the establishment and management of working teams and to fund raising. A plea is made for more co-ordinated initiatives for research proposals, at international but also at national level.

INTRODUCTION

In my capacity as Co-ordinator of Flora Malesiana (FM) I have, for many years, had to repeat the same message again and again: that FM needs to be speeded up and for that we need more collaborators — and thus more money. The present paper will restate the message once more, but maybe some new 'slant' will emerge from a comparison of the consultations I made for the Annual Reports over the last couple of years (see Flora Malesiana Bulletin).

This paper will deal mainly with the future prospects of FM and especially the following two questions:

— what needs to be done to achieve a reliable, complete FM in a relatively short period?
— how optimistic can we be to accomplish that goal?

Fig. 1 may give a rather optimistic view on the progress made so far and expected in the near future. It shows that almost 60% of the total number of taxa has been published so far in FM and that realistically we can expect this to be even 70% or more by the turn of the century. We do not need a great deal of optimism to believe that with some extra effort completion of FM in 20 years is feasible.

However, these figures pertain to the family level and we all know that species numbers are not evenly distributed over the various families. When looking at the same kind of data for species (Fig. 2), the picture changes dramatically. Now it becomes clear what an enormous task we still have to face, when aiming at a timely completion of FM.

[1] Rijksherbarium/Hortus Botanicus, Postbus 9514, 2300 RA LEIDEN, The Netherlands.

Flora Malesiana Series I&II
State of affairs: families

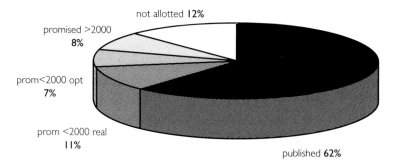

FIG. 1. State of affairs regarding families: prom<2000 real(istic) = submission of manuscript is pretty sure before the year 2000, prom<2000 opt(imistic) = submission before 2000 is dubious.

Flora Malesiana Series I&II
State of affairs: species

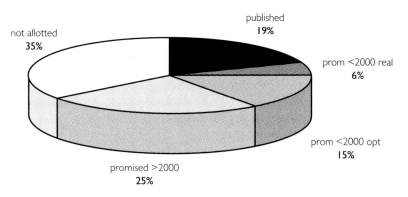

FIG. 2. State of affairs regarding species for Series I & II.

Obviously we need a great deal of optimism to even think of completing the Flora at all, not to speak of completion within the next two decades or so.

MALESIAN PLANT DIVERSITY

Let us recapitulate what we are talking about when dealing with FM.

Malesia is a major hotspot of global biodiversity. By combining estimates by specialists it appears that the Malesian vascular flora comprises over 40,000 species. A comparison of the plant diversity of Malesia and Europe (Table 1) shows how this tremendous botanical diversity relates to a temperate flora.

TABLE 1. Plant diversity in Malesia and Europe.

	MALESIA AREA 3.10⁶ KM² COAST LINE 275.10³ KM		EUROPE AREA 9.10⁶ KM² COAST LINE 150.10³ KM	
VASCULAR PLANTS				
Species (% endemism)	42,000 (70%)		11,000 (30%)	
genera (% endemism)	3,200 (40%)		1,550 (?)	
families	310		200	
% world flora	14		4	
no. spec/100 km²	14		1.2	
% land surface	2		6	
MAJOR TAXA	no. species (% total number)	no. spec/ 1000 km²	no. species (% total number)	no. spec/ 1000 km²
Seaweeds	1000 – 1500 (15 – 20%)	4 – 5 (1000 km coast)	1500 (20%)	10 (1000 km coast)
Bryophytes	3,500 (15 – 20%)	1.2	1,500 (8%)	0.2
Pteridophytes	4,400 (29%)	1.5	142 (1%)	0.0
Dicots	25,500 (9%)	8.5	8,850 (4%)	1.0
Monocots	11,700 (17%)	3.9	2,000 (3%)	0.2

Present input and mission

Even to think of achieving a complete inventory, an extensive network of specialists is needed. At present the FM network comprises 14 participating institutes and over 130 authors from c. 30 countries. They all participate to fulfil the mission of the FM project (Table 2).

— providing a complete scientific inventory of Malesian botanical diversity through taxonomic research, including user-friendly identification tools;
— training of young taxonomists, particularly from the Flora Malesiana region.

TABLE 2. Mission of the Flora Malesiana project.

Research	a complete scientific inventory of Malesian botanical diversity through taxonomic research, including a classification, descriptions and user-friendly identification tools
Education	training of young taxonomists, particularly from the FM region
Participants	14 institutes (4 European)
	>130 specialists (± 50% European) from 30 countries (11 European)
Plan of Action	**completion of Flora Malesiana in 20 years, in view of alarming decline of biodiversity in Malesia**

Challenge

Systematic information and expertise is urgently needed. Policies for nature conservation and sustainable utilisation of the rich natural resources depend on comprehensive biodiversity knowledge. For the plants of a large region like Malesia, only a critical Flora, based on the study of the whole area, can provide this knowledge; existing knowledge of the taxonomy of most taxa is still incomplete or outdated and inaccessible in numerous and scattered publications.

For the accurate generation and application of systematic information we need a sufficient number of active taxonomic experts. Regrettably, there is an acute and even further increasing shortage of tenured professional taxonomists, especially in the Malesian countries.

Consequently, it is correspondingly difficult to complete Flora Malesiana, not to speak of a timely completion, however much effort the present collaborators may put in (and that is much).

Urgency

How urgent is a timely completion of FM? This can be gauged from the figures of the present deforestation in the region (Table 3). And what can we say about the estimated extinction rate (the figures pertain to the tropics worldwide, but there are no indications that estimations for Malesia should be less gloomy)?

In view of the biodiversity crisis, the Board of the Foundation Flora Malesiana has made a concrete interpretation of 'timely' and adopted a very ambitious Action Plan to complete Flora Malesiana in 20 years time. From Figs 1 & 2 it is obvious how ambitious this goal is and there are good friends of FM who regard it as utterly unrealistic.

TABLE 3. Deforestation in the Malesian region and extinction rate world wide

MALESIA	
3000×10^3 km²	land surface
2800×10^3 km²	original forest area 93% of total area
1800×10^3 km²	forest (1990) 60% of total area 64% of original forest area
1300×10^3 km²	± undisturbed forest (1980) 43% of total area 72% of (1990) forest area
17×10^3 km²	deforestation/year (1985 – 1990) 0.9% of forest area
EXTINCTION RATE WORLDWIDE	
25,000 vascular plant species (8 – 10%) in 1988 – 2000 0.5% of the total number of species per year	

To carry out this challenge successfully requires a tremendous amount of work. We need a far bigger network of specialists than we have at present. Unconventional (for taxonomic tradition) coordination and management is needed; and most of all: money.

State of affairs

To keep an overview of progress and the remaining revision work, I consult all participants at least once a year, with the inevitable question (among others) about the expected date of completion of the manuscripts.

Flora Malesiana has to deal with some 42,000 species (Fig. 3). Series I (Volumes I, IV – XI) & II (Volumes I & II) and Orchid Monographs published so far cover 17% of the documentation (7000 species). Because of the exceptional size of the *Orchidaceae* (6500 species), the Orchid Monographs have been set up as a separate series to cover their treatment.

To strike an optimistic note first, the past period has been quite successful with the publication of Volume XI and, just now, of the first part of Volume XII, dealing with the *Mimosoideae, Rosaceae, Sapindaceae* and *Meliaceae* and some other families, comprising c. 900 species in total; an addition of 15% in the space of three years. The authors are to be complimented on their major contribution to the progress of the FM project.

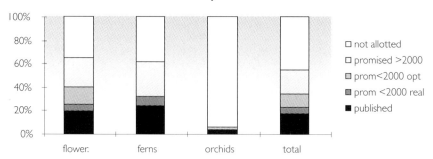

FIG. 3. State of affairs regarding species for Series I & II and Orchid monographs (for legend see Fig. 1).

The next instalment in Series I, comprising the treatments of *Caesalpiniaceae* and some small-sized families, is in the final editorial phase, whereas several manuscripts (e.g. *Boraginaceae, Loranthaceae*) are being edited now. Thus, the publication of treatments of another 550 species will follow soon.

For Series II, the manuscripts of five families covering 50 species are being edited, but we await some more mss to be able to prepare a substantial instalment.

The forthcoming Volume of Orchid Monographs is in the final editorial phase and due to be published by the end of this year, covering some 70 Malesian species.

Series I & II

Now comes the other side of the picture. To give a view on the progress, I restrict myself to Series I and II which comprise some 300 families and 35,000 species — 30,500 seed plants and 4500 ferns. Over 170 genera and some 6850 species have been published so far (Fig. 4).

Although these editorial developments do not sound too bad, they are less than the first questionnaire suggested (Fig. 5). In 1992, 2600 species were promised

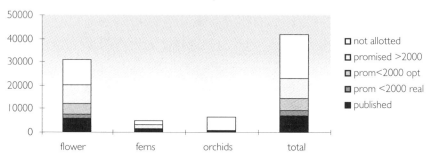

FIG. 4. State of affairs regarding species for Series I & II and Orchid Monographs (for legends see Fig. 1)

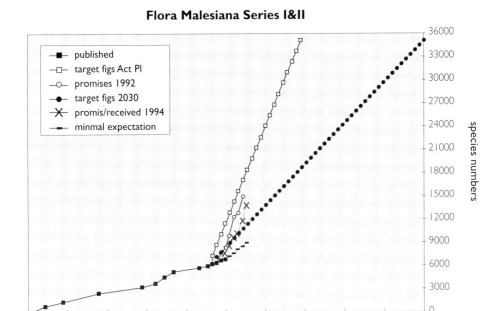

FIG. 5. Progress in publication of treatments, expressed in species covered; promises made in 1992 and 1994, realisation up to 1995 and target figures for completion in 2012 (Action Plan) and 2040 ('realistic scheme').

before the end of 1995. We actually received 1450 species (including those published) between then and now, with another 450 species 'about to be submitted', at least before the end of this year. This means that almost $^3/_4$ of the original promises for 1992 – 1995 will be realised (Fig. 6). But it is also means that we should beware of over-optimism. Experience teaches us that treatments are virtually never finished *earlier* than promised; the final phase of completion always takes more time than expected.

Progress

Realistically, we must anticipate that around the year 2000 not more than $^1/_4$ of the species will have been covered, i.e. a minimal expectation counting only the mss received and those known to be almost finished (Figs 4 & 6). This indeed implies an acceleration, but far less than the increase required for a completion of Flora Malesiana Series I and II in 20 years (i.e. in 2012). This minimal expectation is also less than what I have called a realistic scheme that I discussed in FM Bulletin last year, i.e. completion of FM around the year 2040 (Fig. 5). This would be somewhat disappointing, but we should not be too pessimistic as there may be a chance that more treatments promised will indeed be completed before the year 2000.

Flora Malesiana Series I&II

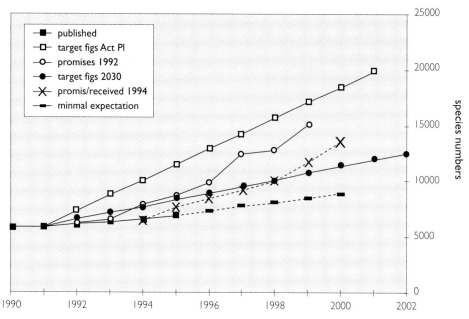

FIG. 6. Detail of the precent decade (promises made in 1992 compared to realisation in 1995 and promises made in 1994).

Prospects

The commitments to Flora Malesiana so far cover almost $^2/_3$ of the total number of species (Figs 2 & 4). The progress in most of the family treatments is slow to moderate. In several families hardly any notable progress has been made the last few years; sometimes we were forewarned of this, sometimes not. Mostly, rapid progress is hampered by other commitments (e.g. research — other Flora projects, fieldwork; teaching; administration) or by lack of funds, especially to attract collaborators to establish a working team. Also, we have to deal with unexpected and unwelcome events like the regretted deaths of Ben Stone (*Rutaceae* p.p., *Pandanaceae*), Karl Kramer (*Pteridaceae*), Ru Hoogland (*Cunoniaceae*) and Frank White (*Ebenaceae*), most of whom we commemorated during the symposium. And we all know that when specialists die before finishing their manuscripts so much information is lost that continuity of progress is unlikely.

Comparison of Figs 1 & 2 clearly illustrates that mainly relatively small families have been completed so far. This is also clear from the list of the largest families in the region, of which only three have been published, and from a similar list of the largest genera (Tables 4 & 5).

Fortunately, in recent years several international teams have been established to cope with large families and develop a research and management model for intensive collaboration to accelerate the production of Flora treatments. To mention some examples which are really gaining momentum: the *Araceae* group on which Alistair Hay reported in the first spring issue of FM Bulletin and which just published a checklist and bibliography for the Malesian region, and *Euphorbiaceae*, with a network of 25 botanists centered on Leiden. Other teams on e.g. *Apocynaceae*, *Papilionaceae* and *Polypodiaceae* have taken up an accelerated approach, which also seems to work.

TABLE 4. Large-sized families in the Flora Malesiana area.

FAMILIES (>500 MALESIAN SPECIES)	NO. SPECIES	COMPLETION / PUBLICATION
Orchidaceae	6500	
Rubiaceae	2000	
Euphorbiaceae	1000	2003
Melastomataceae	1000	
Arecaceae	975	2003
Gesneriaceae	900	
Annonaceae	875	
Poaceae	850	1999 (p.p.)
Ericaceae	750	1967
Araceae	725	1997, 1999
Myrtaceae	725	2000
Dryopteridaceae	700	
Lauraceae	700	
Zingiberaceae	700	
Acanthaceae	625	
Araliaceae	600	
Papilionaceae	575	1998
Moraceae	575	1998
TOTAL	20,775	
Cyperaceae	400	1974, 1979
Dipterocarpaceae	390	1982

TABLE 5. Large-sized genera in FM area.

GENERA (>200 MALESIAN SPECIES)	NO. SPECIES	COMPLETION / PUBLICATION
Bulbophyllum (*Orch*)	1000	1993 p.p.
Dendrobium (*Orch*)	700	
Selaginella	500	
Syzygium (*Myrt*)	500	allotted
Schefflera (*Aral*)	490	checklist
Ficus (*Mor*)	475	allotted
Asplenium	400	allotted
Cyrtandra (*Gesn*)	400	
Pandanus	400	
Diplazium (*Dryopt*)	300	
Grammitis	300	allotted
Rhododendron (*Eric*)	290	1967
Calamus (*Arec*)	280	allotted
Diospyros (*Eben*)	250	
Eria (*Orch*)	250	
Memecylon (*Melast*)	250	allotted
Vaccinium (*Eric*)	240	1967
Ardisia (*Myrs*)	220	allotted
TOTAL	7245	

I am convinced that teams like these are the only way to proceed and successfully attack such families. Besides, working teams may give better possibilities to take over unfinished treatments in case of the unexpected death of experts.

Remaining families

Although some specialists have recently opted to act as co-ordinating authors, those families (Fig. 12) still not allotted include a number of important ones like the *Gesneriaceae* and *Lauraceae*. Furthermore, although people have taken up the challenge to co-ordinate the revision of huge families like the *Rubiaceae*, many collaborators are needed for a completion of the treatment. For all these families an adequate working team has to be formed as soon as possible, otherwise serious problems will arise in the long run for continuous progress towards a complete Flora.

TABLE 6. Selection of remaining families — authors wanted.

Family	Size	Coordinating Author	Collaborators
Series I			
Acanthaceae	54 genera / 610 species	Hansen (C)	more wanted
Annonaceae	56 / 850	—	more wanted
Asclepiadaceae	55 / 500	De Koning (L)/ Wilcock (ABD)	more wanted
Asteraceae	150 /400	—	—
Commelinaceae	19 / ?	Conran (AD)	wanted
Gesneriaceae	35 / 900	—	—
Lauraceae	21 / 700	—	—
Myrsinaceae	16 / 500	Hu Chi Ming (IBSC)	wanted
Rubiaceae	150 / 2000	Puttock (CANB)	more wanted
Santalaceae	9 / ?	—	—
Scrophulariaceae	45 / ?	—	—
Urticaceae	26 / ?	Conn (NSW/K)	more wanted
Series II			
Dryopteridaceae	20 / 465	—	—
Lycopodiaceae	3 / 140	—	—
Selaginella	1 / 500	Camus (BM)	wanted
Vittaria	9 / 150	—	—

WHAT SHOULD WE DO

Although the picture I present may not give an optimistic view of the prospects, we should not be dismayed.

But, what can be done to meet our ambitious goal?

Work force and training

First of all, manifestly, only when more people start working on Flora Malesiana treatments, can the goal of a complete Flora come into sight.

In the last few years, it has been possible to attract a number of new authors. The recent initiatives of especially our Australian colleagues to increase their efforts to treat Malesian taxa is much appreciated. But this very welcome development is not enough to increase the momentum of the Flora sufficiently.

The present work force (taxonomic, editorial and secretarial) will have to be substantially expanded. At least twice the present number of taxonomists is needed, say 20 full-time equivalents per year for the next 20 years, much more than can be recruited from the existing staff of systematic institutions with research programmes on SE Asia. We are approaching the limits to the number of volunteer taxonomists we can recruit. Most of the specialists appointed at systematic institutions and dedicated to research programmes pertaining to Malesian plants are already involved.

Moreover, systematists (including those working on Malesian botany!) comprise a steadily ageing body of scientists, with many experts now close to or past retirement. For example, the average age of the permanent staff of the Rijksherbarium is over 50. Table 7 shows that $^3/_4$ of the botanists working in herbaria with major Borneo collections were over 40 years old in 1990, and that $^1/_5$ had retired.

TABLE 7. Age classes (after Burley)

COLLABORATORS IN HERBARIA WITH MAJOR BORNEO COLLECTIONS (1990)				
age	no.	%	SE Asian vs. non-SE Asian	% non-SE Asian / age class
<30	4	1.0	1 – 3	75
31 – 40	87	23.1	18 – 69	79
41 – 50	115	30.6	17 – 98	85
51 – 60	101	26.9	21 – 80	79
>60	69	18.4	4 – 65	94
Total	376	100.0	61 – 315	84

It is therefore necessary to raise funds to be able to contract additional (young) botanists especially for Flora Malesiana work.

However, even when we have the money there are not enough experienced taxonomists who can be contracted to carry out the work. We need to raise new generations of taxonomists. Training programmes (individual; at undergraduate, MSc, and PhD level), particularly for young taxonomists in the Flora Malesiana region, should be integrated in our efforts. At the same time we should make clear that the time that specialists spend on training they cannot spend on research and Flora-writing. Therefore, training needs substantial additional funding, especially to contract good teachers, thus allowing specialists do as much of what they are most good at as possible — work on their specialist family.

Furthermore, present expertise on Malesian plant diversity is concentrated in Europe. Table 7 shows that almost 85% of the staff members of herbaria with major Borneo collections are not Southeast Asian botanists. Table 8 clearly illustrates that by far the greater part of the authors of Flora Malesiana revisions are European.

This is profoundly unfortunate and very bad news. Taxonomic expertise within the Malesian region itself is very greatly needed, not only to contribute to Flora Malesiana, but especially to ensure the optimal use of systematic knowledge in all fields of botanical enquiry, conservation, and sustainable development. Therefore, we have to develop exchange training programmes and make pleas for more chairs in plant taxonomy at local universities.

TABLE 8. Background of authors of FM treatments.

172 families published by 80 different authors	
European:	70 authors — 157 families / 5750 species
Asian:	5 — 10 / 750
Other nationalities:	5 — 5 / 500

Efficiency

Acceleration of the revisionary work of each participating taxonomist can have some effect and should not be neglected; it is possible to increase efficiency within the resources available.

We have to keep a sharp focus on what we are aiming at, i.e. producing family treatments at high speed as well as disseminating taxonomic information. This requires continuous self reflection, evaluation of research goals, criteria, format, looking for a practical consensus between our scientific integrity and the demands of society for comprehensive biodiversity information.

Floristic approach

People involved in taxonomic inventories always face a conflict of interest: on the one hand the monographic approach which is scientifically required and on the other hand the geographically restricted taxonomic-floristic approach.

A Flora is in my opinion not meant to solve systematic problems, but it should primarily serve to present the best available overview of taxonomic diversity and make existing information accessible to non-specialists. The biodiversity crisis makes accelerated progress in taxonomic inventories most urgent. I call all participants to adopt a pragmatic, i.e. a restricted taxonomic approach. Please, make a realistic planning and keep your promises. I know that you have many duties, that you often are forced to do too many things and take up tasks for several projects at the same time (as you may be the only specialist available on particular taxa!). I fully understand that you are attracted by interesting fundamental questions, and that whatever brings in money will get priority.

But, write down what you know!

I am not advocating that we have to abandon fundamental phylogenetic and biogeographic studies. On the contrary, I only want to say that when you as experts agree to collaborate, the most efficient and productive way to proceed is to compile and check in a quick and clean way the existing knowledge, leaving systematic problems for future research.

We cannot wait until all taxonomic problems have been solved. Existing information, however incomplete for Malesian taxa, is already quite substantial; much is already available in numerous scattered publications. Without doing too much damage to our scientific integrity and quality, the publication of the present knowledge on particular taxa already will mean a great step forwards towards a reliable, complete FM.

Format

To adopt a simple approach means avoiding irrelevant details. Nomenclatural and complex taxonomic problems should be dealt with pragmatically. Only information relevant to the users should be presented. Species delimitation and particularly identification should focus on macroscopic characters which are easy to observe. A more unified format, allowing for differences in format according to the specific nature of the various families, for all ongoing major Flora projects will avoid duplication in the work of each of you.

Computerisation

Efficiency can be further improved by computerisation of the taxonomic 'manual labour' (descriptions, bibliography, etc.). A computerised checklist for the Flora Malesiana region (presenting species names with synonymy, author and literature reference, and geographical distribution) will provide a good starting point for future taxonomic work and for the planning of the whole treatment, and as such will speed up the Flora treatments of the remaining taxa. For some families this has been done, like *Araceae* (Blumea Supplement vol. 8) and *Annonaceae* (provisional).

Collaboration

Important improvements can be made by more intensive collaboration and co-ordination in international working teams. I already have stressed the importance of more working teams for the treatment of the remaining large-sized and complex families to secure future progress. But this holds for all ongoing revisions — division of labour, mutual stimulation, combining data bases, to mention a few of the advantages. I sincerely hope that this symposium will stimulate specialists on the remaining families to come together and establish working teams to treat their respective family for FM.

Dissemination

To increase transfer of systematic knowledge on tropical biodiversity and cope with the demands of policy makers, land use planners, foresters, conservationists and all other non-systematists for more taxonomic information, existing and new knowledge should be readily available, i.e. in a fast, accessible, user-friendly format.

Most experts and libraries are available in Western countries only. Digging out information is often very time-consuming and its distribution is still a costly business. These problems can be tackled by further developing interconnected computerised information systems. Increased computerisation of the data will enable much faster retrieval of information which can be continually updated and expanded.

Present the diagnostic information in as multi-user-friendly a way as possible; include multi-entry and separate keys using characters of the fruits, flowers, or vegetative parts only and keys per main islands and geographical areas. Furthermore, in cooperation with expert centres like Expert Centre for Taxonomic Identification (ETI, Amsterdam), develop presentations on CD-ROM and other multi-media.

Management

Intense management, unusual for taxonomists, is required to secure accelerated progress and a timely completion of inventory projects. Clear goals have to be set and end-products defined. A realistic time schedule and publication scheme based on accurate calculation of the workforce needed, is indispensable for successfully running the project and for fund raising. Again, please be realistic in your promises, and keep them!

Funding

As always, money is the key issue. Funding is needed for contracting capable botanists, for training talented students, exchange programmes, etc.

It turns out that it is extremely difficult to raise funds for the FM project as a whole, and for straightforward taxonomic projects. However, we should be able to make use of the fact that since the 1992 Earth Summit biodiversity is a real political issue. Also the current high profile of biodiversity has focused attention upon how little is known of tropical plant diversity and highlighted the unacceptably slow progress of tropical Flora projects.

I call all of you to try to approach your national funding agencies, be it science foundations or overseas aid agencies, and submit proposals of well-defined and concrete projects, as much as possible jointly with our Malesian sister institutes, e.g. for training through research and mobility, fieldwork with collateral taxonomic studies.

International agencies can be approached the same way. Our network provides ample opportunities to find partners.

There will be unexpected possibilities, as the successful application for a grant for a network on Botanical Diversity in the Indo-Pacific from the European Union Human Capital & Mobility Programme proves. We should not be discouraged when only one out of five applications will be successful.

CONCLUSIONS

Now I come back to the questions I posed at the beginning.

How optimistic can we be? In 1992 when I had just started my job and had prepared the Action Plan, I was very optimistic. Now I am much more realistic, but I see positive signs. We can be proud of the results achieved so far and the number of international programmes that have been funded since 1992. We can be confident about the many family revisions expected in the near future and that there will be accelerated production of instalments. With the present input, completion of FM in 45 years seems feasible. However, a more timely completion of FM is still far-fetched. Nevertheless, I am still convinced that more is possible, because it is so urgently needed and we can argue that.

It is up to all of us here to determine how far optimism is justified and what news I shall be bringing to the fourth Flora Malesiana Symposium in 1998. I am looking forward to it.

Disturbance and its significance for forest succession and diversification on the Krakatau Islands, Indonesia

SUSANNE F. SCHMITT[1] & TUKIRIN PARTOMIHARDJO[2]

Summary. Recolonisation of the Krakatau islands (Rakata, Sertung, Panjang) commenced after the cessation of the 1883 eruptions, with forest closure taking place by around 1930, co-incidental with the emergence of the new volcanic island Anak Krakatau. This paper reports on disturbance phenomena operating within the islands during a phase of renewed volcanic activity, and demonstrates that the group is characterised by a dynamic and variable physical environment, and examines the relevance of these environmental forces to succession and diversification of the inland forests. The islands are characterised by disturbance events of varying frequency, size and intensity, from simple treefalls via landslides to large-scale volcanic ash-falls. For instance, the 1992-94 eruptions appear particularly to have caused elevated levels of gap formation on Sertung. Quantification by transects at the meso-scale (c. 2 ha area) revealed markedly lower cumulative species totals and lower diversities for the canopy and lower canopy for Panjang and Sertung than for comparable areas of Rakata. Thus, Anak Krakatau's eruptions over the past 60 years appear to have had a significant impact on vegetation pattern at this scale. At the patch-scale (16 m²) the species composition of the regeneration layer of gaps was shown to be most dissimilar from the neighbouring understorey for those gap sites with evidence of previous disturbance or that suffered severe disturbance (e.g. a landslide site). It has been shown that disturbance is an important ecological determinant of diversification, and succession for the forests of this island group. Yet, prediction of future successional development is difficult since disturbance and its impact vary in space and time, and its influence is critically dependent on a complex array of interacting factors such as seed availability, dispersal and extreme climatic events.

INTRODUCTION

The Krakatau islands (6°06'S, 105°25'E; Fig. 1) lie in the middle of the Sunda Straits about equi-distant between Java and Sumatra some 45 km from the mainland of Java and Sumatra. The archipelago consists of the islands Rakata (c.750 m a.s.l.), Sertung (187m a.s.l.), Panjang (147 m a.s.l.) and Anak Krakatau (during eruptions between 1992 – August 1993 the island increased in height from 180 to c.280 m a.s.l.; Thornton *et al.*, 1994).

Plant recolonisation of the Krakatau islands began swiftly after the cessation of the sterilising eruptions of 1883. A strandline flora, consisting of such species as *Barringtonia asiatica, Terminalia catappa,* and *Hibiscus tiliaceus,* was very quickly established, and the inland succession progressed from an early phase of fern and grassland cover to a grassland stage with scattered clumps of trees and shrubs, typically *Macaranga tanarius, Ficus septica,* and *Ficus fulva.* These clumps eventually coalesced so that by around 1930 a species-poor closed forest covered each island.

[1] School of Geography, University of Oxford, Mansfield Road, Oxford OX1 3TB.
[2] Herbarium Bogoriense, Jalan Ir. H. Juanda 22, Bogor 16122, Indonesia.

247

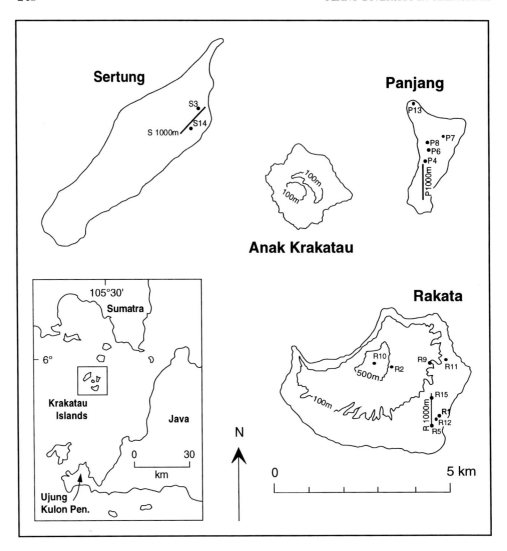

FIG. 1. Map of the Krakatau islands with locations of the 20 × 1000 m belt-transects for species-area enumeration, and gap site locations.

On Rakata large areas became dominated by the early successional, anemochorous tree species, *Neonauclea calycina* (Ernst 1907; Doctors van Leeuwen 1936; Whittaker *et. al.* 1989).

Tagawa *et. al.* (1985) were the first to notice differences in forest types between the islands. Sertung and Panjang were found to be dominated by *Timonius compressicaulis* and *Dysoxylum gaudichaudianum*, and the present pattern of distribution of these species were attributed to the order of arrival and edaphic factors (Tagawa *et al.* 1985; Tagawa 1992; Thornton 1996). However, Tagawa's group failed to recognise the possible influence of volcanism which resumed c. 1930 with the emergence of Anak Krakatau in the centre of the island group. Disturbance from intermittent volcanic activity has since then been an integral part of the Krakatau islands' disturbance regime. Ash-fall and, during severe eruptions (e.g., 1931 – 5 and 1952/3), blast and gas damage (Doctors van Leeuwen 1936; Borssum Waalkes 1954; Whittaker *et al.* 1989) has directly affected Panjang and Sertung but not Rakata. A history of these episodes can be reconstructed from numerous ash deposits in the soil profiles of Sertung and Panjang (Whittaker *et al.*, 1992a). Forest divergence has been argued by Whittaker and colleagues to have occurred, at least in part, as a result. Thus the current forest dominance by *Timonius compressicaulis* and *Dysoxylum gaudichaudianum* on Panjang and Sertung, rather than *Neonauclea calycina* on Rakata, has been suggested to be due in part to a deflection of succession by volcanic disturbance to the former (Bush *et al.* 1992; Whittaker *et al.* 1989).

The Krakatau islands provide a unique long term data-series for a primary successional sequence in the tropics. Overall the forests are still early successional in nature, as indicated by their general pattern of dominance by a small number of early seral tree species such as *Neonauclea calycina, Dysoxylum gaudichaudianum,* or *Timonius compressicaulis* (typically >50% basal area; Whittaker *et al.* 1989; Bush *et al.* 1992). Surveys over the last 17 years suggest that the flora continues to increase in size, and change in terms of the balance of different ecological and taxonomic groups (Bush & Whittaker 1993; Whittaker *et al.* 1992b; Whittaker & Jones 1994). In order to understand the recolonisation and succession several key variables have to be considered: dispersal; the mainland source-pool for propagules; the degree of isolation; total area and habitat availability (e.g. altitudinal range), and the characteristics of the disturbance regime. This paper will concentrate on the latter, but an evaluation in the context of the other main variables will be attempted.

While there is evidence — some historical and much circumstantial — in support of the disturbance-driven successional model (Bush *et al.* 1992; Whittaker *et al.* 1989; Whittaker *et al.* 1992a), most of the evidence indicating forest divergence and reduced diversity has come from plot-based vegetation and general floristic surveys. These have therefore concentrated on quantification at the small scale (400 – 2500 m^2) or the whole island-scale (3 – 17 km^2). For the first time this paper presents data on diversity and species composition at the meso-scale, and for gap regeneration on the patch-scale. Moreover, this is the first study that has recorded in detail the impacts of damage from volcanic activity, due to the coincidence of the study period (1992 – 1994) with renewed activity of Anak Krakatau. Thus the results of this study should not only help to assess the relevance of the disturbance-driven successional model, but also help to answer the following questions:

1. How is the differential impact of disturbance reflected in the forest mosaic structure and species-richness of each island at the meso-scale?
2. Are the inferences drawn from the plot based studies with regard to successional pathways corroborated at the meso-scale?
3. At the patch-scale, can the difference in the composition of the regeneration layer between gaps, and between the gap and the adjacent understorey be explained by the type and scale of disturbance?

METHODS

A) The meso-scale investigation: transect surveys of species presence/absence, and of disturbance.

Surveys (trees and shrubs), and quantification of disturbance were carried out in 1993 and 1994 on Rakata, Sertung, and Panjang.

The survey method reported below provides for the examination of species distributions, β-diversity, and turnover along a spatially contiguous transect. Its advantages include greater objectivity, speed, and relevance, in comparison to a plot-based method, in an attempt to answer questions 1 and 2 posed above.

Species-area curves and β-diversity:

Three 1000 m transects (one for each island) were enumerated in lowland areas (see Fig. 1 for location). Presence/absence scoring of woody species in the understorey (c.0.5 – 5 m), lower canopy (c.5 – 15 m, depending on total height of canopy) and canopy (>15 m) was carried out every 10 m covering a 'field of view' of c.10 m left and right. Over a transect length of 1000 m this yielded an approximate survey area of 2 ha. Although this is a fairly crude way of surveying, this method is appropriate for a fast comparison of alpha- and β-diversity between islands, and for use as evidence for inter-island differences in impact of disturbance at the meso-scale.

The species-area relationship for each transect was established by calculating cumulative species totals from both ends of each transect, for all strata combined, and for 0.1 ha sections. These sections were derived by lumping five 10×20 m sections together.

Whittaker's β-diversity index was calculated according to the following formula: b = S/α-1 (S = cumulative species total for the whole transect; α = average of the cumulative species total of all 0.1 ha sections). Whittaker's index has been found to be the best index available for presence/absence data (Magurran 1988).

Frequency of disturbance:

Disturbance surveys enumerated the number of recent disturbances in an attempt to document if the volcanic activity during the study period has had a significant effect on the number of disturbances experienced, and if inter-island differences are again discernible. Sertung, which historically has received the greatest amount of ash-fall (Whittaker et al. 1992a), and which has the most early successional appearance, was assumed to be most severely affected.

A disturbance event was recorded when evidence of past treefall or large branchfall was noted. Landslides were also recorded. Disturbances were divided into four age-classes according to the state of decay of the trunk, and/or the stage of regeneration (Table 1). Estimation of treefall age above four years was not possible.

The surveys involved recording disturbances encountered along a total of nine gully transects, varying in length from 450 to 1050 m (length surveyed: 3790 m on Rakata; 1850 m on Panjang; 2000 m on Sertung), together with quantification of all disturbances found along a total of six 'cross-country' transects of 450 to 1000 m length (length surveyed: 1950 m on Rakata; 1600 m on Panjang; 1000 m on

Sertung). Three of the 1000 m transects of the latter type were also used to enumerate species presence/absence for the calculation of species-area curves (see above). The difference between the two transect types was that the 'cross-country' transects ran from a chosen starting point along a compass bearing, whereas the gully transects followed gully bottoms, mainly for ease of survey in the restricted time available. Percentage frequency needed to be used as transects were of different length.

TABLE 1

Age-class (range: 1 – 4) determination from the decay stage of fallen trees, and the stage of regeneration. For landslides the stage of revegetation of the landslide scar was used to assess the age-class of the disturbance

Age-class with years since disturbance	scoring criteria
1 ≤ 2 years	fallen tree mostly intact, still with twigs or even leaves; bark intact; new regeneration small.
2 = 2 – 3 years	trunk still firm but beginning of break-down; bark starting to flake/already flaked; only major branches left; regeneration of pioneers >2 m depending on size and type of disturbance.
3 = 3 – 4 years	trunk softening and breaking up fast; no bark left; major branches broken off and very decayed; regeneration >3 m depending on size and type of disturbance.
4 >4 years	trunk breaking apart; all debris decayed; gap filled by regeneration forming low canopy in medium to large gaps; gap-closure from ingrowth and regeneration in small gaps.

B) The patch-scale investigation: species composition and diversity of the regeneration layer in young gaps.

In 15 gap sites (Fig. 1; Table 2), species composition and abundance of the sapling layer (>50 cm <5 m height) were recorded in pairs of 4 × 4 m plots; one in the gap centre and the other in the adjacent understorey. Eight gaps were located on Rakata, five on Panjang and two on Sertung, spanning the spectra from small to large, and from light (e.g. branchfalls) to severely disturbed sites (e.g. landslides).

TABLE 2

Gap sites over-view table showing severity of disturbance and size score, forest type, and whether previously disturbed

gap code ◊	severity score^	size score~	size in m²	disturbance type+	forest type*	previously disturbed#
R1	1	1	361	LNG? (STD)	Nc/Dg	N
R2lng	3	4	3016	LNG	Nc/Va	N
R2tf	2	4	3016	MTF/LNG	Nc/Va	N
S3	2	4	2297	TF	Dg	N
P4	2	4	2209	MTF	Tcl	N
R5	1	2	830	LBF	Dg	Y
P6	2	2	829	MTF	Tcl/Nc	N
P7	3	2	447	LS	Nc/Dg	N
P8	2	2	724	MTF	Nc/Tcl	Y
R9	2	2	486	MTF	Nc	Y
R10	2	3	1084	MTF/LNG	Sn/Va	N
R11	2	1	368	TF	Nc/Dg	Y
R12	1	3	1492	LBF	Dg	Y
P13	1	2	721	LBF	Dg	N
S14	1	3	1487	LNG	Tcl	Y
R15	1	2	485	TF (STD)	Nc	Y

◊: R = Rakata; S = Sertung; P = Panjang; R2 has two gap plots, R2tfg = R2 treefall gap plot, and R2lng = R2 lightning gap plot.
^: score 1 = low severity; score 2 = medium severity; score 3 = high severity.
~: score 1 = small = <400 m²; score 2 = medium = 400 – 1000 m²; score 3 = large = 1000 – 2000 m2 ; score 4= very large = >2000 m2.
+: LNG = lightning strike; STD = standing dead; TF = treefall; LBF = large branchfall; MTF = multiple treefall; LS = landslide.
*: Dg = Dysoxylum gaudichaudianum; Nc = Neonauclea calycina, Tcl = Timonius compressicaulis; Sn = Saurauia nudiflora; Va = Vernona arborea.
#: Y = yes; N = no.

Time constraints and inaccessibility of large areas of the islands prevented the enumeration of more sites.

Gap sizes ranged from 361 – 3016 m², whereby the gap area was approximated to that of an ellipse. The end points of the ellipse's diameters, i.e. the approximate gap edges, were determined as the points were photosynthetically active radiation (PAR) levels dropped to the understorey equivalent for more than two readings (readings taken every 2.5 m using a Sunfleck Ceptometer; AT Delta-T Devices).

Detrended Correspondence Analysis (DCA) and Two-Way-Indicator-Species Analysis (TWINSPAN):

Species compositional data of 4 × 4 m gap and understorey plots were subject to ordination by detrended correspondence analysis (CANOCO; Ter Braak 1988), and divisive classification by Two Way Indicator Species Analysis (TWINSPAN; Hill 1979). For TWINSPAN, pseudo-species cut levels were set at 0, 2, 5, 10, and 20, otherwise default settings were used (Kent & Coker 1992).

The assumption is that if type and scale of disturbance are more important than the island context and the surrounding forest type in influencing diversity and species composition of the gap plots, then the latter should be separated in ordination space and/or placed into different main TWINSPAN groups (determined by the second level of the division) from their understorey plots. The differences in composition should be the greater the larger the scale of disturbance, as that would, according to Bazzaz (1991), result in an increasing component of recruitment from the seed bank and seed immigration rather than advanced regeneration.

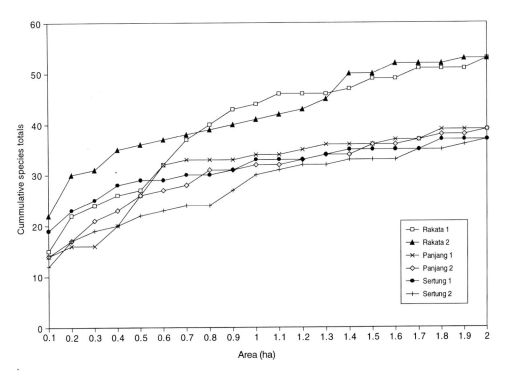

FIG. 2. Species-area curves for woody species (excluding climbers) of all strata. One 1000 m transect was enumerated on Rakata, Sertung, and Panjang respectively. Approximate area covered is 20 × 1000 m (2 ha). The alternative forms of the curves were derived by calculating the cumulative species totals in turn from each end of the three transects. The great variation in the shape of the 0-0.6 ha sections of the alternative versions of the same transect demonstrate the danger of relying on very small plots for species richness comparisons.

RESULTS

A) The meso-scale investigation

Species-area curves and β-diversity:

The Rakata transect has a higher cumulative species total for all strata. It also has the highest β-diversity when canopy and lower canopy strata are considered separately, which, however, becomes very similar to that of the other islands if the understorey is included. The species-area curves for Sertung and Panjang are significantly different from those for Rakata (Mann-Whitney-U test; p<0.01; n=20), but not from each other (Fig. 2; Table 3). The early flattening of the Sertung and Panjang species-area curves is indicative of these islands' comparatively reduced species richness. This is the first time this has been shown at the meso-scale.

Frequency of recent disturbance:

In accordance with casual observations, and with the historical pattern of ash-fall within the islands (Whittaker et al., 1992a), Sertung was found empirically to have the highest proportion of recent disturbances (class 1; 40.9%; Fig. 3A) of all transect surveys combined, while Panjang was intermediate with 30.2%, and Rakata lowest with 28.3%. When comparing the full age-class distributions for 'gully' transects only, Sertung still had the highest average percentage (37.7%) of recent (class 1) disturbances (Fig. 3B). However, Panjang's distribution of age-classes of disturbance, contrary to expectation, has a lower frequency of recent disturbance (20.9%) than both Sertung and Rakata (26.6%), and overall the majority of Panjang disturbances are older (class 3 and 4).

TABLE 3

Arboreal species richness (trees and shrubs) and R. H. Whittaker's β-diversity indices for lower canopy and canopy, and for all strata combined (2 ha area). Comparison between 1000m species presence/absence ('cross-country') transects on Rakata, Sertung, and Panjang. Note that b-diversity values should be compared across islands but within the same category only. Then lower values indicate less community diversity.

	Rakata		Panjang		Sertung	
	species richness	β - diversity	species richness	β - diversity	species richness	β - diversity
lower canopy & canopy	41	4.41	32	3.69	28	3.35
understorey, lower canopy & canopy	54	2.78	39	2.63	37	2.73

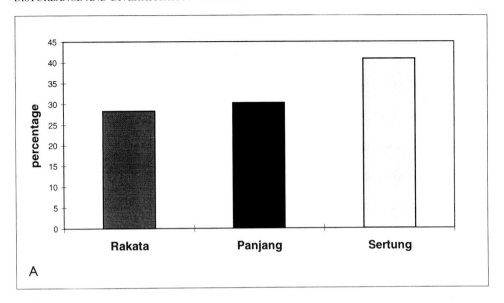

A

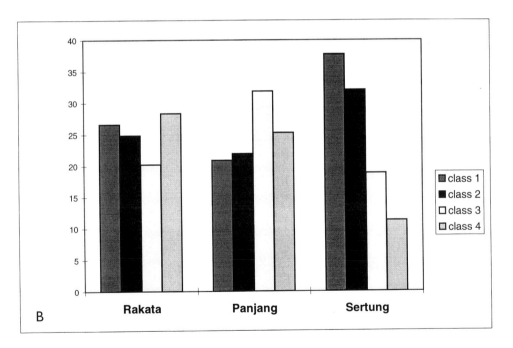

B

Fig. 3. A) Histogram depicting class 1 (<2 years old) disturbances as percentage of the total number of disturbances recorded for transect surveys on Rakata, Panjang and Sertung ('gully' and 'cross-country' transects combined). B) Age-class distribution of disturbances for gully transects on Rakata, Panjang and Sertung expressed in percent of total number of disturbances recorded.

B) The patch-scale investigation

DCA ordination and TWINSPAN classification:

After applying the DCA ordination, TWINSPAN divisive classification to the second level of the division, placed all plots into four main groups. Sub-groups of plots are formed at the third level of the division (Fig. 4). All results are summarised in Table 4A, and the preferential species for each sub-group of plots are indicated.

The main groups (I-IV) appear largely characterised by altitude (for Rakata plots), forest type and/or island affiliation. The type and scale of the disturbance only appeared important in the most extreme case of the P7g landslide site which forms its own main group. For seven out of 15 gaps a separation of the understorey and gap plots took place at the second level of the division, which means their gap and understorey composition is very dissimilar. Except for P7g, a common feature of these gap plots is that they are multiple aged gaps. This result is also supported by the finding that Sorensen similarity indices of gap and understorey sites of previously disturbed sites are significantly lower (Table 4B) than for sites with no evidence of previous disturbance. Thus, multiple gap opening events might have led to several recruitment waves of species often likely to be different from the advanced regeneration component. The repeated damage to and eventual death of some or

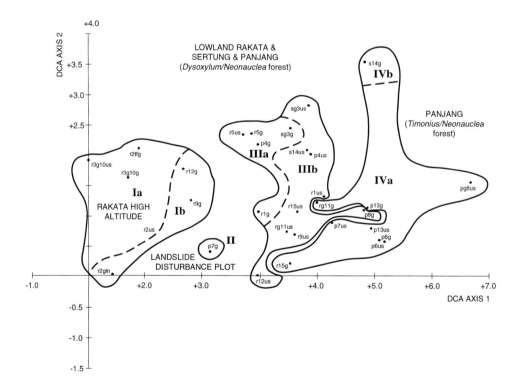

FIG. 4. DCA ordination and TWINSPAN classification of the sapling regeneration layer of pairs of 4 × 4 m gap and understorey plots from Rakata, Sertung, and Panjang.

TABLE 4

A) Lists of plots for second and third level TWINSPAN groupings of 4 × 4 m gap and understorey plots of 15 gap sites.

2nd level main groups	I		II	III		IV	
3rd level sub-groups	Ia	Ib		IIIa	IIIb	IVa	IVb
preferential species	Cyrtandra sulcata 1 (+)	Ficus fistulosa 2 and 3 (-)	Ficus pubinervis 1 (+)	Pipturus argenteus 1 (+); Macaranga tanarius 2 (+)	Buchanania arborescens 1 and 2 (-) Antidesma montanum 2 (-)	Antidesma montanum 1(+)	Bridelia monoica 1, 2, and 3 (-)
plot codes	R3G10g R3G10us R2tfg R2us	R12g R9g R2lng	P7g	R5us R1g R5g P4g S3g	R9us R15us P8g R11us R12us S3us S14us R1us P4us	R15g P13g P13us R11g P6g P6us P7us P8us	S14g

B) Comparison of Sorensen similarity indices of gap vs. understorey plots for all gap sites. A one-tailed Mann-Whitney-U test is applied to test whether gap/understorey similarity of previously disturbed gaps is significantly less similar than of previously undisturbed gap sites.

previously disturbed sites	Sorensen index	previously undisturbed sites	Sorensen index	α significance level
R5	0.16	R1	0.46	
P8	0.10	R2lng	0.34	
R9	0.19	R2tf	0.26	
R11	0.38	S3	0.47	
R12	0.12	P4	0.57	
S14	0.12	P6	0.71	
R15	0.32	P7	0.13	
		R10	0.92	
		P13	0.88	0.0055

Note: R = Rakata; S = Sertung; P = Panjang; g = gap plot; us = understorey plot; gap R2 has two gap plots, R2tf = R2 treefall gap plot, and R2lng = R2 lightning gap plot;

In part A) numbers and (+) and (-) signs behind the species names describe the pseudo-species (range from 1-5) and negative or positive placement in the classification dichotomy of the preferential species; for all 'a' sub-groups these are so-called 'indicator species', i.e., they are exclusively found on either the (+) or (-) side of the dichotomy ; see Kent and Coker (1992) for a detailed description.

most saplings of this component is also likely to have contributed to the greater differences. The finding of significantly lower sapling species numbers in previously disturbed gap plots (t-test; p<0.01) supports this suggestion. In contrast some of the single-disturbance, and very lightly disturbed gaps, such as, for instance, branchfalls or simple treefalls (e.g., PG6 and PG13 in sub-group IVa), even retain both the gap and the understorey plots in the same sub-group, and close together in ordination space.

DISCUSSION

A) The meso-scale investigation

Species-area-curves and β-diversity:

Plot-based and whole-island analyses have previously suggested greater inequitability and lower species richness for the ash-affected islands of Sertung and Panjang (Bush *et al.* 1986; Whittaker *et al.* 1989; Whittaker & Jones 1993). The findings in the present study of lower species richness and β-diversity at the meso-scale for the forests of Sertung and Panjang, appear to add further support to the disturbance-driven successional model for the Krakatau islands (Bush *et al.* 1992; Whittaker *et al.* 1989; Whittaker *et al.* 1992a). The b-diversity results for the canopy and lower canopy layer can be interpreted as showing the matrix to be composed of large patches of a generally similar nature on Sertung and Panjang, relative to Rakata. Multivariate analysis reported elsewhere supports this interpretation (Schmitt & Whittaker, in press). However, if all strata are considered, thus including both saplings of canopy trees and true understorey species, the β-diversity values are similar across the three islands. This might be taken as indicative of a trend towards greater similarity in the degree of mosaic patterning of the canopy (β-diversity) on the three islands in time. This interpretation, however, assumes high transition probabilities of the potential canopy species in the understorey, no severe volcanic disturbance, and several other unknowns.

Prior to Anak Krakatau's emergence, more species were known from the forests of Sertung than from Panjang, but fewer than from Rakata (whole island scale; see Whittaker *et al.* 1989 Table 4). Although differences in survey efforts have to be borne in mind (Whittaker *et al.* 1989 Table 2) simple species-area relationships could explain that difference, as Sertung (13 km^2) is more than three times the size of Panjang (3 km^2), but smaller than Rakata (17 km^2). Additionally, the greater height of Rakata added a high altitude component to its flora.

Similarly, successional development of Sertung and Panjang up to c.1930, was in line with compositional changes in lowland Rakata, although perhaps a little slower and there were differences in commonness of particular species. For instance, *Timonius compressicaulis* was found to be more widespread on Panjang than on Rakata (there mostly found in the North-west), and at that time had not been found on Sertung. *Neonauclea calycina* was already quite frequent in the northern flat and hilly parts of Sertung, and was frequently found on Panjang by 1930, but appeared generally not as common as in the upper reaches of Rakata. Yet, no evidence of significant forest divergence in the lowland associations was remarked upon by Doctors van Leeuwen (1936).

p. 256

"One will have noticed that in the three islands the associations in the lower parts and the various
successions are alike. The vegetation of the higher parts of Mt Rakata is different, and is absent in
the other two islands, since these are no higher than 180 m and 147 m respectively."

It was from its early strongholds in the uplands of Rakata that *Neonauclea calycina*
subsequently spread downwards to dominate in the lowlands (Borssum Waalkes 1960;
Whittaker *et al.* 1984). The possession of upland territory and the early establishment
of *Neonauclea* in it may thus be an important element of the developmental process
(cf. Tagawa 1992). In addition to the undoubted importance of differing patterns of
initial establishment of species populations on the islands, an important role has also
been played by the widespread occurrence of volcanic disturbance. Evidence for this
is to be found in the present, very different forest dominance patterns and
physiognomy, the very uniform age-structure in *Timonius* forests reported upon
c.1983, (Bush *et al.* 1992; Whittaker *et al.* 1989), and as further established in this
meso-scale study, a significantly lower species richness and β-diversity (less mosaic
nature) of Sertung and Panjang's forests compared to Rakata. These observations are
suggestive of the ideas of deflection and successional set-back. It can be argued that
early arrival alone would not have led to such extreme dominance patterns and
uniformity, but rather that establishment prior to large-scale volcanic disturbance may
well have given *Timonius compressicaulis* and *Dysoxylum gaudichaudianum* the
opportunity for rapid invasion of large areas of Panjang and Sertung.

Frequency of recent disturbance:

   The results of the disturbance surveys suggest that during the recent volcanic
activity (1992-1994) Sertung's forests have been more dynamic than those of the other
two islands. Although the scale and severity of the recent disturbances do not appear
to have been as great as the 1930 – 35 and 1952/53 eruptions (Borssum Waalkes 1954;
Borssum Waalkes 1960; Doctors van Leeuwen 1936; Whittaker *et al.* 1992a), these
results lend further support to the hypothesis that volcanic disturbance has a
significant impact on turnover and succession through the creation of gaps. Ash-
affected Panjang, which was also expected to show an increase in recent disturbance,
however, showed somewhat conflicting results. On the one hand, in the combined
analysis (Fig. 3a), it was found to have a slightly higher proportion of recent
disturbances than Rakata. On the other hand, the analysis of only the gully transects,
shows no evidence of increased disturbance since the beginning of the eruptions in
1992 (Fig. 3b). Within island differences in disturbance dynamics have been observed
to be important, and part of the explanation might lie in the fact that Panjang is not
necessarily affected by the eruptions at the same time, and not to the same degree as
Sertung. The differences in the thickness and, partly the chronology of the ash-
profiles as shown by Whittaker *et al.* (1992a), and casual observations on differences of
ash-thickness during the recent eruptive phase (c. 8-10 cm for Panjang and 10-15 cm
for Sertung) support this argument. This may largely be explained by the timing and
duration of eruptions in relation to dominant seasonal wind directions; whereas,
Sertung lies down-wind of the active vents during the dry season (easterly winds),
Panjang receives most ash during the wet season (westerly winds). Furthermore, ash-
fall received by Panjang during the wet season might result in less volcanicity related

disturbance, as a build-up of heavy ash-loads is avoided due to regular rainfall. Thus risk from branch and stem snapping, as was observed on Sertung, is mostly prevented.

B) The patch-scale investigation: interpretation of DCA and TWINSPAN results

The main TWINSPAN groupings (I-IV) of the 4 × 4 m gap and understorey plots predominantly appeared to reflect differences in forest types, which often coincide with island affinity (particularly in the case of group I and IV). The type and scale of the gap-creating disturbance was not, as one might have expected, crucial in differentiating the gap and understorey composition, except in the extreme landslide case. Instead, previous disturbance of the gap site (i.e. frequency of disturbance) appeared more important, as it could have led to progressive damage and death of advanced regeneration, and replacement with newly recruited species. The greater survival of advanced regeneration plus some additional recruitment of new species might explain the significantly higher number of species in previously undisturbed gap plots. This suggestion must be advanced tentatively, however, as some of the difference between gap and understorey may be inherent in the often patchy nature of understorey regeneration prior to gap opening (Brown & Whitmore 1992).

Nevertheless, the DCA and TWINSPAN analysis have shown that, in the majority of the recently created gaps, advanced regeneration appeared to play an important role (Uhl *et al.* 1988) at least in the initial phase of gap fill since otherwise more differentiation of gap and understorey plots in ordination space and TWINSPAN groupings would have been apparent. However, monitoring of survival and growth of the advanced regeneration component, and any gap recruitment, would be necessary to assess whether special gap adaptation (Denslow 1980a) is of importance for gap regeneration on Krakatau. In any event, in the primary successional and island context, there is a lack of a full successional spectrum of species (sensu Swaine & Whitmore 1988). With exception of the extreme pioneers (e.g. *Macaranga tanarius, Neonauclea calycina, Timonius compressicaulis,* etc.), Krakatau's woody flora is dominated by secondary species (e.g. *Dysoxylum gaudichaudianum, Ficus pubinervis, Buchanania arborescens,* etc.) with an apparently broad tolerance of environmental factors (Richards 1996), and their presence or absence, and continued survival in gap regeneration is much more likely to be due to influences such dispersal, chance and, as suggested from this study, site history (e.g. previous disturbance). The latter is strongly influenced by the occurrence of extreme events, such as periods of volcanic activity, droughts or storms.

The above suggests that early composition and diversity of the gap regeneration layer is likely to differ strongly from the advanced regeneration layer of the surrounding forest if the disturbance creating the gap was severe, and/or repeat disturbance to same area has taken place. Cursory analysis of two years of monitoring of gap regeneration (work in progress), suggests that the extreme event of the 1994 El Niño drought was the dominant factor influencing species growth, survival and recruitment. This was particularly so in conjunction with ash-stress, and does not seem to be related to the physical characteristics of the gap. Future canopy composition is, thus, difficult to predict from looking at the species composition and diversity patterns of gap regeneration at the early stages only.

CONCLUSIONS

At the meso-scale, lower diversity and greater uniformity of the canopy layer composition has been found for the forests on Panjang and Sertung in comparison to those of lowland Rakata. This finding supports existing evidence at the plot-, and whole-island scale, and lends support to the emphasis on disturbance in the analyses of Whittaker and co-workers (Whittaker *et al.* 1989,1992a; Bush *et al.* 1992).

At the patch-scale, some indication was found that severe disturbance, and repeat disturbance in particular, has a significant impact on species composition and diversity of the regeneration layer at the early stages of gap regeneration by reducing the importance of advanced regeneration (cf. Bazzaz 1991). Prediction of gap fill from the physical characteristics of the gap, and/or the species composition of the regeneration layer in young gaps, however, is not feasible. This is particularly the case in forests as dynamic as those of the Krakatau islands, and where extreme climatic events (e.g. El Niño droughts) might have an overriding effect on regeneration dynamics in gaps.

The results of this study have highlighted the important impact of disturbance on both scales of the investigation, particularly if it is of severe and/or frequent nature. With spatial scale, severity, and timing of disturbance being extremely variable, predictions of successional development on the Krakatau islands are difficult. Furthermore, this study suggests that gaps as windows of regeneration, and potential loci of colonisation by new species, are not lacking. Nevertheless, both those gap sites monitored and more generally gaps observed by the author were not found to be 'hot-spots' of diversification, as, for instance, species still relatively rare on the islands were only occasionally found in gaps. Instead it was regularly noted that rare species, which were already long established, seemed to radiate from their first point of colonisation only slowly and were apparently unrelated to habitat preferences. Such species can fail to spread for a long time within and between the islands, and one might suggest some dispersal limitation (e.g. in the case of *Semecarpus heterophylla*). Other less common species, however, seem to be increasing in importance and have started to become locally frequent in the understorey (e.g. *Syzygium polyanthum*, and *Planchonella duclitan* in lowland Rakata). Therefore, with the increasing senescence of, for instance, the canopy dominant *Neonauclea calycina* on Rakata, these and other species are likely to play a more important role in the future. Nevertheless, the diversification process can not be expected to occur very fast, and is always likely — at least on Sertung and Panjang — to be subject to further volcanic disturbance.

ACKNOWLEDGEMENTS

We thank the Indonesian Institute of Sciences (LIPI), and the Indonesian Directorate General of Forestry and Nature Conservation (PHPA) of the Ministry of Forestry for permission to work on the Krakatau islands. Particular thanks are due to Dr Soetikno Wirjoatmodjo, Dr Dedy Darnedi, Dr Johanis Mogea and Paul Jepson and to all those who helped with fieldwork in 1993 and 1994. We are very grateful to Dr R.J. Whittaker, Dr B.D. Turner, Dr D. Sheil, Dr G. Vieira, and an anonymous reviewer for comments and suggestions on the manuscript. This project was funded by a NERC studentship (GT4/92/30/L) to S. Schmitt, and additional funds were provided by the Royal Society's South-east Asia Rain Forest Research Programme. This is Krakatau Research Project Publication No. 49 which is an affiliated project of the Royal Society South-east Asia Rain Forest Research Programme (Programme Publication No. A\142).

REFERENCES

Bazzaz, F. A. (1991). Regeneration of tropical forests: physiological responses of pioneer and secondary species. In: A. Gomez-Pompa, T. C. Whitmore & M. Hadley (eds), Rain forest regeneration and management. pp. 91 – 117. Paris: UNESCO.

Borssum Waalkes, J. van (1954). The Krakatau islands after the eruption of October 1952. Penggemar Alam 34: 97 – 104.

Borssum Waalkes, J. van (1960). Botanical observations on the Krakatau Islands in 1951 and 1952. Ann. Bogor. 4: 1 – 64.

Brown, N. D. & Whitmore, T. C. (1992). Do dipterocarp seedlings really partition tropical rain forest gaps? Phil. Trans. R. Soc. Lond. B 335: 369 – 378.

Bush, M., Jones, P. & Richards, K. (eds) (1986). The Krakatoa centenary expedition 1983: final report. Miscellaneous Series Number 33. 267pp. Hull, England: Department of Geography, University of Hull.

Bush, M.B. & Whittaker, R.J. (1993). Non-equilibration in island theory of Krakatau. J. Biogeogr. 20: 453 – 457.

Bush, M. B. & Whittaker, R. J. (1995). Colonisation and succession on Krakatau: an analysis of the guild of vining plants. Biotropica 27: 355 – 372.

Bush, M. B., Whittaker, R. J. & Partomihardjo, T. (1992). Forest development on Rakata, Panjang and Sertung: Contemporary dynamics (1979 – 1989). Geojournal 28: 185 – 199.

Denslow, J. S. (1980a). Gap partitioning among tropical rainforest trees. Biotropica 12: 47 – 55.

Doctors van Leeuwen, W. M. (1936). Krakatau 1883-1933. Ann. Jard. Bot. Buitenz. 46 – 47.

Ernst, A. (1907). Die neue Flora der Vulkaninsel Krakatau. 77pp. Zürich: Faesi & Beer.

Hill, M. O. (1979). TWINSPAN — A FORTRAN program for arranging multivariate data in an ordered two-way table by classification of individuals and attributes. Ithaca, N.Y.: Cornell University.

Kent, M. & Coker, P. (1992). Vegetation description and analysis: a practical approach. 363pp. London: Belhaven Press.

Magurran, A. E. (1988). Ecological diversity and its measurement. 179pp. London: Croom Helm.

Partomihardjo, T. (1995). Studies on the ecological succession of plants and their associated insects on the Krakatau islands, Indonesia. 236pp. D.Phil. thesis. University of Kagoshima.

Schmitt, S.F. & Whittaker, R.J. (in press). Disturbance and succession on the Krakatau islands, Indonesia. In D.M. Newbery, H.N.T. Prins, N.D. Brown (eds),

Population and community dynamics in the tropics. British Ecological Society Symposium Volume.

Swaine, M. & Whitmore, T. C. (1988). On the definition of ecological species groups in tropical rain forests. Vegetatio 75: 81 – 86.

Tagawa, H. (1992). Primary succession and the effect of first arrivals on subsequent development of forest types. Geojournal 28: 175 – 183.

Tagawa, H., Suzuki, E., Partomihardjo, T. & Suriadarma, A. (1985). Vegetation succession on the Krakatau Islands, Indonesia. Vegetatio 60: 131 – 145.

Ter Braak, C. J. F. (1988). CANOCO — a FORTRAN program for canonical community ordination by [partial] [detrended] [canonical] correspondence analysis, principal components analysis and redundancy analysis. Wageningen: ITI – TNO.

Thornton, I.W.B. (1996). Krakatau: the destruction and reassembly of an island ecosystem. Cambridge, Mass.: Harvard University Press

Thornton, I. W. B., Partomihardjo, T. & Yukawa, J. (1994). Observations on the effects, up to July 1993, of the current eruptive episode of Anak Krakatau. Global Ecol. Biogeog. Lett. 4: 88 – 94.

Uhl, C., Clark, K., Dezzeo, N. & Maquirino, P. (1988). Vegetation dynamics in Amazonian treefall gaps. Ecology 69: 751 – 763.

Whittaker, R. J. & Jones, S. H. (1994). The role of frugivorous bats and birds in the rebuilding of a tropical forest ecosystem, Krakatau, Indonesia. J. Biogeogr. 21: 245 – 258.

Whittaker, R. J., Walden, J. & Hill, J. (1992a). Post-1883 ash fall on Panjang and Sertung and its ecological impact. Geojournal 28:153 – 171.

Whittaker, R. J., Bush, M. B., Partomihardjo, T., Asquith, N. M. & Richards, K. (1992b). Ecological Aspects of plant colonisation of the Krakatau islands. Geojournal 28: 201 – 211.

Whittaker, R. J., Bush, M. B. & Richards, K. (1989). Plant recolonisation and vegetation succession on the Krakatau islands, Indonesia. Ecol. Monogr. 59: 59 – 123.

Whittaker, R.J., Richards, K., Wiriadinata, H. & Flenley, J.R. (1984). Krakatau 1883 to 1983: a biogeographical assessment. Progr. Phys. Geog. 8: 61 – 81.

Systematics of *Aporosa* (*Euphorbiaceae*)

ANNE SCHOT[1]

Summary. Aporosa is a genus of small dioecious trees from the South East Asian rain forests. The 82 species are delimited and grouped on a combination of macromorphological characters. Different characters are informative at different levels. Inflorescence type and position of foliar glands are used for grouping at the highest level. Flower and leaf characters discern groups on a lower level, and indumentum and size differences delimit at species level. Various phylogenetic analyses were carried out with Hennig86 and resulted in many thousands of equally parsimonious cladograms. All patterns agree on three monophyletic groups and a paraphyletic remnant. This remnant can be divided geographically: one contains all New Guinean species and the other the remaining Sundanese species. To get a better insight into the processes that cause the multiple patterns, analyses were conducted using selected subsets of the taxa. Hypothetically, the contradictions in the paraphyletic part of the tree regarding the Sundanese species might be caused by extinction of ancestors, and those regarding the New Guinean species by extensive hybrid speciation.

INTRODUCTION

Aporosa Blume is one of the more common Euphorbiaceous genera of the tropical lowland rain forests of the Malay Archipelago. It consists of 82 species of small to medium-sized dioecious trees. They occur commonly in the tropical rain forests of Southeast Asia. Their most westerly distribution is in the tropical forests of Sri Lanka and Kerala (India) with six species; then from the moist parts of the Himalaya southwards into Indo-China and the tropics of the Malay Peninsula up to New Guinea, where one species reaches the Solomon Islands (Fig. 1). Borneo, with 30 species, of which ten are endemic, and New Guinea, with probably up to 20 endemic species, many of which are still poorly known, are the main centres of diversity for *Aporosa*.

Aporosa belongs to subfam. *Phyllanthoideae*, the possibly paraphyletic part of the *Euphorbiaceae* (Webster, 1994). This subfamily is mostly characterized by primitive, unspecialized characters such as dioecy, simple trichomes, 3 – 5 carpels, and dehiscent fruits. *Aporosa* itself is a good example. It has no laticifers, carunculus, cyathia, lepidote hairs, or any other such derived characters. The leaves are simple, alternate, with small, unspecialized glands. The inflorescences are racemose and unisexual. The flowers consist of sepals and stamens or pistil only. The fruit is often a tardily dehiscing capsule and the seed is covered by a thin aril.

Aporosa was classified by Webster (1994) with seven other genera into the subtribe *Scepinae*. Of these seven genera, *Baccaurea* is closest. The dioecious simple axillary racemose inflorescences and tardily dehiscing 2- or 3-locular fruits are also found in some species of *Baccaurea*, which is regarded as sister genus of *Aporosa*. *Aporosa* differs

[1]Rijksherbarium/Hortus Botanicus, PO Box 9514, 2300 RA Leiden, Netherlands

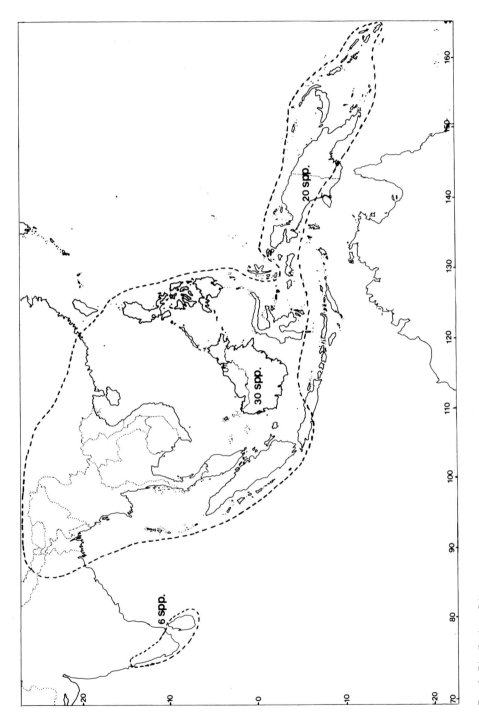

FIG. 1. Distribution of *Aporosa*.

from *Baccaurea* in the small sessile staminate flowers (mean length c. 0.5 mm) that only have a minute pistillode or none at all. The species are hard to delimit. As seen in other members of the *Phyllanthoideae*, specialized characters are rare. With only relatively few characters that vary in a systematically informative fashion, species are mostly delimited on unique combinations of characters. Therefore, the species are often rather similar in appearance and may be linked by intermediates.

This paper is in two parts. In the first an overview is given of the most important macromorphological characters used for delimiting and classifying the species, i.e., inflorescence type, position of the foliar glands, and some lesser characters. The second part contains a summary of the main patterns found by a phylogenetic analysis. A more exhaustive treatment of both will be presented elsewhere (Schot in prep.).

MACROMORPHOLOGY AND SPECIES DELIMITATION

Inflorescence types

A range of different inflorescence types is seen in *Aporosa*. Staminate plants have compact packages (glomerules) of 5 – 20 staminate flowers, each glomerule being surrounded by one bract. Pistillate plants have one flower in each bract. Both staminate and pistillate inflorescences show variation in the arrangement of the glomerules or pistillate flowers along the rachis (Figs. 2 & 3).

The staminate inflorescences range from glomerules densely set and completely covering the rachis (Fig. 2A, B) to individual glomerules fully separate (Fig. 2L, M). Intermediate types, with the glomerules spaced at the base of the rachis and clustered apically are found in many different configurations (Fig. 2 C – K). Of these, two are particularly striking. In one the rachis broadens apically and forms a more-or-less distinct head (Fig. 2J, K); in the other the flowers occur only in the apical part of the inflorescence (Fig. 2C). The latter type is also distinct because the stamens are strongly exserted.

Pistillate inflorescences show variation analogous to that of the staminate inflorescences. Some have densely set flowers that completely cover the rachis (Fig. 3A, B), whereas others have laxly arranged flowers (Fig. 3K, L). A series of intermediates is also present (Fig. 3C – J). An exceptional type of pistillate inflorescence has only a single flower (Fig. 3M, N); only the bud in the topmost bract develops the flower, while the other bracts have dormant buds.

Foliar glands

Three kinds of glands are present on *Aporosa* leaves, viz. marginal, disc-like and basal. They are anatomically identical.

Marginal glands are present in almost all *Aporosa* species and are taxonomically unimportant.

Disc-like glands appear as small black or brownish dots on the lower surface of the leaf (Fig. 4). Their presence/absence and position is generally consistent with groups based on inflorescence types. They may be few and scattered at the base (Fig. 4A), scattered (Fig. 4B) or more numerous (Fig. 4C) around the margins, or regularly scattered inside the marginal arches of the secondary nerves (Fig. 4D).

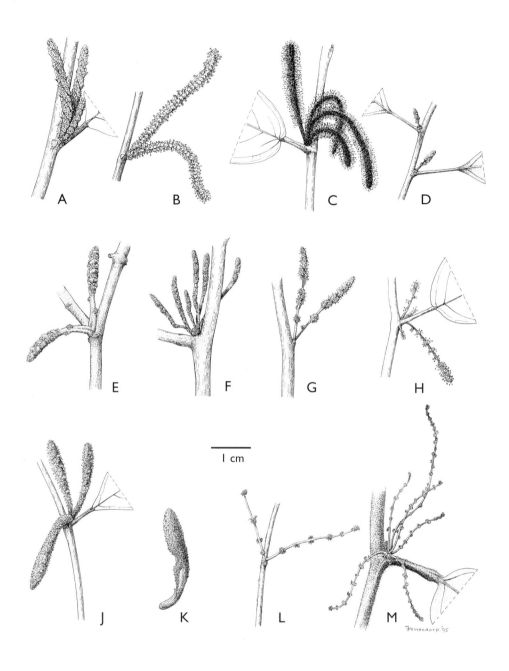

Fig. 2. Staminate inflorescence types. Differences in arrangement of glomerules along the rachis. **A** *A. octandra* (Meijer 6707). **B** *A. yunnanensis* (Put 3529). **C** *A. falcifera* (Hose 40). **D** *A. granularis* (AA 157). **E** *A. subcaudata* (S. 23379). **F** *A. stellifera* (KEP-FRI 7981). **G** *A. banahaensis* (FB 7279). **H** *A. lagenocarpa* (Jacobs 5102). **J** *A. frutescens* (Maxwell 82-35). **K** *A. nitida* (Jacobs 5196). **L** *A. prainiana* (KLU 1183). **M** *A. elmeri* (SAN 29685).

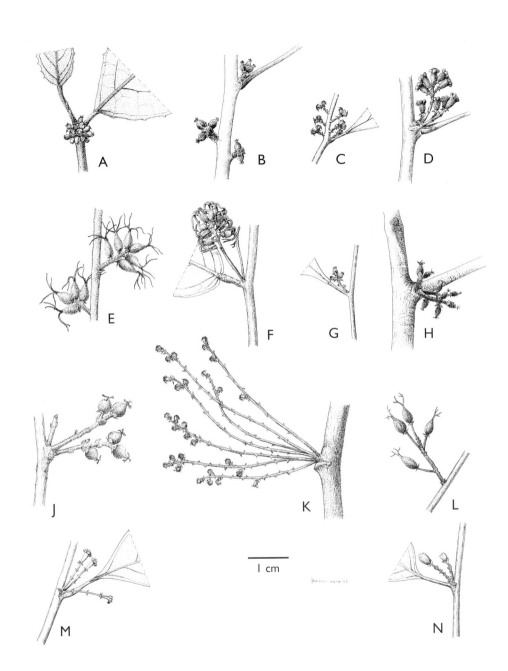

FIG. 3. Pistillate inflorescence types. Differences in arrangement of flowers along the rachis. **A** *A. serrata* (Kerr 8776). **B** *A. aurea* (KEP 105136). **C** *A. granularis* (Jacobs 5193). **D** *A. stellifera* (KEP-FRI 14705). **E** *A. lagenocarpa* (SAN 49765). **F** *A. falcifera* (Hose 92). **G** *A. maingayi* (KEP 105120). **H** *A. nervosa* (Maxwell 86-277). **J** *A. leytensis* (Wenzel 614). **K** *A. elmeri* (Ridsdale 2067). **L** *A. antennifera* (de Wilde & de Wilde-Duyfjes 15619). **M** *A. sarawakensis* (S 14730). **N** *A. frutescens* (Maxwell 82-191).

Basal glands are large and distinct in one small group of related species. Occasionally, small and indistinct basal glands occur in few other species.

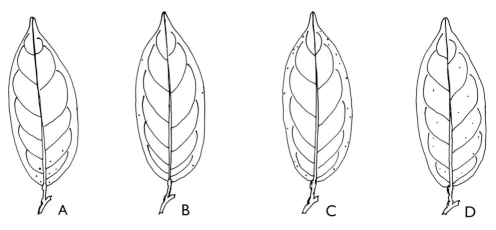

FIG. 4. Different positions of disc-like foliar glands on the lower side of the leaf.

Other characters

Other characters group the species within the groups formed by variation in inflorescence type and glands. These include variation in the staminate and pistillate flowers, colour of the leaf on drying, and shape of the stipules.

Useful characters from the staminate flowers include the degree of fusion of the sepals, the exsertion of the stamens, and the size of the anthers (Fig. 5). Pistillate flowers can vary in the length of the pedicel and the length, nature, and degree of division of the styles (see Fig. 3). Furthermore, in the whole genus there are about as many 2-locular as 3-locular species, while a few species even have 4 or 5 locules. The septae and columella are puberulous in one group within the species with densely-flowered inflorescences (Fig. 3A, B).

Aporosa is known to be an aluminum accumulator (Chenery, 1948). This is indicated by the colour of the dried leaves. When the aluminum concentration reaches a certain level the herbarium specimens, independent of their collecting

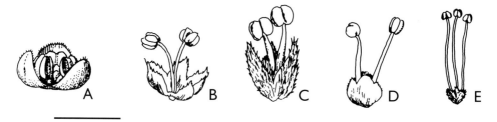

1 mm

FIG. 5. Variation in staminate flowers. Note the differences in sepal fusion and the relative size of stamens and anthers. **A** *A. symplocifolia* (redrawn from Pax & Hoffmann, 1922). **B** *A. lagenocarpa* (redrawn from Airy Shaw, 1974). **C** *A. bourdillonii* (redrawn from Stapf, 1894). **D** *A. campanulata* (redrawn from Smith, 1909). **E** *A. falcifera* (redrawn from Pax, 1890).

and drying method, have yellowish-green or greyish-green leaves. Species with low concentrations of aluminum in their leaves often dry brownish and intermediate colour shades can also be observed. This coloration is a useful feature for species recognition in herbarium specimens. From a systematic viewpoint, the greenish colour is characteristic for one group of species with laxly-flowered inflorescences, and a slightly yellower shade of green delimits one subset from the species with densely-flowered inflorescences.

Stipules are caducous or persistent. There are three main categories: mostly caducous (narrowly) ovate; shortly persistent, narrowly ovate when young and becoming more-or-less oblique with age; persistent, falcate.

Occasionally, the complete set of the above-mentioned characters of inflorescences, glands, flowers, leaf colour, and stipule shape is identical in closely related species. Species are then separated on differences in size and indumentum of leaf and inflorescence.

Species delimitation and species concepts

Species are recognized by specific combinations of characters. Thus *A. frutescens* and *A. globifera* have almost identical vegetative characters, but different inflorescences. On the other hand, *A. nitida* and *A. confusa* both have the one-flowered pistillate inflorescences and clavate-shaped staminate inflorescences, but differ in indumentum of the vegetative parts.

The use of combinations of characters instead of unique characters results in a rather narrow species concept. Two characters, e.g. dense or lax inflorescences and glabrous or puberulous leaves, can define four species on the basis of the character combinations dense and glabrous, dense and puberulous, lax and glabrous and lax and puberulous. When delimiting species on combinations of characters, intermediates might be assigned specific status, especially if the intermediate combination holds throughout this species, e.g. inflorescence intermediate & leaves only puberulous below. Synonymising such intermediates with their extremes means that only one very variable species is defined. This also effects the circumscription of related species, because the boundaries between them also become obscure.

Usually, species are merged on the number of intermediates present: many intermediates call for uniting species, few for separating species. This is a decision that is not based on the characters. If in another species-complex the same characters were present without intermediates, they would be kept apart. In the case of *Aporosa*, where there are so few characters but so much variation, the blurring of species boundaries by uniting species of intermediate combinations would confuse about half of the species. Therefore, I strictly use recognizability for species delimitation, independent of whether the formed character combinations are based on extreme or intermediate states.

A good example is delimitation in the species-complex of the widespread Asiatic *A. octandra*. *A. octandra* occurs from the Himalaya to South China and into the Malay Archipelago up to Sulawesi. The species is constant and recognizable on the periphery of its distribution, but in Indo-China and North Thailand it appears in many different forms. It even merges gradually into another species, *A. villosa*, that is separated from *A. octandra* on the tomentose and broad leaves with cordate base (the lamina is puberulous, narrowly elliptic, and with a cuneate base in *A. octandra*). The fact that specimens exist that have tomentose, elliptic leaves with a rounded base does not make *A. villosa* less recognizable. In the northern part of its range *A. octandra* is sympatric with another related species, *A. wallichii*, that is easily separated from *A. octandra* by the glabrous, green drying leaves and glabrous ovary. Several specimens from Indo-China are present that combine puberulous leaves and ovaries with green drying leaves, or glabrous ovaries with brown drying leaves. Again, this does not take away the fact that *A. wallichii* is still distinct from *A. octandra*, even though the individual characters that make the specific character combination vary. I have thus decided to keep both *A. villosa* and *A. wallichii* separate from *A. octandra*. The specimens with the intermediate combination are classified as *A. octandra* - which is anyway quite variable, keeping the extreme species strictly delimited.

Phylogeny reconstruction

Summary of the analyses

Multiple phylogenetic analyses were run with Hennig86 (Farris 1988). The data-matrix contained 82 multistate macromorphological characters and 103 taxa of *Aporosa*, consisting of the 82 species, eight varieties, six forms, and seven incompletely known specimens. Seven species of the sister genus, *Baccaurea*, were used as the outgroup. Because of the large number of taxa, the analysis met with a memory overflow at 875 equally parsimonious trees, indicating that possibly suboptimal trees are being retained. Various options were tried, but all resulted in a memory overflow with trees of different topologies. Analyses with variously weighted characters did not clarify the situation. The characters, data-set, and results of the various analyses are treated in detail by Schot (in prep.). Here, I will describe the general pattern seen over all analyses and explore possible explanations for the lack of consensus.

General consensus pattern

I have depicted the general consensus pattern that can be seen over all the different topologies in a combined diagram (Fig. 6). Different shadings represent the 11 species groups that can be delimited with the aid of the combinations of the macromorphological characters given in the first part. The species belonging to these 11 groups and their main characteristics are summarized in Table 1. I have chosen to give a combined diagram in the format of Fig. 6 rather than putting all 110 taxa in a consensus cladogram. In this way the interactions of the members of the groups in the various analyses are easier to explain.

The consensus diagram in Fig. 6 shows how the 11 groups were distributed in the various analyses. Groups 1, 2, 3, 7, and 8 were always present and they are given as distinct clades. However, group 7 interacts with group 6 and is only partly distinct. Therefore, these two are shown as joined. In the analyses this means that taxa out of group 6 can sometimes be placed between the taxa of group 7, and vice-versa. The same is true for members of other groups that are drawn adjacently in the diagram. Thus taxa out of the intermediate groups 5 and 11 can have one or more members appearing between members of almost any other group.

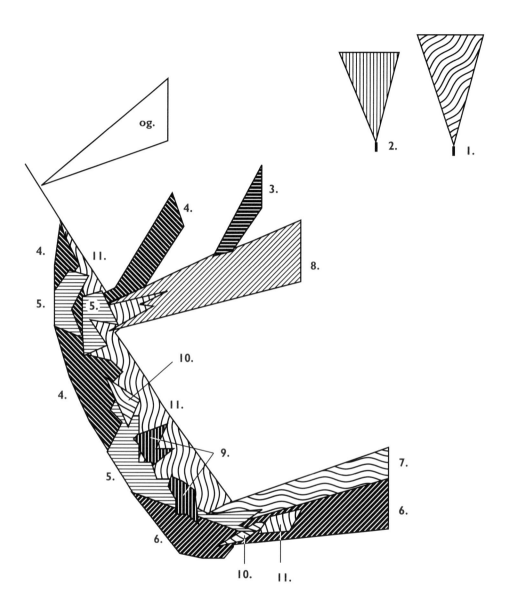

FIG. 6. Compilation of the consensus cladograms of all phylogenetic analyses, showing how the 11 groups delimited on macromorphological characters are placed. See Table 1 for the characterization of the groups. 1. monophyletic group around *A. octandra*. 2. monophyletic group around *A. frutescens* and *A. nitida*. 3. monophyletic group around *A. benthamiana*. 4. paraphyletic Sundanese group with *A. arborea*. 5. paraphyletic Sundanese group with *A. nervosa*. 6. Sundanese species around *A. maingayi*. 7. Sundanese species around *A. subcaudata*. 8. New Guinean group around *A. papuana*. 9. New Guinean group around *A. sclerophylla*. 10. New Guinean species around *A. nigropunctata*. 11. Intermediate New Guinean species. og. = outgroup.

TABLE 1. The species belonging to the groups in Figs. 6, 7, 9, and 11 and their most important macromorphological and distributional characteristics.

Group	species belonging	characteristics
1.	*acuminata, aurea, bourdillonii, cardiosperma, ficifolia, fusiformis, globifera, lanceolata, latifolia lucida, macrophylla, nigricans, octandra, planchoniana, pseudoficifolia, serrata, symplocifolia, tetrapleura, villosa, wallichii, yunnanensis*	Inflorescences densely flowered, septae & columella puberulous, flowers sessile, mostly 2-locular, disc-like glands scattered; Mainland of Asia, two widespread in the Malay Archipelago
2.	*banahaensis, chondroneura, confusa, frutescens, fulvovittata, nitida, prainiana, quadrilocularis, sarawakensis, stellifera, stenostachys, symplocoides, whitmorei, specE*	Inflorescences laxly flowered, some uniflorous, leaves drying green, papery, inflorescences drying yellow; Sundaland and Philippines
3.	*benthamiana, bullatissima, falcifera, lunata, sylvestri*	brush-like staminate inflorescences, pistillate flowers with long smooth stigmas, 3-locular, stipules large, falcate, disc-like glands distinct; Sundaland
4.	*arborea, grandistipula, illustris*	inflorescences laxly flowered, 3-locular, stipules large, falcate; Sundaland
5.	*basilanensis, caloneura, dendroidea, nervosa, sphaeridiophora, specG*	intermediate; Sundaland
6.	*alia, antennifera, ?duthieana, granularis, lagenocarpa, maingayi, microstachya*	inflorescences short, densely flowered, flowers sessile, 2-locular, leaves smallish; Sundaland
7.	*elmeri, penangensis, rhacostyla, selangorica, subcaudata, specF*	as 6, but larger leaves
8.	*brassii, decipiens, lamellata, leytensis, papuana, petiolaris, vagans, specD*	large leaves, long thick inflorescences, largish flowers, 3-locular, stipules large falcate; New Guinea
9.	*brevicaudata, egregia, parvula, sclerophylla*	small inflorescences and leaves, 2-locular, montane; New Guinea
10.	*longicaudata, nigropunctata*	as 9, 3-locular; lowland New Guinea
11.	*annulata, carrii, flexuosa, hermaphrodita, heterodoxa, latifolia, ledermanniana, leptochrysandra, misimana, reticulata, specA,B,C*	intermediate; New Guinea

It may seem from the diagram that almost all taxa "jump" groups. However, three groups (1, 2, and 3) do not show this pattern of exchange and are monophyletic in all the analyses. They are firmly characterized by a unique combination of characters. Their monophyly was tested by inactivating various characters from the matrix. In almost all cases they remained present, indicating high confidence in their monophyly.

Group 1 contains mainly Asiatic species with dense inflorescences (Fig. 2A, B, 3A, B), sessile flowers, minute staminate flowers with free sepals and short stamens (Fig. 5A), and scattered foliar glands (Fig. 4D). Furthermore, most have 2-locular ovaries and all share the puberulous septae and columella. Species belonging to this group are, among others, the species around *A. octandra* (all Sri Lankan and Indian species) and species such as *A. aurea* with thick shiny yellow leaves when dry.

Group 2 is distributed on Sunda and is best known from the common species, *A. frutescens* (the type of the genus), the Bornean *A. nitida*, and the Sundanese *A. symplocoides*. They are characterized by lax inflorescences (Fig. 2L, cf. Fig. 3L), but in some species these are clavate (Fig. 2J, K) or uniflorous (Fig. 3M, N). Foliar glands have the plesiomorphic configuration (Fig. 4B), and the stipules are often shortly persistent and strongly oblique. Species of this group are easily recognized when dry by the papery greyish-green or yellowish-green leaves and the yellowish inflorescences.

These two monophyletic groups do not have a consistent position in the cladograms. They can root in groups 4, 5, 6, or 11 and are therefore shown separate from the rest of the diagram. Group 3 is the smallest, made up of the five species of the *A. benthamiana - A. lunata* affinity. These have flowers densely set at the apex of the inflorescence (Fig. 2C, fig 3F); the staminate flowers have stamens about three times the length of the sepals (Fig. 5E), and the pistillate flowers have rather long, smooth stigmas (Fig. 3F). The stipules are falcate and often persistent. Many disc-like glands ornament the margins of the leaves (Fig. 4C). The species have a Sundanese distribution, and two are endemic on Borneo. This group usually roots at the top or near the base of group 8.

After taking out these three monophyletic clades we are left with a paraphyletic residue. Three parts are still distinct and present, shown as end-clades in the diagram. These are groups 4, 8, and 6+7. However, their basal relationships are unclear. They are mostly characterized by plesiomorphic features and thus cannot be strictly delimited on macromorphological grounds.

The paraphyletic residue can be geographically split into a Sundanese part (groups 4 to 7) and a New Guinean part (groups 8 to 11). The Sundanese group contains either species with intermediate inflorescences (e.g. *A. subcaudata* (Fig. 2E) or *A. nervosa* (Fig. 3H)) and other plesiomorphic characters, or species with characters that only partly fit into one of the three monophyletic groups. For example, the common Bornean species *A. grandistipula* has elongated inflorescences (cf. Fig. 2L, 3L) in common with the species around *A. frutescens* but lacks the greenish drying leaves. The New Guinean species are rather difficult to characterize. They combine derived characters with primitive ones, and thus do not fit into any of the other groups.

We have now a general view of the consistent relationships of *Aporosa* in the various analyses. It would be easy to stop here and assume that we cannot obtain more from the present data. But there must be more to be derived from the phylogenetic analysis. In fact there are only two points that are inconsistent in the analyses: the variable positions of the monophyletic groups with respect to the paraphyletic residue, and the lack of resolution and stability of the basal relationships within this residue (remembering that a couple of distinct, but paraphyletic lineages were present next to the ever changing groups 4/5 and

9/10/11). These problems may even be correlated, i.e., the same individual taxa with particular character state combinations may cause the ambiguity. If these species are removed, the data may be re-analysed to try and obtain some insight into and possible causes of the basic branching pattern in the genus. Thus the next step is to find which species are causing the difficulties and re-run the analyses in various configurations without these species.

One of the first problems is the ambiguous placement of the monophyletic clades in groups 4, 5, 6 or 11. Of these possibilities the relationship with group 11 can be discarded. The specific taxa of group 11 that are depicted in the sister-group position are those which have about half of their characters scored as "?". Parsimony analysis tends to avoid optimizing long phyletic lines characterized by many state changes at the same node. Thus, it is more parsimonious to root strongly apomorphic groups on taxa that are incompletely known, because then the development of the synapomorphies can be more gradually proposed. We can thus safely assume that these rootings are wrong. The remaining possibilities are group 4, 5 or 6. But the basal relationships of the taxa within these groups are not resolved either. This means that the problem in rooting the monophyletic group is correlated to the ambiguity of the relationships within these groups and we can treat them together with the problem of these groups.

Thus the complete ambiguity has boiled down to the unravelling of the paraphyletic part. Geographically speaking, the paraphyletic remnant consists of two distinct entities: the Sundanese species, consisting of the small groups 4, 5, 6 and 7, and the New Guinea species, also separated into 4 groups, group 8 to 11. The Philippine/ Sulawesian *A. leytensis* falls under the New Guinean species group 8, on base of its characters and its consistent placement there. The consensus diagram in Fig. 6 shows close relationships between all taxa of group 11 and all other taxa, and close relationships of groups 8, 9 and 10 with parts of the Sundanese groups. When translating such mixtures back to evolutionary pathways, this would mean that various taxa colonized frequently New Guinea and recolonized Sundaland in the same high numbers. This is highly unlikely considering what we know of colonization possibilities. It is more likely that only one or two lineages have colonized New Guinea. Considering this, we can start by dividing the group into Sundanese and New Guinean species and analyse them separately.

Sundanese species

In an analysis of only the Sundanese remnant the inflorescence type and foliar glands have intermediate states and do not indicate lineages. The species mostly vary in the size and indumentum of the leaves and inflorescences and can be arranged along an axis that represents variation in these characters (Fig. 7). The large-leaved, long-inflorescenced, 3-locular species, e.g. *A. arborea* or *A. grandistipula* (group 4), are at one end of the spectrum; the smaller-sized, 2-locular ones like *A. maingayi* or *A. antennifera* (group 6) at the other. In between lie *A. nervosa* and related species (group 5) and midway between these last are the large-leaved species around *A. subcaudata* with short inflorescences and 2-locular ovaries (group 7). All groups overlap: there is at least one species that can be placed in two groups at the same time, and this is not the same species in all the analyses. Thus it is seen that

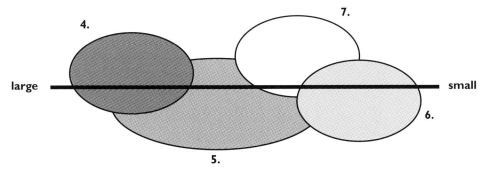

FIG. 7. The remaining Sundanese species-groups 4 – 7 arranged along a size & indumentum axis. Left are the large-leaved and long-inflorescenced, puberulous species; at the right side the smaller-leaved and shorter-inflorescenced species. See for a short characterization of the groups table 1.

these are four small groups that are always somehow distinct, but in different configurations in the various analyses.

A theoretical model is plausible that explains these "jumping" species: the extinct ancestor model. Table 2 gives a hypothetical data-matrix of ten species and eight characters. Analysis of the complete matrix with an all-zero outgroup results in one tree (Fig. 8a) with 13 steps and a CI-value of 0.61. A couple of parallelisms and some reversals are present, but the synapomorphies are strong enough to balance the homoplasies. Now, let it be assumed that for some reason, e.g. a changing climate - which would be quite natural because of ice ages — the possession of character 1 becomes disadvantageous. All species with this character (species A, C, D, E and F) fail to adapt to the change and die out. Re-analysis of the matrix in table 1 with only the surviving species B, G, H, I, and K, and characters 2 – 6 and 8, gives 2 equally

TABLE 2. Data-matrix for ten hypothetical species with eight characters. Analysis of this matrix with an all-zero outgroup results in the cladograms shown in Fig. 8.

	1	2	3	4	5	6	7	8
A	1	0	0	1	1	0	0	0
B	0	0	0	1	1	1	0	0
C	1	0	0	1	1	1	0	0
D	1	0	0	1	0	0	0	0
E	1	1	0	0	0	0	0	0
F	1	1	1	0	0	0	1	0
G	0	0	1	0	0	0	0	1
H	0	1	1	0	0	0	0	1
I	0	1	0	0	1	0	0	0
K	0	1	1	0	1	0	0	0

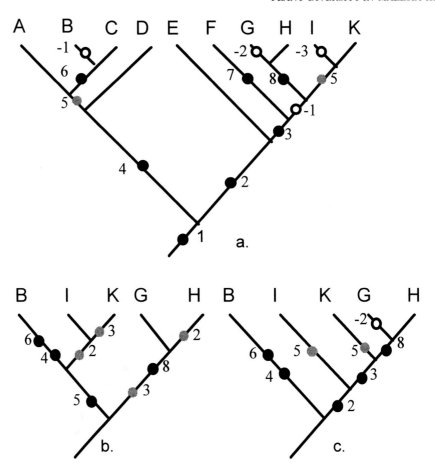

FIG. 8. Cladograms derived from the data matrix in Table 1. a. Cladogram found when all species are included; length 13, CI 0.61. b, c. The two cladograms found with a decimated data-matrix after an hypothetical extinction of all species with character 1 present (A, C, D, E, F); length 8, CI 0.75.

parsimonious trees with 8 steps and CI 0.75 (Fig. 8bc). The connection of the "middle" species G and H, that have evolved a synapomorphy, is the only one that is stable. The relations of species I and K, whose synapomorphy had a parallel with species B, have become ambiguous. The homoplasious character 5 cannot be balanced any longer by the character distribution of characters 2, 3 and 4 in species D, E, and F and the position of species I and K becomes inaccurate.

The above is a theoretical example. However, it has also been shown with real data-sets that extinct taxa — in the form of fossils — are neccessary to obtain the true phylogeny (e.g. Donoghue *et al.* 1989). When fossils are not known the phylogeny of a genus that is presumably ancient has a high chance of being incorrect (Doyle & Donoghue, 1992).

Looking back at *Aporosa*, we can see analogies in the "jumping" species of the Sundanese remnant. The paraphyletic group around *A. maingayi* (group 6) and *A.*

subcaudata (group 7) might be like species I and K, i.e. monophyletic but without strong synapomorphies and therefore sometimes depicted as paraphyletic. Others species, such as *A. antennifera* might be like species B, i.e. survivors that are jumping in or out of the group I – K. And finally, the monophyletic groups might be as G – H, without a consistent placement.

Rerunning the analysis without the various categories of "jumping" species gives no better results. This might be seen as indication that the supposed analogy with loss of consensus through exclusion of extinct taxa holds for the Sundanese *Aporosa*. Since no fossils are known, it may not be possible to find an accurate dichotomous estimate of the phylogeny, which makes a polytomous depiction of the basal relationships of the Sundanese species acceptable.

New Guinea species

The New Guinean species present a different phenomenon. The characters that cause the lack of solution within the 1000 cladograms is not caused by their states, but by their combinations. Many New Guinean species combine character states found only in the derived Sundanese groups with the plesiomorphic character states of the Sundanese intermediates. The exact combination is slightly different for every species, but based on the same characters. Thus every species can be placed near only a few others, but all these positions are equally parsimonious.

These problems are, however, only present when the New Guinean species are compared with the Sundanese species. Taking only the New Guinean species as an unrooted network, they can be arranged like the Sundanese group, along an axis of size differences. This results in the schematic representation of Fig. 9. There are four groups. Of the three distinct ones (8, 9, 10) the large-leaved thick-inflorescenced group 8 contains the species around the widespread *A. papuana*. Group 9 and 10 are two small-leaved short-inflorescenced groups of which 9 has the lowland, 3-locular species *A. nigropunctata* and *A. longicaudata* and group 10 its four vicariant, mountainous, 2-locular relatives like *A. sclerophylla*. The last and largest group, 11, consists of the species that have the "wrong" combination of characters to belong in any of the other groups. Species with exceptional characters, such as *A. hermaphrodita*, are also placed here. This is mainly the group that causes difficulties in the complete analysis.

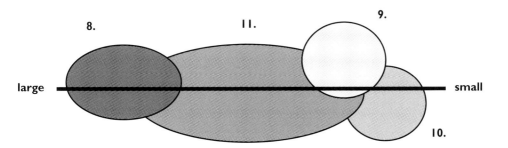

FIG. 9. The New Guinea species-groups 8 – 11 arranged along a size & indumentum axis as in figure 7. See for a short characterization of the groups table 1.

Re-analysis of the complete data-set without this troublesome group 11 gives an immediate reduction in the number of trees found and thus a better resolution in the consensus tree. The remaining three New Guinean groups settle down as sister-groups to two particular groups of the Sundanese cladogram, namely group 8 with group 4 and group 9 and 10 in the overlap-zone of group 5 and 6.

The question arises why omitting the species of group 11 has this effect on the number of possible solutions. Logic dictates that it must be the "wrong" combinations of characters states shown by them that disrupts the phylogenetic pattern. Why do they have such combinations of characters? One possible explanation is that we are dealing with hybrids. According to Rieseberg (1995) hybrids are a mosaic of parental, intermediate, and novel characters that show no coherence. McDade (1992) showed how inclusion of hybrids between distantly related parents disrupts the patterns in the cladogram.

Fig. 10 and Table 3 give a theoretical example of such a hybrid pattern. The analysis of the true species A – E with an all-zero outgroup is a straightforward case (Fig. 10a: one tree, length 5, CI 100). But suppose that species B and E hybridize and produce species F. F inherits one derived character from E. Analysing the complete matrix of Table 3 results in six trees (length 6, CI 83) and the solution of the phylogenetic analysis is lost (Fig. 10b: strict consensus). F is either placed at the ancestral node of D – E (Fig. 10c), or between B and C (Fig. 10d). The ambiguity lies in the optimization of character 3 and 4 respectively. Depicting it as a parallelism or reversal causes collapse of one of the nodes marked by an asterisk in the figure, resulting in the six possible trees. Funk (1985) shows how simple hybrids might be identified on the basis of such character conflicts and depicted in a reticulation in the cladogram.

TABLE 3. Data-matrix for five hypothetical species A – E and a hybrid species F with five characters. Analysis of this matrix with an all-zero outgroup gives no solution for the position of the hybrid F. The cladograms are depicted in Fig. 10.

	1	2	3	4	5
A	1	0	0	0	0
B	1	1	0	0	0
C	1	1	1	0	0
D	1	1	1	1	0
E	1	1	1	1	1
F	1	1	0	1	0

Looking back at group 11 of *Aporosa*, the analogy is easily made. The New Guinean species of this group bring only conflicts in the character combinations of the Sundanese species, without adding information in the form of new synapomorphies.

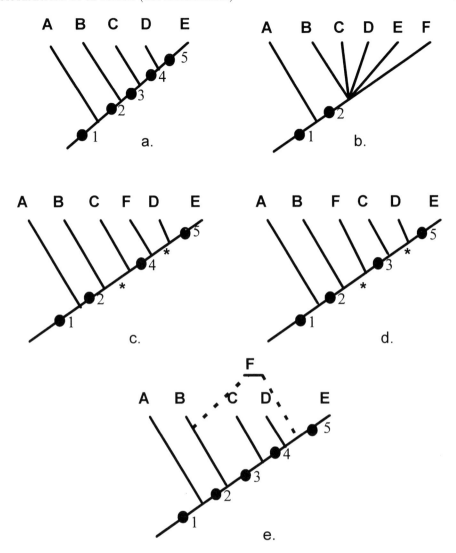

FIG. 10. Cladograms derived from the the data matrix in Table 3. a. Cladogram with only "true" species A-E; length 5, CI 100. b. Strict consensus of the analysis including hybrid F; 6 trees, length 6, CI 0.83. c & d. Two of the six cladograms showing the ambiguous position of F. e. cladogram with A – E and the hybrid F depicted in a reticulation.

Reconstructed pattern

The separate subanalyses of the Sundanese and the New Guinean species have cleared some of the ambiguity of the consensus pattern of Fig. 6. The main conclusions of the foregoing paragraphs were:

(a) the putative New Guinean hybrids — group 11 — introduce conflict and should be kept separate.

(b) Without them, the rest of the New Guinean species settled down in two lineages, group 8 with 4, and 9/10 at the 5/6 transition.

(c) The Sundanese polytomy at the base, formed by groups 4, 5 and part of 6 can not be resolved. This has to remain a polytomy. However, coupled to the fact that the two New Guinean clades separate 4 from 5/6, a main distinction between two lineages can be made. Part of the ambiguous taxa remain in the basal polytomy, others function as ancestral to group 8 and 9/10 respectively.

(d) Monophyletic group 1 was rooted, like the New Guinean lineages 9/10, in the 5/6 overlap-zone.

(e) Monophyletic group 2 mainly connected to one species of group 4, *A. grandistipula.*

(f) Monophyletic group 3 was in all cases placed at the base or the top of New Guinean clade 8.

(g) Group 7, that showed a paraphyletic connection with group 6, possibly forms a monophyletic subclade of the same group 6, like the species I-K of the theoretical example.

(h) Other "jumping" taxa of groups 4, 5, and 6 may represent relict species of once species-rich clades and are best placed in a polytomy. A consensus diagram with these new patterns might look something like Fig. 11.

At the base is a polytomy of the Sundanese groups 4 and 5. Out of this polytomy arise two distinct lineages. The first clade (a) contains the species with inflorescences with laxly set flowers, mostly 3-locular ovaries, and often large falcate stipules. This clade comprises one species of group 4, *A. grandistipula*, from whose ancestor the monophyletic group 1, the species around *A. frutescens,* arose. The New Guinean group 8 is also nested in this clade. Its most primitive species is the Philippine/Sulawesian *A. leytensis*, which also represents the ancestral line that gave rise to the highly apomorphic monophyletic group 3 (Fig. 11, clade a).

The second lineage is formed by species with more densely set flowers, mostly 2-locular ovaries, and smaller, slightly oblique to narrowly ovate stipules (Fig. 11, clade b). Groups 5 and 6 form the plesiomorphic base of this lineage. The relationships at the root of this second clade could not be clarified. It was found that monophyletic group 1 nested here together with the New Guinean groups 9 and 10. This pattern was explained by hypothesizing extinction. Of group 5 and 6 a few isolated species have survived little changed, e.g. *A. caloneura* of group 5 and *A. duthieana* and *A. antennifera* of group 6. Others are now derived, e.g. the monophyletic group 1, *A. octandra* and allies, group 9, 10, and the rest of group 5 and 6. The direct ancestors of all these groups are now extinct, and the relationships of their extant descendants cannot be clarified. Therefore, this clade has a polytomy at the base from which arise these three groups and the extant parts of group 5 and 6. Group 7 has only a connection with group 6.

The excluded group 11 was hypothesized to contain hybrid species. The reconstructed pattern in figure 11 shows that two clades colonized New Guinean out of Sundaland (Fig. 11: group 8 and 9/10 respectively). This supports the hypothesis that hybrids were formed on reaching New Guinea by secondary contact of two distantly related lineages. The hybridization event is shown as a reticulation in figure 11. It gave rise to the hybrid swarm, group 11.

The three monophyletic groups, the paraphyletic Sundanese group, and the New Guinea-group will be described as sections in the forthcoming monograph on *Aporosa* (Schot, in prep.).

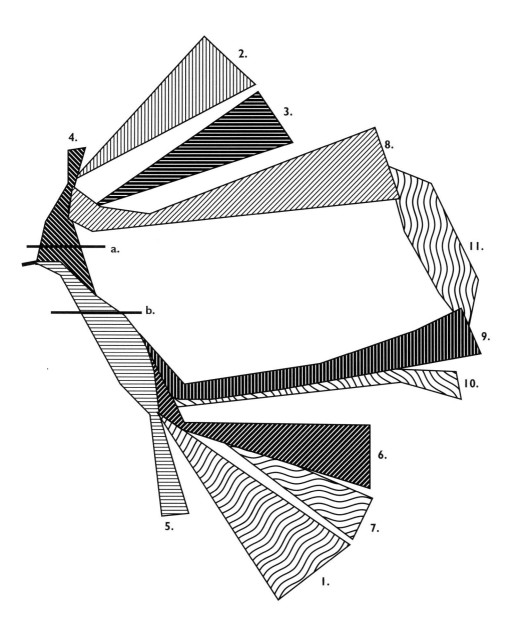

FIG. 11. Reconstructed phylogenetic consensus diagram with the 11 groups as reached after separate analysis of the Sundanese and the New Guinean species. Two lineages are present: a and b. Group 11 is depicted in a reticulation. See for a short characterization of the groups table 1 and figure 6.

CONCLUSION

In many cases when phylogenetic analysis yields many equally parsimonious trees, it is tempting to find fault with characters or taxa and start manipulating them to find a better solution. If no solution is found, the group under study is often regarded as unresolved. However, it is equally likely that the multiple patterns have natural causes. The ambiguity is then itself the result of the phylogeny reconstruction.

As I have shown here, it is important to have some idea why the analysis does not yield unambiguous results. Knowing the limits of phylogenetic analysis — and herbarium techniques — we might point out the elements causing trouble and re-analyse the data to find indications to proposed hypotheses of speciation. Polytomies can be quite natural, as already proposed by Bremer & Wanntorp (1979) and should be acceptable when theoretically explainable by, for example, a specific speciation process. Even partly resolved cladograms give insight into the process of evolution, and that is what we are aiming for.

ACKNOWLEDGEMENTS

The manuscript was read and commented on by Dr M. C. Roos and Prof. P. Baas. Mr. J. J. Wessendorp prepared the drawings.

REFERENCES

Airy Shaw, H. K. (1974). *Aporosa lagenocarpa.* Hook. Ic. Pl.: t. 3701.

Bremer, K. & Wanntorp, H. -E. (1979). Hierarchy and reticulations in systematics. Syst. Zool. 28: 624 – 627.

Chenery, E. M. (1948). Aluminium in the plant world I. Kew Bull. 2: 173-183.

Donoghue, M. J., Doyle, J. A., Gauthier, J., Kluge, A. G., & Rowe, T. (1989). The importance of fossils in phylogeny reconstruction. Ann. Rev. Ecol. Syst. 20: 431 – 460.

Doyle, J. A. & Donoghue, M. J. (1992). Fossils and seed plant phylogeny reanalysed. Brittonia 44: 89 – 106.

Funk, V. A. (1985). Phylogenetic patterns and hybridization. Ann. Miss. Bot. Gard. 72: 681 – 715.

McDade, L. A. (1992). Hybrids and phylogenetic systematics II. The impact of hybrids on cladistic analysis. Evolution 46: 1329 – 1346.

Pax, F. (1890). *Euphorbiaceae — Aporosa.* In: A. Engler & G. Prantl, Pflanzenf. 3, 5: 29 – 30.

Pax, F. & Hoffmann, K.(1922). *Euphorbiaceae - Phyllanthoideae - Phyllantheae - Aporosa.* In: A. Engler, Pflanzenr. 4, 147, 15: 80 – 105.

Rieseberg, L. H. (1995). The role of hybridization in evolution: old wine in new skins. Am. J. Bot. 82: 944 – 953.

Schot, A. M. (in prep.). Taxonomy and phylogeny of *Aporosa* Blume (Euph.). Blumea Suppl.

Smith, J. J. (1909) *Aporosa campanulata.* Ic. Bogor. 3: t. 229.

Stapf, O. (1894). *Aporosa bourdillonii.* Hook. Ic. Plant.: t. 2204.

Webster, G. L. (1994). Classification of the *Euphorbiaceae.* Ann. Miss. Bot. Gard. 81: 3 – 32.

Tree Flora of Sabah and Sarawak project — a botanical inventory of tree resources

E. Soepadmo[1] & R. C. K. Chung[1]

Summary. The Tree Flora of Sabah and Sarawak Project, which was officially launched on the 18th of November 1991, is a joint effort undertaken by the Forest Research Institute Malaysia (FRIM), Sabah Forestry Department (FD – Sabah) and Sarawak Forestry Department (FD – Sarawak). This ten-year project is jointly funded by the Government of Malaysia, the United Kingdom Overseas Development Administration (ODA), and the International Tropical Timber Organization (ITTO). The main objectives of the project are (1) to document and update the taxonomic status of all tree species that are native to Sabah and Sarawak; (2) to produce up-to-date accounts of the Tree Flora of Sabah and Sarawak in eight volumes within ten years; (3) to upgrade Malaysian capability and expertise in plant taxonomic research; and (4) to strengthen the curation and management ability of the herbaria of the three main participating institutions (FRIM, FD – Sabah and FD – Sarawak). The first volume of the Tree Flora containing 310 species in 31 families has been published. Thirty-six Malaysian and 21 overseas botanists are involved in the project. To achieve objectives 3 and 4, workshops on Principles and Methods of Flora Writing, the BRAHMS Program, DELTA Software, and Botanical Illustration have been organised. Two Malaysian botanists have completed their postgraduate training in U.K., and another three are currently doing Ph.D. programmes in U.K. and Malaysia. Three major collecting expeditions were made to the Lanjak Entimau Wildlife Sanctuary in Sarawak, Bukit Tawai in Sabah and Ulu Belaga area in Sarawak. To date, data from 5,500 specimens in 23 families have been computerised, and 4,110 entries/articles of taxonomic and ecological references were added to our literature database. The project also receives close collaboration from botanists attached to overseas herbaria and universities from Austria, Denmark, Germany, Hong Kong, the Netherlands, U.K. and U.S.A.

INTRODUCTION

Borneo, the third largest island in the world, of which Sabah and Sarawak are parts, has been frequently acknowledged as one of the most important centres of plant diversity in the world. The island, which occupies a total land area of approximately 740,000 sq. km, is conservatively estimated to harbour 10,000 – 12,000 species of flowering plants, representing about 5 – 6% of the world total (Merrill 1950; van Steenis 1950; Kiew 1984; Mat-Salleh *et al.* 1992). Of these, 40 – 50% are endemic to the island, and up to 80% of the endemic species in Borneo occur in Sabah and Sarawak.

In certain localities in Sabah and Sarawak, where botanical exploration has been carried out more intensively, the species diversity is indeed extremely high. Beaman & Beaman (1990), for instance, have found that the flora of the Mt Kinabalu Park, Sabah, encompassing an area of about 700 sq. km, contains not less than 4,000 species of vascular plants in 180 families and 980 genera.

---

[1] Forest Research Institute Malaysia, Kepong, 52109 Kuala Lumpur, Malaysia

For a number of economically important families and genera of flowering plants, Borneo (Sabah and Sarawak in particular) is also known as the centre of distribution and species diversity. For example, of the 386 known species of the *Dipterocarpaceae* 291 species or about 75% are recorded from Borneo, of which 257 species or about 66% occur in Sabah and Sarawak. Of the 291 species occurring in Borneo 156 or about 54% are endemic, of which 59 species or about 20% are restricted to Sabah and Sarawak (Ashton 1982). In the genus *Durio* (*Bombacaceae*), of the 30 species which have been described to date 20 species or about 66% are recorded from Borneo, with 16 species or about 53% occurring in Sabah and Sarawak (Kostermans 1958; 1990). Nine or about 45% of the 20 species known from Borneo are endemic and all occur in Sabah and Sarawak.

For non-woody components of the flora, *e.g.*, orchids and rafflesias, Borneo (and Sabah and Sarawak in particular), also have the most number of species in the Malesian phytogeographic region. In this region, the orchid family or the Orchidaceae is known by 3,000 – 4,000 species, representing 12 – 16% of the entire flora (Chan *et al.* 1994). Of these, Lamb (1991) has estimated that 2,500 – 3,000 species are found in Borneo, equivalent to about 75% of the Malesian orchid flora. A great number of the Bornean species have been recorded from Sabah and Sarawak. The genus *Rafflesia* (*Rafflesiaceae*) with the largest flower in the world, also has its centre of species diversity in Borneo where 5 – 6 of the 14 species known to date have been recorded, and all occur in Sabah and Sarawak (Meijer 1984; Mat-Salleh 1991; Nais 1992).

Other Old World tropical families which have their centres of distribution and species diversity in Borneo include the *Anacardiaceae*, *Burseraceae*, *Celastraceae*, *Clusiaceae* (*Guttiferae*; *Calophyllum*), *Euphorbiaceae*, *Fagaceae*, *Myrtaceae* (*Syzygium*), *Rhizophoraceae*, and several others (Airy Shaw 1975; Ding Hou 1958, 1962, 1978; Kochummen 1995; Leenhouts 1956; Merrill & Perry 1939; Soepadmo 1972; Stevens 1980).

The presence of high species diversity in the natural forests of Sabah and Sarawak also means that there is a wealth of forest products to be harvested. There is no doubt that, in the past few decades, the harvesting and utilisation of these forest products, *e.g.*, tropical hardwood timbers and rattans, has contributed significantly toward the socio-economic development of these two eastern Malaysian states. In recognising that of late, the exploitation and conservation of biodiversity in Sabah and Sarawak has become the focus of international attention and scrutiny, and the need to strike an acceptable balance between development and conservation of natural resources in these two states, it is imperative that up-to-date botanical inventories should be carried out without further delay. Such basic information is of paramount importance to the understanding of the availability, distribution, ecological and conservation requirements, and economic potential of the plant resources. Without such information, it will be extremely difficult if not impossible to develop and manage the available resources on a sustained basis.

The need for a Flora of Borneo

The flora of a given region provides an inventory of plant species occurring in that region, facilitates a means of species identification and provides a source of information pertaining to the up-to-date taxonomic and conservation status, distribution, ecological amplitude, and economic potential of the treated species. It is, therefore, unfortunate to note that despite the widespread national and international recognition of the great economic and conservation value of Bornean rain forests, and that botanical exploration and collection in Borneo begun as early as 1822, with botanical accounts of its flora appearing since 1894 (Wong 1995), this species-rich island has neither a comprehensive flora of any kind nor even a concise checklist of plant species. For this reason Kiew (1984) highlighted the need and urgency of producing a simple "Flora of Borneo" in order to facilitate the implementation of sustainable management practices of forest resources of this species-rich island. She argued that by using Merrill's (1921) and Masamune's (1942) enumerations as a basis, and by adopting a pragmatic approach, a concise "Identification Flora of Borneo" could be completed within 10 years. However, due mainly to the lack of or insufficient institutional and financial support, as well as a shortage of qualified man-power, the proposed project never materialised.

The Tree Flora of Sabah and Sarawak project

In recognising the urgent need for producing a "Flora of Sabah and Sarawak" of any kind, that recent collections (more than 200,000 numbers from Sabah and Sarawak up to 1990) necessitate revision and updating of taxonomic accounts in the Flora Malesiana itself and that trained scientific personnel are now available at various national institutions, the Tree Flora of Sabah and Sarawak Project was officially launched in November 1991 and was fully operational by April 1993.

Organization. To execute the project smoothly, a Steering Committee comprising the Director General of Forest Research Institute of Malaysia (FRIM, as Chairman), the Director of Sabah Forestry Department, the Director of Sarawak Forestry Department and representatives of the funding agencies (ODA – UK and ITTO) has been established. Its main function is to provide advice and guidance to the Project Coordinator and Editorial Board in matters pertaining to administrative and financial policies (Fig. 1). To coordinate all aspects of the project activities and to monitor progress in flora writing, a Coordinating Committee consisting of the Project Coordinator (Chairman) and Representatives of FRIM, Sabah Forestry Department, and Sarawak Forestry Department has also been formed. To ensure that the publications of the Tree Flora meet international standards, an Editorial Committee/Board comprising Chief Editor, Co-Editor(s) and Advisors has been appointed. FRIM, being the leading executing institution, also acts as the Secretariat for the project.

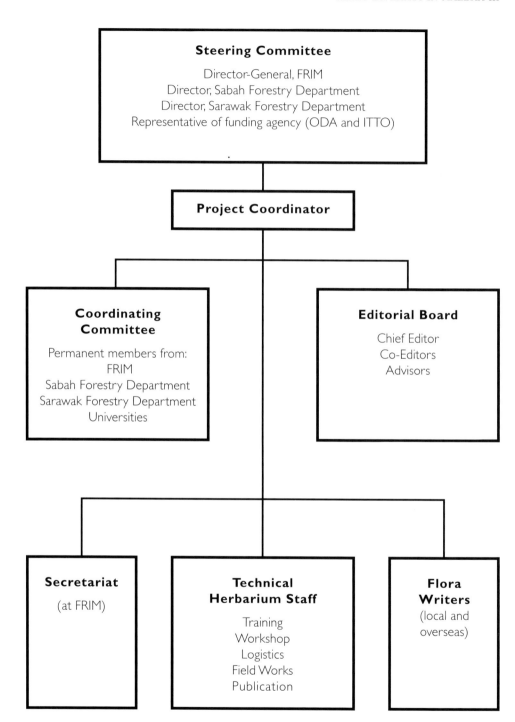

FIG. 1. Project Organisation

Objectives. This modest ten-year project is executed jointly by the Forest Research Institute Malaysia (FRIM), the Sabah Forestry Department (FD – Sabah), and the Sarawak Forestry Department (FD – Sarawak) with the collaboration of other research institutions and universities. The main objectives of the project are:

- To document and update the taxonomic status of all tree species native to Sabah and Sarawak.

- To publish eight volumes (each containing 300 – 400 species) of a concise Tree Flora of Sabah and Sarawak within ten years.

- To upgrade Malaysian capability and expertise in plant taxonomic research and the survey and documentation of tree diversity.

- To develop and strengthen the management capability of herbaria and their related data bases in the three participating institutions (FRIM, FD – Sabah and FD – Sarawak).

Funding. For the first five years, the project is being jointly funded by the Malaysian Government, the Overseas Development Administration (ODA) of the United Kingdom, and the International Tropical Timber Organization (ITTO).

Botanists Involved. Botanists of the following national and international institutions are taking part:

MALAYSIA — Forest Research Institute Malaysia, FD – Sabah, FD – Sarawak, Universiti Kebangsaan Malaysia (UKM), Universiti Malaya (UM), Universiti Malaysia Sarawak (UNIMAS), Universiti Pertanian Malaysia (UPM), Mt Kinabalu Park, Malaysian Agricultural Research and Development Institute (MARDI), and the WWF – Malaysia. In all 36 botanists are involved.

OVERSEAS — Institute of Botany, University of Vienna, Austria; University of Brunei Darussalam, Brunei; University of Aarhus, Denmark; Botanischer Garten und Botanisches Museum, Berlin, Germany; University of Hong Kong, Hong Kong; Rijksherbarium Leiden, the Netherlands; Singapore Botanic Garden, Singapore; Oxford Forestry Institute, Oxford, United Kingdom; Royal Botanic Gardens, Kew, United Kingdom; Arnold Arboretum, Harvard University, USA; University of Florida, USA; and Everglades National Park, Florida, USA. A total of 21 botanists are involved.

Coverage. Being an identification type of Flora intended to be "user-friendly" to non-specialist readers, the Tree Flora will be written in simple and easy-to-understand English. The use of highly technical botanical terms is, therefore, being avoided.

In the Tree Flora, all dicotyledonous trees, here defined as woody plants with the main upright stems measuring not less than 5 m tall and 10 cm diameter, will be treated in full. Non-tree and introduced tree species known only in cultivation will be given cursory treatment and annotated in the keys only. For each tree-taxon (family, genus, species), treatment will be confined to the following aspects — correct/accepted scientific name, major references, description of diagnostic characters, vernacular names (if applicable), distribution, ecology, uses, notes on taxonomy (if applicable), and key(s) to lower rank taxa. For each genus, at least one line-drawing depicting important vegetative and reproductive characters will be provided.

Format. To standardise the format of manuscripts to be published in the Tree Flora, a set of guidelines has been prepared by the Editors (Soepadmo & Wong 1995). This guideline is obtainable from the project Secretariat at FRIM.

Other Project Activities. Apart from preparing and publishing eight volumes of the Tree Flora of Sabah and Sarawak, the project also carries out a number of activities relevant to its overall objectives. These activities include:

• Organisation of collecting expeditions to a number of botanically little-known localities in Sabah and Sarawak.

• Establishment of a database for specimens and taxonomic references using BRAHMS and related software.

• Conducting workshops and specialised training on Flora writing, documentation software, botanical illustration, editing, curation and management of herbarium specimens, etc.

• Postgraduate training to upgrade Malaysian capability in plant taxonomic research and the inventory and documentation of plant/tree diversity.

• Collaboration with other institutions and agencies with similar aims in the Malesian region.

Work Plan. The Tree Flora of Sabah and Sarawak is anticipated to contain about 3,000 species in 110 families, and will be published in eight volumes (Fig. 2). Volume 1 containing 310 species in 100 genera and 31 families was published in 1995, while volume 2 with about 360 species in 30 families is scheduled to be published in 1996. Volume 3 (with about 360 species in 15 families) and volume 4 (with about 360 species in 12 families) are planned to be issued in 1997 and 1998, respectively. Volume 5 and 6 containing about 720 species in 15 families are to be published by the middle of year 2000, and volumes 7 and 8 with about 890 species in 7 families by the year 2001/2002.

PROGRESS TO DATE

Tree Flora Volumes. Immediately following the publication of volume 1, manuscripts for volume 2 are currently being edited. Revisions of various families to be published in the subsequent volumes have been initiated by various authors.

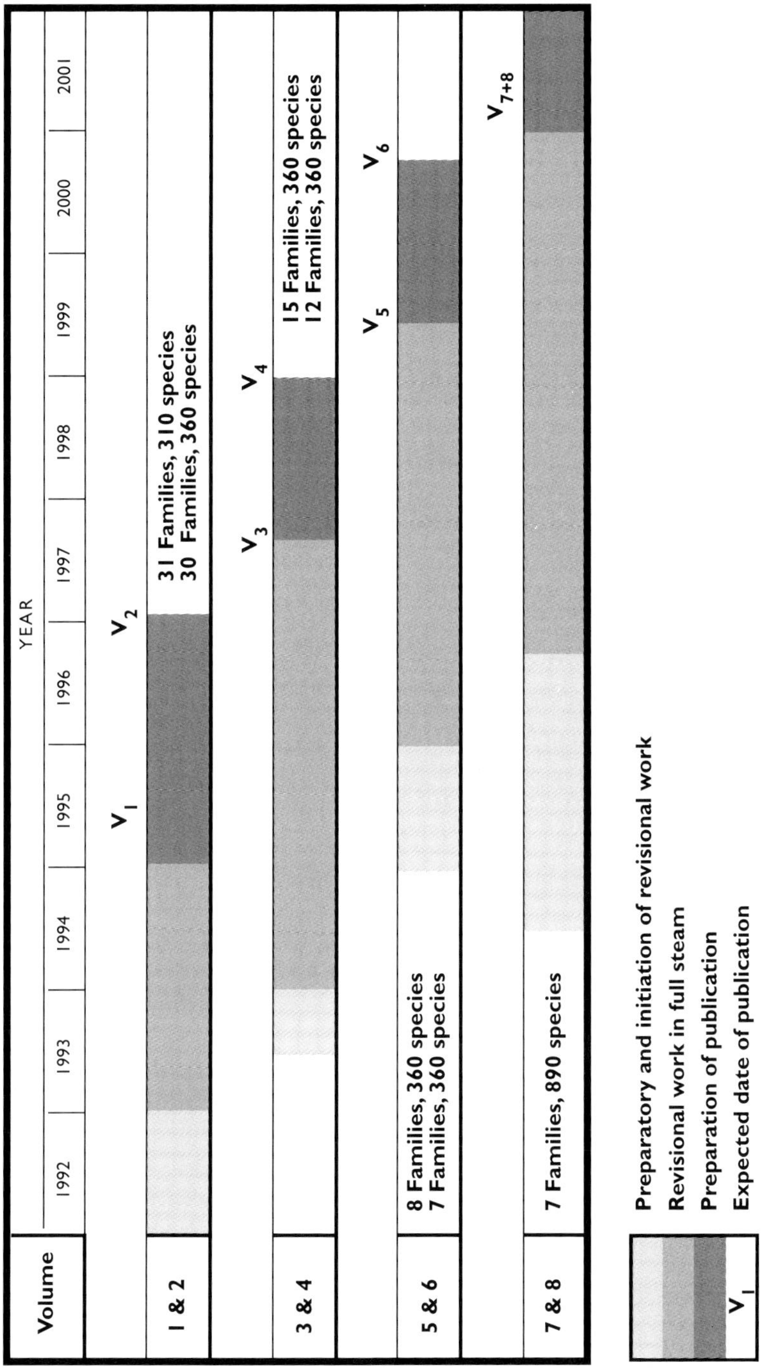

FIG. 2. Work Plan

Collecting Expeditions. Three major expeditions to botanically little-known localities in Sabah and Sarawak have been organised, *i.e.* to the newly established Lanjak Entimau Wildlife Sanctuary Reserves in Sarawak, Bukit Tawai in Sabah, and the Ulu Belaga area in Sarawak. In these expeditions, a total of 2,660 collections numbers were made.

Workshop/Training Course. To date four workshops/training courses, *viz.* Principles and Methods of Flora Writing, the BRAHMS Programme, DELTA Software, and Technique of Botanical Illustration, have been conducted.

Postgraduate Training. Under the Tree Flora Project's full scholarship, three young Malaysian botanists have undergone postgraduate training (Ph.D and M.Sc.) at the University of Reading, England. In addition, the Tree Flora Project partially sponsors three other postgraduate candidates (two Ph.Ds and one M.Sc.) at local universities. Financial support for short/specialised training lasting for a few weeks to about three months has also been given to a number of young Malaysian botanists.

Specimen and Reference Database. Up to June 1995, data of 5,500 specimens of 23 families have been computerised using RDE/BRAHMS software. Concurrently, a total of 4110 entries of taxonomic references have been filed on our literature database using Notebook software.

Herbarium Visits. Twenty-five senior and junior botanists involved in the Tree Flora project have been provided with travel grants to conduct taxonomic research in Herbaria (Malaysia and overseas) other than their own institutions.

Specimen Loan. The Secretariat at FRIM has handled loans (overseas and within Malaysia) of about 3,000 specimens to various botanists involved in the project and working in different herbaria.

Liaison Officers. To assist Malaysian botanists in solving various nomenclatural and taxonomic problems, each year two senior Malaysian botanists are sent to Kew Herbarium and Rijksherbarium for periods ranging from a few weeks to 3 months.

ACKNOWLEDGEMENTS

The authors wish to express their deep gratitude to Dato' Dr. Salleh Mohd. Nor, the Director General of FRIM, for his continuous guidance and encouragement, and to the Malaysian Government, ODA – UK and ITTO for the financial support. We would like to thank the Organisers of the Third International Flora Malesiana Symposium for their cordial hospitality, and for the excellent facilities accorded to us during the symposium.

REFERENCES

Airy Shaw, H. K. (1975). The *Euphorbiaceae* of Borneo. Kew Bull. Add. Series IV. 245p.

Ashton, P. S. (1982). *Dipterocarpaceae*. Flora Malesiana 1, 9(2): 237 – 552.

Beaman, J. H. & Beaman, R. S. (1990). Diversity and distribution patterns in the flora of Mount Kinabalu. In: P. Baas, K. Kalkman & R. Geesink (eds.), The Plant Diversity of Malesia, pp. 147 – 160. Kluwer Academic Publisher, Dordrecht, the Netherlands.

Chan, C. L., A. Lamb, Shim, P. S. & Wood, J. J. (1994). Orchids of Borneo. Vol. 1. Introduction and a Selection of Species. Sabah Society & Royal Botanic Gardens, Kew. xviii + 402 p.

Ding Hou (1958). *Rhizophoraceae*. Flora Malesiana 1, 5(4): 429 – 493.

—— (1962). *Celastraceae* I. Flora Malesiana 1, 6(2): 277 – 291.

—— (1978). *Anacardiaceae*. Flora Malesiana 1, 8(3): 395 – 548.

Kiew, R. (1984). Towards a Flora of Borneo. In: Ismail Sahid, Zainal Abidin A. Hasan, A. Latiff Mohamed & A. Salam Babji (eds.), Research Priorities in Malaysian Biology, pp. 73 – 80. Penerbit Universiti Kebangsaan Malaysian, Bangi, Malaysia.

Kochummen, K. M. (1995). *Burseraceae*. In: E. Soepadmo & K. M. Wong (eds.), Tree Flora of Sabah and Sarawak, Vol. 1: 45 – 106. Celastraceae, *ibid.*: 107 – 154.

Kostermans, A. J. G. H. (1958). The genus *Durio* Adans. (*Bombacaceae*). Reinwardtia 4(3): 47 – 153.

—— (1990). *Durio bukitrayaensis* Kosterm. (*Bombacaceae*), a new species from Borneo. Botanica Helvetica 100/1: 29 – 31.

Lamb, A. (1991). Orchids of Sabah and Sarawak. In: R. Kiew (ed.), The State of Nature Conservation in Malaysia, pp. 78 – 88. Malayan Nature Society & IDRC, Canada.

Leenhouts, P. W. (1956). *Burseraceae*. Flora Malesiana 1, 5 (2): 209 – 296.

Masamune, G. (1942). *Enumeratio Phanerogamarum Bornearum*. 739 p.

Mat-Salleh, K. (1991). *Rafflesia* — Magnificent Flower of Sabah, 49 p. Borneo Publishing Company, Kota Kinabalu, Sabah.

——, Beaman, J. H. & Beaman, H. (1992). Specimen database and their utilization for the Flora of Borneo. In: Ghazally Ismail, Murtedza Mohamed & Siraj Omar (eds.), Forest biology and Conservation in Borneo, pp. 117 – 137. Center for Borneo Studies, Publ. No. 2.

Meijer, W. (1984). New species of *Rafflesia* (*Rafflesiaceae*). Blumea 30: 209 – 215.

Merrill, E. D. (1921). A Bibliographic Enumeration of Bornean Plants. J. Str. Br. Roy. As. Soc., Special Number. 637 p.

—— (1950). A brief survey of the present status of Bornean botany. Webbia 7: 309 – 324.

—— & Perry, L. M. (1939). The Myrtaceous genus *Syzygium* Gaertn. in Borneo. Mem. Am. Acad. Arts & Sci. 18(3): 135 – 202.

Nais, J. (1992). Distribution, dispersal and some notes on *Rafflesia* around Kinabalu, Malaysia. In: Ghazally Ismail, Murtedza Mohamed & Siraj Omar (eds.), Forest Biology and Conservation in Borneo, pp. 97 – 108. Center for Borneo Studies, Publ. No. 2.

Soepadmo, E. (1972). *Fagaceae.* Flora Malesiana 1, 7(2): 165 – 403.

Soepadmo, E. & Wong, K. M. (1995). Guide to Preparing and Editing Manuscripts. 31p. Forest Research Institute Malaysia, FD Sabah and FD Sarawak, Malaysia.

Steenis, C. G. G. J. van (1950). The delimitation of Malaysia and its main geographical division. Flora Malesiana 1, 1: LXX – LXXV.

Stevens, P. F. (1980). A revision of the Old World species of *Calophyllum* (*Guttiferae*). J. Arn. Arb. 61 (2 & 3): 117 – 699.

Wong, K. M. (1995). A brief history of botanical collecting and documentation in Borneo. In: E. Soepadmo & K. M. Wong (eds.), Tree Flora of Sabah and Sarawak, 1: XXI – XLI.

What kind of classification should the practising taxonomist use to be saved?

P. F. STEVENS[1]

Summary. Changing limits of higher taxa (mainly genus level and above) in the Malesian flora and their effects on our understanding of evolution and biodiversity are discussed. The three main ways of naming organisms — evolutionary, cladistic and phylogenetic — are compared. Phylogenetic naming is not currently a viable option. The limits (to a large extent) and ranks of higher taxa are assigned by humans, not nature, and findings in cognitive science are used to help clarify how groups are circumscribed and ranked in evolutionary and cladistic classifications. The aspects of evolutionary classifications discussed are the nature of taxa, the multiple uses, stability and convenience of classifications, and their relationship to "nature". The criteria used to delimit and rank groups in evolutionary classifications are shown to have little force, and shifting ideas of convenience and utility often shape taxon limits. In cladistic classifications convenience is subordinated to the need for groups to be monophyletic; within this constraint, convenience can guide taxon rank. Two possible confounding factors, hybridisation and paraphyly, are evaluated. Hybridisation compromises the monophyly of genera, but genera that hybridise can be combined. Paraphyly is a complex issue. Establishing the monophyly of higher taxa will entail some familiar names disappearing and some taxa greatly changing size, but this will clarify our ideas of evolution and diversification. Concerns that persisting ancestral and hence paraphyletic species will confound cladistic classification seem to be misguided, both because of extinction (in its absence, even evolutionary classification is impossible) and because the concept of paraphyly is not applicable to processes and patterns at the level of divergence and individuation of species. It is recommended that those working on plants from the *Flora Malesiana* region make their higher taxa monophyletic, with the nomenclatural changes that this entails being listed in the final volume.

INTRODUCTION

> I am more and more against voting a genus a good one merely because it has a few good characters, but to judge of its value really one must take a view of the allied ones. Where a large group is natural and well defined I do not like lopping off a few species here and there on account of some remarkable exceptional character, for it is that which carried to excess has produced the present lamentable chaos in our botanical system (George Bentham to Asa Gray, Febr 1/[18]61).

This spring term I taught Flowering Plant Families, a course which focuses on plants of the eastern United States. You may well wonder what connection that course has to do with a meeting on the Malesian flora. The connection is what I can teach, why, and the students' reaction. Every year I incorporate the findings from morphological and molecular phylogenetic studies appearing the previous year. Each year the course becomes more satisfactory both to teach and to learn as the limits and relationships of more monophyletic groups are clarified. Instead of

---

[1]Harvard University Herbaria, 22 Divinity Avenue, Cambridge, MA 02138, U.S.A.

making vague statements of relationships and offering questionable delimitations of groups, a body of work supported by articulated theory is developing that allows us to discuss in a way hitherto impossible the evolution, biogeography and relationships of in many cases newly circumscribed groups. Thus understanding that monocots probably evolved from a plant that we would call a dicot is very exciting. The student looks at features like the position of the prophyll relative to the axis and the nature of the stele in the stem with new eyes — indeed, students previously would not have looked critically at such features at all — since there is now a context for thinking about character evolution.

The question is, should phylogenetic studies be integral to the Flora Malesiana project? Negative assessments of the cladistic approach still appear about once a year (Hedberg 1995 and Brummitt 1996, are convenient entries into the literature; see especially Cronquist 1987) about the problems caused by parallel evolution, hybridisation, the hubris consequent on over-reliance on computers, etc. I will turn the question around, and ask, are we interested in subjects like diversification, evolutionary ecology, biogeography, and the like? Other papers given here show that it is clear we are, and there is a very large literature showing how such subjects can begin to be answered when we have a phylogeny.

My focus here is on supraspecific groups, mostly on the genus and above. I first discuss recent studies, both cladistic and conventional, on genera and families in the Malesian flora, and show how different circumscriptions of taxa affect a diversity of issues from our understanding of evolution to conservation priorities. I then outline the three main types of classification currently most discussed: evolutionary, cladistic and phylogenetic. I detail briefly some of the claims made for evolutionary classifications, and some of the criteria used for recognising higher taxa in evolutionary and cladistic classifications. My approach is in part cognitive, for example, how do we memorise, and what is the relationship between naming and assessments of similarity? I will suggest that work in cognitive science helps clarify a number of operations entailed in classification. I do not discuss how we detect genealogies, or how hybridisation might affect this, or issues bearing on naming fossils, or naming species.

It might be argued that databases change the relationship between nomenclatural change and optimal structure of classifications, since names used in any one system can readily be related to names used in other systems, as well as to phylogenies. However, names are still the basis of our everyday discussions, and how we use classifications and commit them to memory are still reasonable concerns. Little of what follows is new, but in light of the persisting discussion over the goals and merits of different approaches to classification, clearly bears repeating.

GENERA, FAMILIES AND PHYLOGENY OF MALESIAN PLANTS

There is a fast-growing number of cladistic studies affecting Malesian families. For example, Graham *et al.* (1993) link *Sonneratiaceae* firmly to lythraceous genera, even if the monophyly of the *Lythraceae* s.l. still needs to be confirmed. Judd *et al.* (1994) question the monophyly of Malesian families such as *Sapindaceae*, *Capparidaceae*, and *Apocynaceae*. The traditional concepts of *Simaroubaceae* and *Loganiaceae* have fallen apart (Fernando *et al.* 1995; Struwe *et al.* 1995), and major changes in the relationships and delimitation of groups within the Lamiales, that is, the bulk of the zygomorphic sympetalous families, are in the offing (Olmstead & Reeves 1995).

Such revisions greatly affect how we interpret the diversity of these groups in Malesia and adjacent areas. Thus we can no longer think of the *Ericaceae* as being almost absent from Australia (Anderberg 1993; Kron & Chase 1993) — on the contrary, they have undergone a major diversification there, being represented by what has previously been called *Epacridaceae*. We now need to find the lineage within *Ericaceae* to which epacrids are related (*Prionotes* may not be immediately related to other epacrids — Crayn *et al.* 1995). Similarly, the limits and relationships of *Loganiaceae* and *Simaroubaceae*, clearly difficult groups as discussed in the *Flora Malesiana* (Leenhouts 1962; Nooteboom 1962), are clarified by recent studies. Previously, I fretted over why *Simaroubaceae* were recognised as a family yet were so variable, now distributions of characters such as quassinoid versus limonoid triterpenoids, the nature of seed fats, the presence of stipules, and the degree of ovary fusion, make better sense because several genera have been assigned to related or not so related lineages. *Suriana* may share amphitropous ovules with *Harrisonia* (Noteboom 1962), but this is no longer of particular relevance to *Simaroubaceae* as cladistically delimited since neither remains in the family. The association of *Leitneriaceae* — the only endemic family of the Southeast United States (well, it used to be) — with *Simaroubaceae* makes sense; although *Leitneriaceae* do not have notably bitter bark (*Simaroubaceae* usually do) there are other features, cryptic although they may be, that support this alignment (Fernando *et al.* 1995; Fernando & Quinn 1995).

Clearly, such changes bear on the issue of endemism in the Malesian flora, not so coincidentally the title of a recent paper by Johns (1995). His discussion of endemism, conservation, and biogeography uses conventional generic (and familial names). However, he senses problems, noting that the nature of taxonomic treatments affects data on regional endemism. Conservation interests focus on genera with restricted distributions, so he suggests that if large, widely-dispersed and unwieldy genera such as *Ficus* were divided into smaller genera, these might be of use to such interests.

Genera as currently circumscribed are heterogeneous, and some convey little information about biogeographical or morphological relationships. But if raising sections to genera would change ideas of endemism, so would ensuring that genera are monophyletic. De Wilde (1994) suggested that *Paramyristica sepicana*, a monotypic genus endemic to New Guinea, is closely related to the New Guinean *Myristica marcgraviana*, *M. carrii*, and *M. hooglandii* (although not so similar anatomically). It seems to have evolved from within a lineage of New Guinean *Myristica*, to which *Paramyristica* belongs cladistically. Other endemic Papuasian genera listed by Johns (1995) include *Papuacedrus* — which is included in the southern genus *Libocedrus* by de Laubenfels (1988). Similarly, *Kerigomnia bilabrata*, a New Guinea endemic, was transferred to *Octarrhena* (Kittredge 1985), a genus known from Sri Lanka to Polynesia. However, *O. bilabrata* finds its relatives in a group of New Guinean species — which perchance could be raised to the rank of genus, and there is a New Guinean endemic à la *Paramyristica* again... *Papuaea reticulata*, a monotypic endemic New Guinean ground orchid with non-resupinate flowers, looks much like *Goodyera* in a reconstruction of the plant drawn by Garay from Schlechter's original description, and Schlechter (1919) noted that it had relationships with *Goodyera*. Is it cladistically a *Goodyera*? (It also has beautiful leaves and would be a good greenhouse plant (Schlechter 1919) — and perhaps hence a conservation priority!) Clearly, the limits of genera depend on the different grouping and ranking criteria used in classifications — not a satisfactory situation.

How our understanding of endemism in the *Flora Malesiana* area is affected by our understanding of genealogy is shown by other recent studies. *Fagraea* is paraphyletic, and includes *Anthocleista* and *Potalia* (Struwe *et al.* 1995). The next step is to elucidate relationships within such groups, and Struwe & Albert (1997) have begun the cladistic studies needed for nomenclatural realignments. Our ideas of the evolution and biogeography are changing dramatically; the lineage is pantropical, but *Fagraea* in the old sense is Indo-Malesian. Similarly, *Mammea* (*Clusiaceae*) is pantropical, but most of the Malagasy-Indo-Malesian species are characterised by fibres free in the mesophyll, almost unique in the family. There is one exception in Malesia, the Bornean *M. calciphila* (and also *M. touriga*, from Queensland). Do they form a lineage with the other taxa lacking such fibres (the four species known from the African continent and from America, and perhaps three from Madagascar)? The answer to this question will clarify the diversification of *Mammea* in Malesia and at the same time the evolution of the Malesian flora.

In *Agapetes*, too, the current classification obscures our understanding of diversification and biogeography. *Agapetes* is most diverse in S.E. Asia. It is by all accounts close to species of *Vaccinium* there, but there is one species of *Agapetes* in Malaya, an outlying centre of diversity in New Guinea, and a few species south and east — a unique distributional pattern in Malesian plants. Some years ago, I wanted to resurrect the genus *Paphia* for the the New Guinea-Oceanic species of *Agapetes*. B. L. Burtt suggested that while I had shown that they differed from the mainland species in anatomy and some aspects of morphology, I had not shown clearly to what they were most related and *A. scortechinii*, from Malaya, was something of a problem. The result was that I placed the extra continental SE Asian taxa in *Agapetes* subgenus *Paphia* (Stevens 1972). However, work since then confirms that the complex of distinctive anatomical features of Malesian and Oceanic *Agapetes* (and *Dimorphanthera*) is consistent in those genera, and is found elsewhere in S. American genera. These features include the possession of a deep-seated phellogen; mainland *Agapetes*, all Malesian *Vaccinium* sampled, as well as the great majority of Southeast Asian species of *Vaccinium* have superficial phellogen (the character varies at the sectional level). It is exciting that Atkinson *et al.* (1995) have recently suggested that cytological data support these relationships. Mainland Asian species of *Agapetes* and Malesian *Vaccinium* are predominantly diploid with the odd polyploid; all counts from New Guinean and oceanic *Agapetes* (and *Dimorphanthera*) are polyploid, as are those from neotropical *Vaccinieae* with which they share morphological and anatomical features. (South American *Notopora* was not counted; it has superficial phellogen [pers. obs.], but is otherwise very distinct from S.E. Asian-Malesian taxa with that feature. Atkinson *et al.* [1995] also found that some Malesian non-*Vaccinium Vaccinieae*, including *A. scortechinii*(!), and some tropical

New World *Vaccinieae*, are diploids.) The sampling is still poor, but the concordance of the different kinds of data is striking, and suggests that the 100+ species of large-flowered Vaccinieae — "*Agapetes*" + *Dimorphanthera* — in Papuasia have close relationships with New World taxa, certainly not with the Old World taxa in which they have been placed or with which they are associated geographically (Table 1). Ultimately, combinations of the New Guinean and Oceanic species of *Agapetes* in *Paphia* and of the mainland species of *Agapetes* in *Vaccinium* will probably be required. But this is only a step in the right direction, an overhaul of generic limits in the *Vaccinieae* as a whole.

TABLE 1. The distributions of suites of characters in some *Vaccinieae* (for more details see Stevens 1972, Atkinson *et al.* 1995).

Taxon	Anatomical characters	Ovary characters	Chromosomes
Vaccinium: SE Asia/Malesia	+	+	2x/4x
Agapetes: SE Asia	+	+	2x/4x
Agapetes scortechinii: Malaya	–	intermediate	2x
Agapetes: New Guinea/Oceania	–	–	6x
Dimorphanthera: Papuasia	–	–	6x

There are several recent cladistic studies of individual genera or small groups of genera of the Malesian flora. These have been used to justify the description of new genera (Adema & van der Ham 1993), while studies on *Cupaniopsis* and *Guioa*, sizeable genera in the Sapindaceae, integrate species distributions with relationships, and implicate dispersal in the evolution of the first genus (Adema 1991), more vicariance in the second, but with dispersal and speciation in West Malesia and the island arcs of the Pacific (van Welzen 1989). There are about thirteen species of *Hopea* in New Guinea: what is their history there? *Hopea inexpectata* is the only member of section *Dryobalanoides* in New Guinea, all other species there belonging to section *Hopea* (Ashton 1982); does this suggest different dispersal histories for the two groups? Recent evidence suggests that *Hopea* is sister taxon to part of *Shorea* (*Shorea* is paraphyletic — Dayanandan 1995), and so the biogeography and evolution of *Hopea*, *Shorea*, and other genera must be reinterpreted. Phylogenetic relationships are crucial for any discussion of diversification, biogeography in general, and endemism in particular.

THREE WAYS OF NAMING ORGANISMS

Although phylogeny is the basis for our understanding of nature, there is no agreement as to what of phylogeny a classification should reflect; some even argue that classification and genealogies are incompatible. But as Flora writers, we deal with names, and so how we name organisms is of central importance to us, and others. There are two common ways in which organisms are named, conventional or evolutionary (with the superscript [evol] below) and cladistic ([clad]), while a third way, phylogenetic ([phylo]), has recently been proposed. Systematists seem caught between the devil of evolutionary classifications, the deep blue sea of cladistic classifications, and the wild empyrean of nominalist phylogenetic definitions. Practising systematists may be inclined to stay with the devil they know — some variant of evolutionary classification — and one of the main purposes of this paper is to explore some of the consequences of this attitude.

The differences between these three ways of naming can best be illustrated by examples from *Ericaceae* and *Poaceae*. Evolutionary classifications try to combine genealogical relationships with ideas about the distinctness of groups (see below). Taxa can be paraphyletic, that is, they may include only descendants of a particular ancestor, but need not contain all of them (note that this is called monophyly by evolutionary systematists). Thus both molecular and morphological studies suggest that the largely north temperate, wind-pollinated *Empetraceae* and the largely Australian, sclerophyllous and insect- and bird-pollinated *Epacridaceae* are part of the lineage that also includes the *Ericaceae*[evol], and that *Enkianthus*, a Southeast Asian genus, is sister taxon of the combined group (Fig. 1: e.g. Anderberg 1993; Kron & Chase 1993). *Empetraceae*[evol] and *Epacridaceae*[evol] are kept separate because of their distinctness, with *Enkianthus*, although a distinct enough genus, not a particularly remarkable member of the paraphyletic *Ericaceae*[evol] (note that *Pyrolaceae*, *Monotropaceae*, *Vacciniaceae*, and *Rhodoraceae* have all been segregated from *Ericaceae*[evol] at one time or another). Similarly, recent studies suggest that there are two, very well supported basal lineages of the *Poaceae*[evol], one including some herbaceous bamboos and the other *Pharus* (plus *Leptaspis*); the main lineages within the rest of the family are poorly resolved (Fig. 2: Clark *et al.* 1995).

In cladistic classification conventional rules of nomenclature — typification, priority, etc. — are applied in a cladistic context (for useful summaries, see Wiley 1979, 1981, chapters 6, 7; Forey 1992). Only monophyletic lineages (ancestors and all their descendants) are named, and sister taxa are assigned the same rank (but see below). For example, a monophyletic group results from the addition of *Empetraceae* and *Epacridaceae* to *Ericaceae*[evol]; the name *Ericaceae*[clad] is kept because the other family names are younger than it. That *Enkianthus* is sister taxon to the rest of the *Ericaceae*[clad] has no necessary nomenclatural implications at the family level — *Enkianthus* could be segregated, but it does not have to be, and it could as well remain in *Ericaceae*[clad]. In a cladistic classification, subfamilial rank for *Enkianthus* would be appropriate, and the other *Ericaceae* would be placed in one or more subfamilies. Subfamilial rank would also be appropriate for the two basal lineages of the *Poaceae*[clad], these being added to the subfamilial names *Pooideae*, *Panicoideae*, etc., already in use. This is the practice of phyletic sequencing — successive branches of a lineage are recognised at the same hierarchical rank, so allowing the numbers of taxonomic ranks used to stay within bounds (Wiley 1979, 1981).

In phylogenetic naming (de Queiroz & Gauthier 1994, and references), names are related more directly to lineages. A lineage can be named by specifying a species in both basal branches of that lineage and their common ancestor. Attempts to name an *Ericaceae*[phylo] prior to the studies mentioned above might well have used species in genera like *Rhododendron* and *Vaccinium* as reference points, and the name would then be attached to ("defined" by) the common ancestor of those species and its descendents (see Ghiselin 1995, and de Queiroz 1995, for a discussion of definitions and lineages). The common ancestor of the *Epacridaceae* (whether it is monophyletic or not — see Crayn *et al.* 1995) and of *Empetraceae* include only parts of *Ericaceae*[phylo], and so those names refer to different entities; inclusion of such groups will not affect the name *Ericaceae*[clad]. However, if *Enkianthus* is sister taxon to *Ericaceae*[phylo] as circumscribed above, a new name for the whole

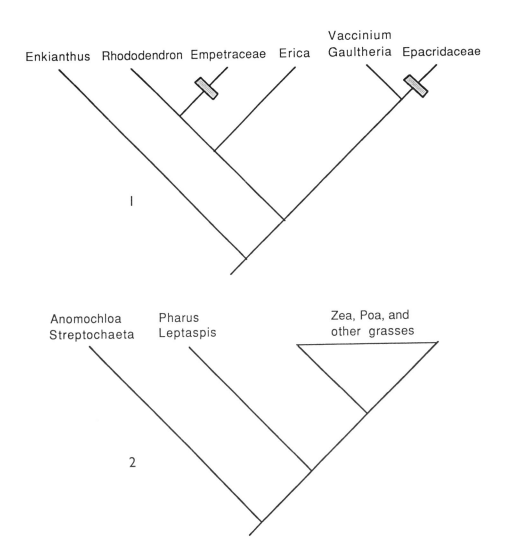

Figs 1 – 2, Genealogy and classification. Fig. 1: Some major lineages in *Ericaceae* (from Anderberg 1993; Kron & Chase 1993). Bars represent especially distinctive characters; the sister group relationships are grossly simplified. Names are those used in classifications[evol]; all the genera in the diagram are included in *Ericaceae*[evol]. Fig. 2: Basal lineages in *Poaceae* (from Clarke *et al.* 1995). Not all members of the *Streptochaeta* and *Pharus* lineages are shown; *Leptaspis* is included in the latter *teste* E. A. Kellogg.

group will be needed, since *Enkianthus* is not a descendant of the common ancestor of *Rhododendron* and *Vaccinium* (Fig. 1). Similarly, to provide a phylogenetic name for the taxa currently included in *Poaceae*, one species in the basal herbaceous bambusoid branch and one in the other branch — species of *Pharus*, *Poa*, or *Phragmites* will all serve equally well — should be mentioned, and reference made to their common ancestor. The name *Poaceae*[phylo] attached to the common ancestor of a species of *Zea* and *Poa* refers to a different unit (Fig. 2).

Other aspects of phylogenetic naming schemes become evident when the initial assessment of relationship of some members of a group is wildly wrong. Take Saxifragaceae as an example. If the *Saxifragaceae*[phylo] had been named with reference to the common ancestor of *Saxifraga* and *Hydrangea*, the well-supported suggestion that *Hydrangea* is to be placed close to *Cornaceae* (Morgan & Soltis 1993) means that *Saxifragaceae*[phylo] would now designate a clade that included half the angiosperms or more. This is because what are named are not classes with properties assigned to included members, but lineages, which are recognisable only in the context of ancestor-descendant relationships. Furthermore, since phylogenetic naming has no rules about terminations of names (cf. Brummitt 1996), *Saxifragaceae*[phylo] could include lineage names with ordinal terminations — or terminations could be entirely rejected. Only names applied to identically circumscribed lineages compete. Thus (a made-up example) *Chrysosplenieae*[phylo], with *Chrysosplenium* and *Philadelphus* as reference points, and *Saxifragaceae*[phylo], as delimited above, refer to the same lineage since they have the same common ancestor; they would compete.

The main argument for phylogenetic naming is that what are being named are lineages, not members of classes, so Linnaean nomenclature, which assumes the existence of a class hierarchy in nature, is conceptually inappropriate — whether in an evolutionary or a cladistic guise. But classification[clad] is the application of nomenclatural rules to genealogical relationships, and is not about taxa being classes in a Linnaean sense, just as the use of type specimens is a way of assigning names to taxa, not about types being typical or taxa having essences. Finally, note that in no naming scheme does a name specify a particular circumscription of a group; changes in content of a group may not entail corresponding changes in its name. Such flexibility is invaluable.

I see at present no compelling argument against classification[clad], and hence no need to rename nature. Nevertheless, the hierarchy in both classifications[evol] and [clad] is easily misinterpreted. It has long been acknowledged (Brown, 1818: see also Davis & Heywood 1963; Stuessy 1990; Stevens 1994b) that members of higher taxonomic ranks are equivalent only by designation. Their treatment as if they were equivalent can cause serious problems (e.g. Vilgalys & Hibbett 1993; Stevens 1997); it often makes little sense to use families and genera as units of comparison. Yet much of the debate about classification centres on the issue of ranking — are familiar families (for example) going to be "lost" when their rank is changed? — even although ranks are entirely human constructs. Hence explaining the role of rank is an important educational matter. The retention of rank in classifications is an aid to learning, since as one learns genealogies, the inclusion relationships implied by terminations can be helpful, but such relationships are all that terminations signify;

rank is relative, not absolute (Forey 1992). Such problems are avoided by the rejection of rank and rank terminations in phylogenetic names. However, the discussion is moot, since there is as yet no formal scheme for phylogenetic naming.

ON EVOLUTIONARY CLASSIFICATIONS

A. The classification itself

Before outlining criteria for delimiting and ranking cladistic taxa, I first discuss isssues in evolutionary classification and some criteria for group circumscription there. This will clarify what classifications in general are all about.

Many of the issues surrounding the supposed merits and demerits of classifications[evol] stem from what is considered to be their main virtue. They attempt to be "synthetic" by incorporating as many kinds of data as possible, or at least cladistic relationships, phenetics, and time of origin (see Stuessy 1990, for a summary), in the one classification. Below I discuss the epistemological status of taxa in classifications[evol], and some of the properties of such classifications.

Nature of taxa: Evolutionary synthesis combines different kinds of data that are incommensurable — what is a common measure for the age of a lineage and the amount of divergence that lineage has undergone since splitting? Even if such a measure were devised, and a decision made that generic rank entailed a given amount of divergence, there would be no way of knowing whether a particular genus was monophyletic or paraphyletic, had separated a long time ago or recently, was phenetically distinct or not, and so on. Furthermore, change in the knowledge of a group that did not affect our understanding of its relationships could lead to a change in rank if the amount of difference between it and other taxa changed. Higher-level taxa in classifications[evol] (including "the new phyletics" — Stuessy 1990) are unlikely to be stable.

Multiple uses: Attempting to produce a single classification that will fulfil optimally all the separate possible uses that have been proposed for classifications is like looking for the pot of gold at the end of the rainbow. General-purpose classifications are unlikely to be optimal for any one purpose (see also Stevens 1990; Humphries 1991). The rank of genus, however conceived, has little value in helping us to understand questions of comparative biology **and** to establish conservation priorities (for ways of relating classifications and phylogenies to the latter, see Humphries & Williams 1994).

Stability: Agreement between taxonomists has often been invoked as a defence of traditional classifications (e.g. Brummitt 1996). However, agreement in the absence of articulated principles is the agreement over a convention (Stevens 1994b) that is of as much intrinsic biological importance as agreement as to how many buttons there should be on the cuff of a blazer, or where the hem-line should be. Even if a classification is designed to be useful to as many people as possible (Morin *et al.* 1989 — interior designers are included), stability may well not ensue, since the user pool changes over time (Hawksworth & Bisby 1988). The recognition of a genus because its distribution is such that the genus is valuable in conservation discussion is a good example. Use in conservation is a new (in this context) use for a classification, and, not surprisingly, would cause nomenclatural instability (to say nothing of systematics becoming a graphic indication of how "science" is driven by social goals, and how that agenda may influence the scientist's interpretation of data). To suggest that the generic concepts of "taxonomic users" should always be preferred over those of "current specialists" seems singularly misguided advice (cf. Gentry 1993, p. 1).

On convenience: The idea that classifications should somehow be generally useful persists (Just 1953; Cronquist 1987; Estes & Tyrl 1987; Brummitt 1996), that is, there is a tendency to think that genera and families can both reflect natural relationships and aid in correct identification by being easy to recognise; taxa should be recognisable by simple inspection. However, as has frequently been pointed out by systematists since Lamarck in 1778 (Bather 1927; Stevens 1994b) natural groups may not be easy to recognise. Lamarck devised a particular format for keys to get to names of an unidentified organism as quickly as possible, while Bentham (1875) promoted the use of keys because characters used in identifying a group might be different than those used in assigning a group its position in the natural (evolutionary) system. Along similar lines, Barkworth (1992) noted that using keys to get to names invokes characters and procedures that have little to do with how those groups were established in the first place.

Use or convenience can be criteria for group recognition when constructing classifications (see below). However, group circumscription in classification[evol] is subject to so many but at the same time such weak controls that usefulness and convenience speedily become ends in themselves. Let us never again hear of systematists "obfuscating" non-botanists' concepts of taxa like *Cruciferae* as if that were a valid argument against changing their circumscription (cf. Brummitt 1996). Paraphyletic taxa should disappear from general systematic use.

On nature: Apparent distinctiveness of a group to a human classifier may not reflect any evolutionary or biological importance of that group. In any case, as Donoghue & Cantino (1988, p. 114) noted, "there may be no single [biological] distinction that is of overriding importance to all biologists, let alone to all people" (see also Hempel 1965, p. 145; Brummitt 1996). We usually know little about the biological or ecological significance of the characters on which genera are based, and this includes those characterising *Paramyristica*, *Papuaea*, *Papuacedrus* and *Kerigomnia*. It has been suggested that ethnobiological classifications, perhaps especially of animals, may approach classifications[evol] (Berlin 1992), with "nature" somehow being captured by both. However, that there is a single nature that can be encapsulated in a classification seems unlikely (Stevens 1994a). Bather (1927, p. lxxi) observed perceptively for the pre-Darwinian period that "not a single naturalist had a clear idea of what he meant by 'natural'. All he knew was that the other fellow's classification was unnatural", a sentiment that is scarcely less true today. The adjective "natural" as applied to classifications should be expunged from our vocabularies unless appropriately qualified each time it is used.

B. Some Guidelines for Circumscribing and Ranking Evolutionary Taxa

Many earlier guidelines for the recognition of taxa at the level of genus and above in evolutionary classifications (e.g. McVaugh 1945; Stuessy 1990, and references therein) concern aspects of cognition. I discuss some of these aspects below, both to clarify what the guidelines entail and to see if their use is appropriate. Cronquist (1988) mentioned the desirability of having taxa subdivided into the same number of subtaxa using similar characters (he recognised this was impossible); the likelihood that large groups would be subdivided by less in the way of differences than smaller groups (see A, below); and what he called McVaugh's rule (e.g. Cronquist 1983), the placing of an intermediate between a small, well-defined segregate group and a more heterogeneous, larger (ancestral) group in the larger group — hence maintaining the distinctness of the small group (B — see Fig. 3). There has also been some discussion about the supposedly dubious virtues of recognising too many families (C). Of course, it is generally hoped that higher taxa are recognisable (D), although it is also conceded that the higher the rank of the taxon the less likely this is (e.g. Stuessy 1990, see also A below).

A. It has often been observed that, other things being equal, the size of groups in classifications[evol] tends to be inversely proportional to their distinctness (see also Davis & Heywood 1963; Mayr & Ashlock 1991; Stuessy 1990). Large groups tend to be subdivided simply because they are large, not because they include much variation. This seems to be a version of the distance/density model in psychology — objects that are scattered are perceived to be more similar than objects that are close together, the latter being more divided into groups despite the small distances separating those groups (Krumsahl 1978). In classifications[evol] this can be rationalised as an expected effect of parallel or convergent evolution (e.g. Kubitzki & Gottlieb 1984), an effect particularly prevalent in related groups (see Vavilov's Law of Parallel Variation and ideas of orthogenesis — Kellogg 1990, for some references).

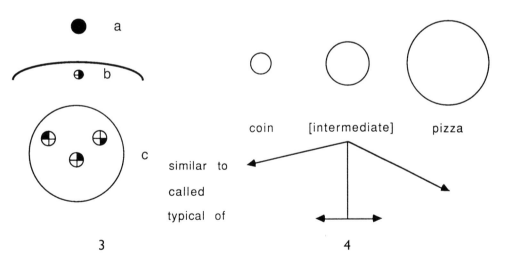

FIGS 3 – 4, Naming and similarity assessments of pizzas and paraphyletic groups. Fig. 3: McVaugh's rule, a = derived genus, b = intermediate genus, c = larger, heterogeneous, ancestral group, the circles inside it being three included genera. The distribution of four characters is shown in the quadrants, filling of quadrants representing the derived states; all four characters have derived states in genus a. Bold line separates the families. Fig. 4: Name, similarity, and typicality of intermediate when compared with a coin and a pizza.

B. McVaugh's rule (Fig. 3) appears to enhance the distinctiveness of segregate groups. This must be in part an illusion, because intermediate taxa cannot be wished out of existence. Rather, this rule seems to relate to how humans group objects, animate or inanimate, and how variation is interpreted. This is clarified by considering some well-known experiments in cognition (Rips 1989) where subjects are shown three figures of the same shape, but of different sizes. Two are referred to categories, of which one is smaller and of a fixed size (a quarter, a 10 penny or 100 ruppiah coin), and another larger, like a pizza, and potentially varying in size (Fig. 4). The third, unnamed, is exactly intermediate in size between the coin and the size of the particular pizza used in the test. Subjects are asked what the intermediate "is" (to which category does it belong), and to which of the categories is it more similar (Rips 1989). Subjects usually give the name of the larger object to the intermediate object — after all, one can make very small pizzas, but a coin of a particular denomination has to be a particular size. The intermediate is considered to be more similar to the smaller object — the potential universe of pizza sizes is such that the intermediate may be very different to some of them. Judgements of typicality (of which two categories is the intermediate object more typical?) tend to be intermediate (Fig. 4).

Systematists' decisions about grouping are far more complex. They involve many more features, few of which are simple size differences of the kind just discussed; however, how many characters are actually evaluated and what elements of complex characters are taken into account by systematists is unclear (Stevens 1990; Donoghue & Sanderson 1994). To paraphrase McVaugh (1945), for a systematist, the retention of the intermediate taxon in the more inclusive taxon will change the limits of the latter slightly, if at all. Individual members of the larger taxon — not necessarily immediately related to the intermediate taxon — may have particular morphological features in common with the intermediate taxon, and this is what is important (Fig. 3). The larger taxon as a whole will then appear more similar to the intermediate than the smaller taxon, because the characters of such taxa are the characters of their included members. Similarity assessments and naming will tend to go together.

There is a further phenomenon to be borne in mind here. Decisions whether to merge groups or keep them separate can be influenced by the order in which objects are compared (Krumsahl 1978, for references), so *Empetraceae* are likely to be considered more similar to *Ericaceae*[evol] than *Ericaceae*[evol] to *Empetraceae*. In such comparisons, features of the first member of the pair are matched with those of the second, and there are more features of at least some *Ericaceae*[evol] that are not found in any Empetraceae than vice versa.

Since McVaugh's rule entails the recognition of paraphyletic groups, it will have no place in classifications[clad]. However, the aspects of cognition mentioned here should be borne in mind as we look at individual features and delimit characters states in the variation we observe (see also Gift & Stevens 1997).

C. Arguments on whether too many taxa are recognised have recently focussed on family-level taxa. Is there any substance to this discussion? Interestingly, individuals frequently utilise the region of 500 "folk generics" in folk taxonomies (Stuessy 1990; Berlin 1992) and a similar number of place names (Hunn 1994; cf. [in part] Kelly 1994). If this is a real constraint of human memory (but see below), and if it could be agreed that having some knowledge of all Angiosperm families was a reasonable goal for systematists — I think it is — then I see no reason not to bear this number in mind as we segregate and merge families.

D. In ethnobiological classifications, the concept of salience — when is a group salient enough to be a candidate for naming and incorporation into the folk taxonomic lexicon — is notoriously elusive (Berlin 1992, for a summary; see also Carter & Gluck 1992; Coley *et al.* 1996). Factors like the size of a group, its internal cohesiveness, its distinctiveness compared with other groups, its overall commoness and importance in the local environment — all elements that we might consider when grouping and ranking taxa in classifications[evol] — all may be a part of this concept. It is hardly surprising that decisions as to when a group "really is" distinctive enough for recognition are inherently problematic. Assertions that taxa or characters have biological importance perhaps can best be seen as attempts to increase the salience of a group for a systematist, although all too often such assertions are poorly supported.

However, this issue returns us to the synapomorphies by which monophyletic groups are recognised, and their hierarchical arrangement. As Cronquist (1988) reasonably observed, groups without characters are not very helpful. Nevertheless, it is often hard to make sense of groups in classifications[evol] in which reticulating relationships are implied or actually represented (note that the issue here is usually not that of hybridisation), and individual characters often have a broad and apparently sporadic distribution (Stevens 1986, 1994b). However, since synapomorphies that underpin cladistic classifications are hierarchically arranged, characters marking lineages at different levels are related in a simple fashion. Even if few features diagnose a lineage, and some may even not be notably common in members of those lineages, there is a rationale for this. It is certainly easier to commit words in hierarchically arranged classifications to at least short-term memory than words in schemes with no obvious hierarchy — rather similar to the relative ease with which people can remember strings of words that make sense when compared with strings that do not make sense (Bower *et al.* 1969; Gelman *et al.* 1989). A great advantage of classifications[clad] is the ease with which existing fragments serve as the foundations of more extensive knowledge as new studies clarify relationships. As Keil (1989, p. 290) suggested, theory links observations and concepts like memory loci did in the great memory systems of antiquity; here theory is genealogy and the algorithms (with their evolutionary assumptions) used to analyse the data; observations are the data we accumulate from our study of organisms; and concepts are classifications.

ON CLADISTIC CLASSIFICATIONS

A. Naming Groups

The guidelines below are similar to those offered by Wagner & Beitel (1992); Wiley (1981) should be consulted for an extended discussion. Taxa at the rank of genus and above should be named only

1, if they are cladistically monophyletic;
2, (ideally) if they are recognisable by characters of external morphology;
3, if nomenclatural and cognitive implications have been taken into account.

1. Monophyly is essential; this is the keystone of classifications[clad] (e.g. Hennig 1966). However, monophyly is not a ranking criterion, and until an absolute ranking criterion such as age is established, ranking remains arbitrary (e.g. Forey 1992). But once we have satisfied the prerequisite of monophyly for the recognition of higher taxa, we can actively seek the classification[clad] that is most convenient for us without fear of comprimising our taxa — a very different situation from the role such convenience plays in classifications[evol].

1A. "Short of the death penalty, there will always be lumpers and splitters. Splitting tends to be accentuated with age, a characteristic of the malaise being that the sufferer seldom admits that he *is* a splitter" (Davis & Heywood 1963: 85). Thus, that monophyletic groups can be recognised within a genus (for example) does not mean that they, too, should be named as genera. As has often been observed in general arguments over lumping and splitting, and can be rephrased here, infrageneric taxa can always be recognised to accommodate these monophyletic groups (Stuessy 1990). Unfortunately, there seems to be difficulty in adopting this course of action! Dividing a monophyletic *Casuarina* into four genera (Johnson & Wilson 1993; they may not all be monophyletic) is not necessitated by any school of classification. More specifically, techniques like phyletic sequencing can be used to reduce the number of ranks used in a highly resolved and complex genealogy (e.g. Wiley 1981, for details).

1B. There is no value in recognising as genera monophyletic segregates from a large, monophyletic genus if their segregation makes that genus paraphyletic (Funk 1985). Indeed, Olmstead & Reeves (1995) discuss the distinction between paraphyly and polyphyly in the context of maintaining a *Scrophulariaceae*[evol], and show how slight the difference is between them. Paraphyletic taxa have no positive characters, that is, no synapomorphies, and are likely to be defined by plesiomorphous characters (unless the synapomorphies of the excised monophyletic group(s) are transformations of the synapomorphies of the combined group — then the argument is whether the excised groups still have the original synapomorphy [Stevens 1984]). Polyphyletic taxa are likely to be defined by plesiomorphies, convergences or parallelisms. There is little reason for recognising segregate taxa in any type of classification, as Bentham's epigraph suggests (see also van Steenis 1978). Furthermore, such segregation tends to be self-reinforcing and so habit-forming behaviour — remove one piece of a group, and there will be another piece that just has to be removed (Clayton 1974; Turner 1985).

2. Other things being equal, it is of course useful if taxa are readily recognisable. Furthermore, there are several suggestions in the literature that gap size between groups at the same hierarchical level should be similar, e.g., that generic limits should be similar throughout a family or part thereof (Orchard 1981; Judd 1982), or more generally throughout flowering plants (Thorne 1976), or throughout all pteridophytes (*Lycopodiaceae*, ferns, etc.: Wagner & Beitel 1992). If genera are made morphologically equivalent, there may be an unfortunate tendency to think that they are equivalent in other ways as well — age, for example. There is no reason to suppose that delimiting groups by morphological gaps of similar sizes in lineages that have been separate for 400 million years (pteridophytes) has any value except for mental book-keeping.

3. If keeping the number of names low or minimising the number of name changes is desireable, then expanding the limits of taxa is an option to be considered. In general, *the number of new genera and families to be described will be small if limits of established genera and families are changed at will.* If an unnamed lineage is sister taxon to a named genus, the limits of the named genus can be extended to include the new taxon just as easily as a new genus described to accommodate it (cf. Figs. 5a, b) — unless its inclusion makes the combined genus excessively heterogeneous (Eriksson 1990). The fact that genera such as *Anomochloa* and *Enkianthus* are basal lineages in the poaceous and ericaceous clades respectively makes them candidates for separate familial status, and *Enkianthus* in particular has morphological features setting it off from other *Ericaceae* (e.g. Anderberg 1993); however, this does not prevent them being kept in *Poaceae* and *Ericaceae.*

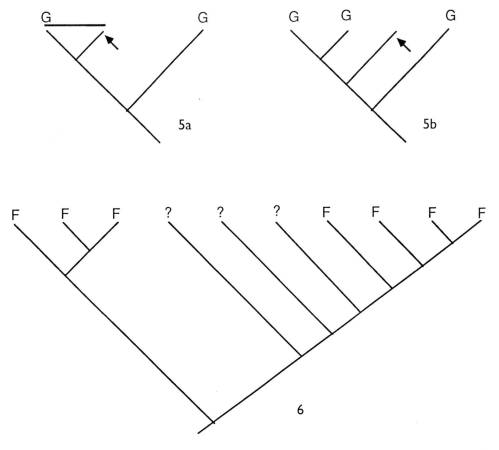

FIGS. 5 – 6. Some aspects of cladistic classification. Fig. 5a: Topology in which addition of unnamed lineage (arrow) does not necessitate a new genus; broad bar is changed circumscription of genus. Fig 5b: Topology in which unnamed lineage either needs a new name, or causes synonymisation of currently-accepted name. Fig 6: Some lineages (f) have family names; if these are to be kept, lineages (?) currently lacking such names will need them.

When describing a genus or family, it should be remembered that branches lower down the tree, i.e. basal to it in a phylogeny, will need names at that rank if they do not already have them (Fig. 6). An example is the odd genus *Triplostegia*, found scattered in the mountains in Malesia (and elsewhere). Peng *et al.* (1995) recognised a *Triplostegiaceae* in the context of the following set of relationships: ((((*Dipsacaceae*) *Morinaceae*) *Triplostegiaceae*) *Valerianaceae*). Recognising these as families will entail recognising at least two other families made up of lineages in the old *Caprifoliaceae* (Judd *et al.* 1994), while because of uncertainties in the relationships between *Viburnum*, *Adoxa*, and *Sambucus* (*Adoxaceae*) it has been suggested that all three be placed in monotypic families (Erbar 1994). The result is that a total of at least ten families will be recognised. This hardly seems desirable.

In this context, the possible cognitive constraints on how, and how many things (see above), we memorise might suggest structure for classifications[clad] — perhaps up to 700 families, grouped and subdivided into quite small units. Similarly, the relative ease with which humans commit to memory short (3 – 5, but not many more) strings of objects of the same general type (numbers, restaurant orders, positions of pieces on a chess board) is also well attested (e.g. Ericsson & Polson 1988. Such studies also show that this constraint does not limit how much we can remember.) Of course, even if these are accepted as reasonable guidelines, it will be difficult to reach agreement as to just which 500 – 700 families should be recognised. (It is of interest that the entire classification in Bentham & Hooker's remarkably successful *Genera plantarum* is reasonably interpreted as attempting to satisfy cognitive constraints. The number of families was fixed at 200, and these families grouped and subdivided into units of fewer than 6, exceptionally 12, included members [Stevens 1997]. The guidelines Bentham & Hooker followed were never published. There may be a moral here!) Moreover, a monographer, a Flora writer, and a herbarium curator may have different memory goals and needs. Indeed, if the issue is simply lumping *versus* splitting, then thinking of memory constraints may be helpful. Otherwise, the whole argument has little content and is very largely a waste of time and energy.

B. Potential problems for cladistic classifications

Two objections to classification[clad] that are still often raised are (1) hybridisation and (2) the persistence of an ancestral species from which other species descended.

1. The issue of hybrids. In evolutionary systematics, if genera produced hybrids, this was grounds to suspect the validity of the generic boundaries (Rollins 1953). Monophyly, defined in terms of inclusion of all descendants of a common ancestor in the one taxon, suggests that all members of a hybridising group should be placed in that lineage. Note that I consider species to be *sui generis* in this matter; they can be considered distinguishable groups of organisms recognisable as reticulating relationships become less general (see below).

A good example of hybridisation is the *Triticeae*. Here, hybridisation between species, sections, and even genera, is almost commonplace, and the generic-level classification[evol] seems in a state of eternal flux. Treatments appearing in the same year may differ by a factor of two in the number of genera recognised, and in one fairly conservative generic treatment, of the 559 species placed in 23 genera (of about 60 names available), 248 of those species were initially described in genera other than those in which they were now included (Barkworth 1992). Furthermore, genome differences, on which considerable reliance has been placed when delimiting genera, may not be associated with morphological differences (or vice versa), or genome similarities with similarities in cytological behaviour (Zhang & Dvorák 1991). Given such inherent uncertainty, placing all *Triticeae* in *Triticum* (e.g. Stebbins 1956), seems to be a course with much to recommend it. I would follow the same course in the *Saxifragaceae*[clad] and *Crassulaceae*, other places where species currently placed in different genera hybridise with some facility (for the *Saxifragaceae*, see Soltis *et al.* 1993). As in other cases where it may be objected that wholesale lumping will never be accepted, I would simply note that the whole of the *Ericeae* — perhaps even more — is in the process of being placed in *Erica* (Oliver 1991) because the latter is paraphyletic (authors in the past having done exactly what Bentham warned us not to do). The *Ericeae*, at some 760 species, are a far larger problem than the *Triticeae*, but an expanded concept of *Erica* will be adopted for Kubitzki's *Families and genera of vascular plants*.

It may be claimed that I have simply moved the problem to the infrageneric level. However, nature is not simple. Hybridisation between species is a carry-over from processes that act at below the species level, and I see no reason to burden the next categorical rank with arguments that can be avoided by fixing that rank beyond the reticulating processes. Attempting to delimit genera when there are or have been reticulating events between species is an even more arbitrary operation than usual. At the infrageneric level, names can be prefixed by "x", denoting hybrid origin, or other conventions used.

At the other end of the spectrum, and perhaps not sharply distinguishable (Skála & Zrzavy 1995), endosymbiosis presents few problems. The "hybridised" organisms form a metaorganism quite unlike its components. The history of the symbionts in the metaorganism can be tracked by studying organelles and their genomes and combining the study with that of the "ancestral" free-living forms. Study of the history of the metaorganism exclude these free-living forms and include the nuclear genome.

As for hybridisation at an intermediate level, yes, there may be challenging cases. However, one possible example, a hybrid origin for *Rosaceae-Maloideae*, remains unconfirmed (Morgan *et al.* 1994).

2. The issue of ancestors. Emphasis on ancestors, whether fossil or extant, may have connections with 18th- and 19th-century world views where nature, as presented to the systematist, seemed to be continuous; emphasis on ancestors and intermediates represents a 20th-century reworking of such ideas (Stevens 1994b). Nevertheless, the possible effect of ancestors on classification continues to exercise many evolutionary systematists, although the issue can be dismissed largely on methodological grounds, the problem of proving that a given taxon is an ancestor (Patterson 1981). The apparent problem posed by living ancestors is the leitmotif of Brummit's defence of paraphyletic taxa, which is in substantial part an attack on the possibility of basing a classification[clad] on monophyletic taxa alone. As Brummitt sees it,

> [I]f the members of one genus can never give rise through evolution to another genus in the way I believe *Tephrosia* has given rise to *Ptycholobium*, how can new genera ever arise? Every genus since the primaeval one must be descended from a member of a different genus, and so must make that genus paraphyletic. .... All the genera in one family would have to be telescoped into the ancestral genus. Applying the principal more broadly, every monophyletic group would sink into its ancestral taxon. The theory is nonsensical, and it seems to me to be not a reflection of evolution; it is a denial of evolution (Brummitt 1996, p. 8).

Although Brummitt does not always clearly distinguish between ancestral species and "ancestral" higher taxa, here he refers to the latter. Such taxa will not be nomenclatural whirlpools, as successively earlier ancestors suck in their paraphyletic descendents (cf. Brummitt 1996). Including asclepiads in the *Apocynaceae* and *Lemnaceae* as part of *Araceae* will not cause either of the first mentioned members of these pairs to fall through the nomenclatural bottom of the other. Rather, some tribe or subfamily of the second will become more speciose than we had thought — which will greatly clarify our ideas of diversification. All monophyletic groups do not leave paraphyletic ancestral taxa behind them (cf. the above quotation) unless they are delimited in the way *Ptycholobium* is delimited. In any event, extinction often solves the problem of ancestors, as Hooker, Darwin, and Wallace noted long ago (e.g. Stevens 1994b); extinction produces the gaps that we use to limit species, genera, families and the like, and without it, classification[evol] also becomes impossible (cf. Figs. 9, 10). Finally, it should be remembered that *humans* decide if a group is a genus or family, and in the case of *Ptycholobium*, first there is the particularly problematical decision that it is separated by an "important" gap from *Tephrosia*.

In a diagram showing "paraphyly" relationships of species, ancestors seem to include their descendents (Figs. 7, 8: Brummitt 1996, Fig. 1a). However, the immediate ancestor of a group will have the synapomorphies of the group and will be a bona fide member of it (Fig. 11). In fact, the situation at the level of species is far more complex than can be dealt with adequately here (e.g. Kornet 1993; Rieseberg & Brouillet 1994; Graybeal 1995 for references). A few remarks are still in order. There are two elements in the concept of monophyly, (1), an ancestor and all its descendents, and (2), a entity whose members are more closely related to one another than to members of any other entities (e.g. de Queiroz & Donoghue 1990); the two do not coincide when relationships are reticulating. Avise (1994, pp. 128, 133) even suggests that for a gene and its alleles the usual course of speciation (here the splitting of a lineage in some fashion) entails a transition polyphyly-paraphyly-monophyly. "Reciprocal monophyly" at the level of a gene (all alleles of the gene in different lineages are there each others' closest relatives), "genome monophyly", and synapomorphies, are different albeit related concepts. If there are balanced polymorphisms in the lineages, genome-level monophyly may be attained only tens of millions of years after splitting of a lineage; cladistic monophyly might seem applicable after any splitting or budding of a lineage (the fact that the split or bud may cease to exist next year is not the point here). Monophyly, and paraphyly in particular, may quite simply be inappropriate concepts at the level of species (e.g. Nixon & Wheeler 1990).

Hence to argue from a poorly-defined notion of species-level paraphyly to the demise of classification[clad] seems a trifle premature. Furthermore, the absence of any examples of such ancestors does not make for a compelling argument.

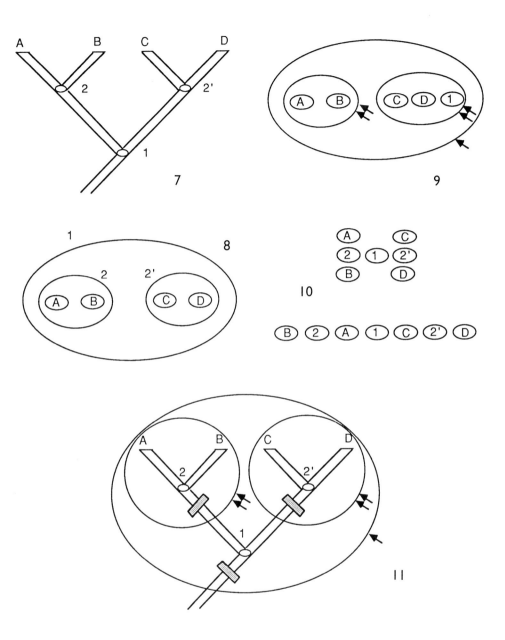

Figs 7 – 11. Ancestors, descendents, and classification. Fig. 7: Phylogeny of four species; 1, common ancestor of the whole group; 2, ancestor of species A and B; 2', ancestor of species C and D. Fig. 8: Paraphyly relationships of ancestors and descendants (see Brummitt 1996, Fig. 1a). Fig. 9: Hierarchical (Linnaean) classification; ellipses represent taxa, single arrow, family; two arrows, genus; an ancestor (1) included in genus with some of its descendants (see Brummitt 1996, cf. Fig. 1b: Brummit does not explain this position). Fig. 10: Representation of species and ancestors in some sort of phenetic space; non-arbitrary grouping compromised. Fig. 11: Ancestors, descendants, and higher taxa, with phylogeny superimposed on a Venn diagram; ellipses, taxa (see Fig. 9); grey bars, synapomorphies (those for terminal branches not shown).

CONCLUSION

Flora writers — and the collaborators in the Flora Malesiana project are a large and active group — do much of the systematic work carried out today, describing genera and families, not to mention hundreds of species. The easiest way for our work to be integrated with that of other biologists is to ensure that our higher taxa are monophyletic, and to interpret these taxa in an appropriate fashion as we make recommendations about conservation, or discuss biogeography or historical ecology. Even accepting monophyly as an absolute criterion for grouping species, genera, etc., *we must learn to live with the absence in nature of any ranking criterion for our higher taxa; the level at which we recognise a particular taxon is a convention*. However, to have monophyly as a grouping criterion is a major advance, although it is asking too much of organisms that the criterion will always be found easy to apply. Of course, we can opt to maintain traditional criteria for group recognition. If we do, the groups we recognise will remain unclear, convenience and expedience will continue to play a subversive role in taxon delimitation, and as far as I am concerned, higher level classification will remain trivialised. In groups not yet revised for Flora Malesiana the names we use should reflect cladistic realignments, and in the final volume changes in familial and generic delimitation that have occurred since the inception of the project can be summarised. Such changes will indicate continuing progress in systematics.

ACKNOWLEDGEMENTS

I am very grateful to S. Davies, T. Eriksson, E. A. Kellogg, S. Madriñán, S. Renner, G. Romero and R. Spangler for discussion on a draft of the manuscript, R. K. Brummitt and M. J. Donoghue for arguments about naming, C. J. Humphries for a helpful review, and to participants of the Third Flora Malesiana Symposium for their comments.

REFERENCES

Adema, F. (1991). *Cupaniopsis* Radlk. (*Sapindaceae*): A Monograph. Rijksherbarium/ Hortus Botanicus, Leiden.

Adema, F., & Ham, R. W. J. M. van der (1993). *Cnesmocarpon* (gen. nov.), *Jagera* and *Trigonachras* (*Sapindaceae-Cupanieae*): Phylogeny & systematics. Blumea 38: 173 – 215.

Anderberg, A. A. (1993). Cladistic interrelationships and major clades of the Ericales. Pl. Syst. Evol. 184: 207 – 231.

Ashton, P. S. (1982). *Dipterocarpaceae*. In: C. G. G. J. van Steenis (ed.), Flora Malesiana, vol. 9(2), pp. 237 – 552, Martinus Nijhoff, The Hague.

Atkinson, R., Jong, K. & Argent, G. (1995). Cytotaxonomic observations in tropical *Vaccinieae* (*Ericaceae*). Bot. J. Linn. Soc. 117: 135 – 145.

Avise, J. C. (1994). Molecular Markers, Natural History & Evolution, xiv + 511 pp. Chapman & Hall, New York & London.

Barkworth, M. E. (1992). Taxonomy of the *Triticeae*: a historical perspective. Hereditas 116: 1 – 14.

Bather, F. A. (1927). Biological classifications: past & future. Quart. J. Geol. Soc. 83: lxii – civ.

Bentham, G. (1875). On the recent progress & present state of systematic botany. Rep. British Assoc. Advancem. Sci. (1874): 27 – 54.

Berlin, B. (1992). Ethnobiological Classification: Principles of Categorization of Plants & Animals in Traditional Societies, xvii + 335 pp., Princeton University Press, Princeton.

Bower, G. H., Clark, M. C., Lesgold, A. M. & Winzenz, D. (1969). Hierarchical retrieval schemes in recall of categorised word lists. J. Verbal Learning & Verbal Behavior 8: 323 – 343.

Brown, R. (1818). Observations, systematical and geographical, on Professor Christian Smith's collection of plants from the vicinity of the river Congo. In: J. K. Tuckey, Narrative of an Expedition to Explore the River Zaire..., pp. 420 – 488, John Murray, London.

Brummitt, R. K. (1996). In defence of paraphyletic taxa. In: L. J. G. van der Maesen *et al.* (eds), The Biodiversity of African Plants, pp. 1 – 14, Kluwer,

Carter, J. E. & Gluck, M. A. (1992). Explaining basic categories: Feature predictability and information. Psychol. Bull. 111: 291 – 303.

Clark, L. G., Zhang, W. & Wendel, J. F. (1995). A phylogeny of the grass family (Poaceae) based on *ndh*F sequence data. Syst. Bot. 20: 436 – 460.

Clayton, W. D. (1974). The logarithmic distribution of angiosperm families. Kew Bull. 29: 271 – 279.

Coley, J. D., Medin, D. L., Atran, S. & Lynch, S. (1996). Does privilege have its rank? Folk biology and induction in two cultures. Cogn. Psychol., submitted.

Crayn, D. M., Kron, K. A., Gadek, P. A. & Quinn, C. J. 1995. Delimitation of *Epacridaceae*: preliminary molecular evidence. Ann. Bot. 77: 317 – 321.

Cronquist, A. (1983). Some realignments in the dicotyledons. Nordic J. Bot. 3: 75 – 83.

Cronquist, A. (1987). A botanical critique of cladism. Bot. Rev. 53: 1 – 52.

Cronquist, A. (1988). The Evolution and Classification of Flowering Plants. Ed. 2, viii + 555 pp, New York Botanical Garden, New York.

Davis, P. H. & Heywood, V. (1963). Principles of Angiosperm Taxonomy, xx + 556 pp., Oliver and Boyd, Edinburgh.

Dayanandan, S. (1995). Phylogeny of the tropical tree family *Dipterocarpaceae* based on nucleotide sequences of the chloroplast *rbc*L gene and morphology. Ph. D. Thesis, Boston University.

Donoghue, M. J. & Cantino, P. D. (1988). Paraphyly, ancestors and the goals of taxonomy: A botanical defense of cladism. Bot. Rev. 54: 107 – 128.

Donoghue, M. J. & Sanderson, M. J. (1994). Complexity and homology in plants. In: B. K. Hall (ed.), Homology: The Hierarchical Basis of Comparative Biology, pp. 393 – 421, Academic Press, San Diego.

Erbar, C. (1994). Contributions to the affinities of *Adoxa* from the viewpoint of phylogenetic development. Bot. Jahrb. Syst. 16: 259 – 283.

Ericsson, K. A. & Polson. P. G. (1988). Cognitive analysis of exceptional memory for restaurant orders. In: M.T.H. Chi, R. Glaser & M. J. Farr (eds), The Nature of Expertise, pp. 23 – 70, Lawrence Erlbaum, Hillsdale, New York.

Eriksson, T. 1990. Reinstatement of the genus *Leucoblepharis* Arnott (*Asteraceae-Heliantheae*). Bot. Jahrb. Syst. 112: 167 – 191.

Estes, J. R. & Tyrl, R. J. (1987). Concepts of taxa in the *Poaceae*. In: T. R. Soderstrom *et al.* (eds), Grass Systematics and Evolution, pp. 325 – 33, Smithsonian Institution, Washington, D. C.

Fernando, E. S., Gadek, P. A. & Quinn, C. J. (1995). *Simaroubaceae*, an artifical construct: evidence from *rbc*L sequence variation. Amer. J. Bot. 82: 92 – 103.

Fernando, E. S. & Quinn, C. J. 1995. *Picramniaceae*, a new family, and a recircumscription of *Simaroubaceae*. Taxon 44: 177 – 181.

Forey, P. L. (1992). Formal classification. In: P. L. Forey *et al.* (eds), Cladistics, a Practical Course in Systematics, pp. 160 – 169, Clarendon Press, Oxford.

Funk, V. A. (1985). Cladistics and generic concepts in the *Compositae*. Taxon 34: 72 – 80.

Gelman, S. A., Wilcox, S. A. & Clark, E. V. 1989. Conceptual and lexical hierarchies in young children. Cogn. Develop. 4: 309-326.

Gentry, A. H. (1993). A Field Guide to the Families and Genera of Woody Plants of Northwest South America..., xxiii + 895 pp., Conservation International, Washington, D. C.

Ghiselin, M. T. (1995). Ostensive definitions of the names of species and clades. Biol. and Phil. 10: 219 – 222.

Gift, N. & Stevens, P. F. (1997). Vagaries in the delimitation of character states in quantitative variation — an experimental study. Syst. Biol., 46: 112 – 125.

Graham, S.A., Crisci, J. V. & Hoch, P. C. (1993). Cladistic analysis of the *Lythraceae sensu lato* based on morphological characters. Bot. J. Linn. Soc. 113: 1 – 33.

Graybeal, A. (1995). Naming species. Syst. Biol. 44: 237 – 250.

Hawksworth, D. L. & Bisby, F. A. (1988). Systematics: The keystone of biology. In: D. L. Hawsworth (ed.), Prospects in Systematics, pp. 3 – 30, Clarendon Press, Oxford.

Hedberg, O. (1995). Cladistics in taxonomic botany — master or servant? Taxon 44: 3 – 10.

Hempel, C. G. (1965). Aspects of Scientific Explanation and Other Essays in the Philosophy of Science, ix + 505 pp, Free Press, New York.

Hennig, W. (1966). Phylogenetic Systematics, 263 pp., University of Illinois Press, Urbana, Illinois.

Humphries, C. J. (1991). The implications of pragmatism in systematics. In: D.L. Hawksworth (ed.), Improving the Stability of Names: Needs and Options, pp. 313 – 322, Koeltz, Königstein.

Humphries, C. J. & Williams, P. H. (1994). Cladograms and trees in biodiversity. In: R.W. Scotland, D. J. Siebert & D. M. Willams (eds), Models in Phylogeny Reconstruction, pp. 335 – 352, Systematics Association special volume 52, Clarendon Press, Oxford.

Hunn, E. (1994). Place names, populaton density, and the magic number 500. Current Anthropology 35: 81 – 85, 232 (erratum).

Johns, R. J. (1995). Endemism in the Malesian flora. Curtis's Bot. Mag. 12: 95 – 110.

Johnson, L. A. S. & Wilson, K. L. (1993). *Casuarinaceae*. In: K. Kubitzki, J. G. Rohwer & V. Bittrich (editors). The Families and Genera of Vascular Plants. II. Flowering Plants — Dicotyledons. Magnoliid, Hamamelid and Caryophyllid Families, pp. 237 – 242, Springer, Berlin.

Judd, W. S. (1982). A taxonomic revision of *Pieris* (*Ericaceae*). J. Arnold Arbor. 63: 103 – 144.

Judd, W. S., Sanders, R. W. & Donoghue, M. J. (1994). Angiosperm family pairs: preliminary phylogenetic analyses. Harvard Papers in Botany 5: 1 – 51.

Just, T. (1953). Generic concepts and modern taxonomy. Chron. Bot. 14: 103 – 114.

Keil, F. C. (1989). Concepts, Kinds and Cognitive Development, xv + 328 pp., M. I. T. Press, Cambridge, Mass.

Kellogg, E. A. (1990). Ontogenetic studies of florets in *Poa* (*Gramineae*): allometry and heterochrony. Evolution 44: 1978 – 1989.

Kelly, K. M. (1994). On the magic number 500: an expostulation. Curr. Anthropol. 35: 435 – 438.

Kittredge, W. (1985). Notes on the orchid flora of New Guinea 1. Bot. Mus. Leafl. 30(2): 95 – 102.

Kornet, D. J. (1993). Reconstructing Species: Demarcations in Genealogical Networks, 120 pp., Instituut voor Theoretische Biologie, Rijksherbarium/Hortus Botanicus, Leiden.

Kron, K. A. & Chase, M. W. (1993). Systematics of the *Ericaceae*, *Empetraceae*, *Epacridaceae* and related taxa based upon *rbc*L sequence data. Ann. Missouri Bot. Gard. 80: 735 – 741.

Krumsahl, C. L. (1978). Concerning the applicability of geometric models to similarity data: the interrelationship between similarity and spatial density. Psychol. Rev. 85: 445 – 463.

Kubitzki, K. & Gottlieb, O. R. (1984). Micromolecular patterns and the evolution and major classification of the angiosperms. Taxon 33: 375 – 391.

Laubenfels, D. J. de. (1988). Coniferales. In: W. J. J. O. de Wilde (editor). Flora malesiana, ser. 1, vol. 10, pp. 337 – 453, Kluwer, Dordrecht.

Leenhouts, P. W. (1962). *Loganiaceae*. In: C. G. G. J. van Steenis (editor). Flora Malesiana, ser. 1, vol. 6, pp. 293 – 387, Walters-Noordhoff, Groningen.

Mayr, E. & Ashlock, P. D. (1991). Principles of Systematic Zoology, ed. 2, xvi + 475 pp., McGraw Hill, New York.

McVaugh, R. (1945). The genus *Tridanis* Rafinesque, and its relationships to *Specularia* and *Campanula*. Wrightia 1: 13 – 52.

Morgan, D. R. & Soltis, D. E. (1993). Phylogenetic relationships among members of Saxifragaceae sensu lato based on *rbc*L sequence data. Ann. Missouri Bot. Gard. 80: 631 – 66.

Morgan, D R., Soltis, D.E. & Robertson, K.R. (1994). Systematic and evolutionary implications of *rbc*L sequence variation in the *Rosaceae*. Amer. J. Bot. 81: 890 – 903.

Morin, N. R., Whetstone, R. D., Wilken, D. & Tomlinson, K. L. (1989). Floristics for the 21st Century, Monographs in Systematic Botany 28, Missouri Botanical Garden, St Louis.

Nixon, K. C. & Wheeler, Q. D. (1990). An amplification of the phylogenetic species concept. Cladistics 6: 211 – 223.

Nooteboom, H. P. (1962). *Simaroubaceae*. In: C. G. G. J. van Steenis (ed.), Flora Malesiana, ser. 1, vol. 6, pp. 192 – 226, Walters-Noordhoff, Groningen.

Oliver, E. G. H. (1991). The *Ericoideae* (*Ericaceae*): A review. Contrib. Bolus Herb. 13: 158 – 208.

Olmstead, R. G. & Reeves, P. A. (1995). Evidence for the polyphyly of the *Scrophulariaceae* based on chloroplast *rbc*L and *ndh*F sequences. Ann. Missouri Bot. Gard. 82: 176 – 193.

Orchard, A. E. (1981). The generic limits of *Ixodia* R. Br. ex Ait. (*Compositae–Inuleae*). Brunonia 4: 185 – 197.

Patterson, C. (1981). Significance of fossils in determining evolutionary relationships. Ann. Rev. Ecol. Syst. 12: 195 – 223.

Peng, C. I., Tobe, H. & Takahashi, M. (1995). Reproductive morphology and relationships in *Triplostegia* (Dipsacales). Bot. Jahrb. Syst. 116: 505 – 516.

Queiroz, K. de. (1995). The definitions of species and clade names: A reply to Ghiselin. Biol. and Phil. 10: 223 – 228.

Queiroz, K. de & Donoghue, M. J. (1990). Phylogenetic systematics or Nelson's version of cladistics? Cladistics 6: 61 – 75.

Queiroz, K. de & Gauthier, J. (1994). Toward a phylogenetic system of biological nomenclature. Trends Ecol. Evol. 9: 27 – 31.

Rieseberg, L. & Brouillet, L. (1994). Are many plant species paraphyletic? Taxon 43: 21 – 32.

Rips, L. J. (1989). Similarity, typicality and categorisation. In: J. Vosniadou & A. Ortony (eds), Similarity and Analogical Reasoning, pp. 21-59, Cambridge University Press, Cambridge.

Rollins, R. C. (1953). Cytogenetical approaches to the study of genera. Chron. Bot. 14: 133 – 139.

Schlechter, R. (1919). *Orchidaceae* novae et criticae. Decas LIX – LXIII. Additamenta ad orchideologiam papuanum II. Rep. Spec. Noc. Regni Veget. 16: 103 – 131.

Skála, Z. & Zrzavy, J. (1995). Phylogenetic reticulations and cladistics: discusssion of methodological concepots. Cladistics 10: 305 – 313.

Soltis, D. E., Morgan, D. R., Grable, A., Soltis, P. S. & Kuzoff, R. (1993). Molecular systematics of *Saxifragaceae* sensu stricto. Amer. J. Bot. 80: 1050 – 1061.

Stebbins, G. L. (1956). Taxonomy and evolution of genera, with special reference to the family *Gramineae*. Evolution 10: 235 – 245.

Steenis, C. G. G. J. van. (1978). On the doubtful virtue of splitting families. Bothalia 12: 425 – 427.

Stevens, P. F. (1972). Notes on the infrageneric classification of *Agapetes*, with four new taxa from New Guinea. Notes Roy. Bot. Gard. Edinburgh 32: 13 – 28.

Stevens, P. F. (1984). Homology and phylogeny: morphology and systematics. Syst. Bot. 9: 395 – 409.

Stevens, P. F. (1986). Evolutionary classification in botany 1960 – 1985. J. Arnold Arbor. 67: 313 – 339.

Stevens, P. F. (1990). Nomenclatural stability, taxonomic instinct, and flora writing — a recipe for disaster? In: P. Baas, K. Kalkman and R. Geesink (editors). The Plant Diversity of Malesia, pp. 387 – 410, Kluwer, Dordrecht.

Stevens, P. F. (1994a). [Review of Berlin, 1992]. Syst. Biol. 43: 293 – 295.

Stevens, P. F. (1994b). The Development of Biological Systematics: Antoine-Laurent de Jussieu, Nature and the Natural System, xxiii + 616 pp., Columbia University Press, New York.

Stevens, P. F. (1997). How to interpret botanical classifications — suggestions from history. BioScience 47: 242 – 250.

Struwe, L. & Albert, V. A. (1995). Phylogeny and classification of *Gentianaceae*. 1. *Anthocleista*, *Fagraea* and *Potalia* (*Potalieae*). Brittonia.

Struwe, L., Albert, V. A. & Bremer, B. (1995). Cladistics and family level classification of the Gentianales. Cladistics 10: 175 – 206.

Stuessy, T. F. (1990). Plant Taxonomy: The Systematic Evaluation of Comparative Data, xvii + 514 pp., Columbia University Press, New York.

Thorne, R. F. (1976). A phylogenetic classification of the Angiospermae. Evol. Biol. 9: 35 – 106.

Turner, B. L. 1985. A summing up. Taxon 34: 85 – 88.

Vilgalys, R. & Hibbett, D. S. (1993). Phylogenetic classification of fungi and our Linnaean heritage. In: D. R. Reynolds & J. W. Taylor (eds), The Fungal Holomorph: Mitotic, Meiotic and Pleomorphic Speciation in Fungal Systematics, pp. 255 – 260, C.A.B. International, Wallingford.

Wagner, W. H. & Beitel, J. M. (1992). Generic classification of modern North American *Lycopodiaceae*. Ann. Missouri Bot. Gard. 79: 676 – 686.

Welzen, P. van. (1989). *Guioa* Cav. (*Sapindaceae*): Taxonomy, Phylogeny and Historical Biogeography. Leiden Botanical Series 12. Rijksherbarium/Hortus Botanicus, Leiden.

Wilde, W. J. J. O. de. (1994). *Paramyristica*, a new genus of *Myristicaceae*. Blumea 39: 341 – 350.

Wiley, E. O. (1979). An annotated Linnaean hierarchy, with comments on natural taxa and competing systems. Syst. Zool. 28: 308 – 337.

Wiley, E. O. (1981). Phylogenetics, the Theory and Practice of Phylogenetic Systematics, xv + 439 pp., John Wiley, New York.

Zhang, H. -B. & Dvorák, J. (1991). The genome origin of tetraploid species of *Leymus* (*Poaceae*: *Triticeae*) inferred from variation in repeated nucleotide sequences. Amer. J. Bot. 78: 871 – 884.

Note added in proof

An undescribed species of *Agapetes* has recently been collected on Bougainville and there is a further undescribed species from Papua New Guinea mainland; when these species are described, the genus *Paphia* will be reinstated.

Floristics, cladistics, and classification: three case studies in Gentianales

Lena Struwe & Victor A. Albert[1]

Summary. Gentianales are a monophyletic group of Asteridae that is well represented in Malesia. Previous classifications of the order have suffered from the circumscription of *Loganiaceae*, which is polyphyletic. *Buddleja*, *Desfontainia*, *Plocosperma*, *Polypremum*, and *Retzia* are misplaced within Gentianales and new positions are suggested. The gentianalean members of *Loganiaceae* are themselves paraphyletic, showing closest relationships to other families of the order. Monophyly is preserved if *Loganiaceae* are divided into four segregate families: *Gelsemiaceae*, *Geniostomaceae*, *Loganiaceae* sensu stricto, and *Strychnaceae*. *Apocynaceae* (including *Asclepiadaceae*), *Gentianaceae* (including *Loganiaceae-Potalieae*), and *Rubiaceae* remain as monophyletic families, and a new familial key is provided. Species relationships within *Strychnaceae* are investigated, and endemic Malesian *Norrisia* is shown to have amphipacific, disjunct relationships to the South American genera *Antonia* and *Bonyunia*. Malesian *Potalieae* show a similar disjunct pattern: paraphyletic *Fagraea* is basal to African/Malagasy *Anthocleista*, which is in turn paraphyletic to American *Potalia*. The results of an *ad hoc* molecular clock experiment place their divergence as contemporaneous with *Strychnaceae*, between c. 45 and 160 million years before present, depending upon the rate assumption used. Although the younger date is coincident with the proposed Tertiary breakup of a boreotropical continuity, the older date, which supports cladogenesis before the breakup of Gondwana, fits the area cladograms better and suggests an ancient (Jurassic) derivation of modern clades. Recognition of phylogenetic patterns within Gentianales, *Strychnaceae*, and *Potalieae* underscores the utility of a cladistic approach to classification, floristics, and biogeography of Malesian plants.

Introduction: floristics and cladistics

The primary components of floristic studies are the plant species of an area. The description and identification of species have been the major issues for most floristic works over the centuries. However, not only are the species in a flora of interest, but the relationships of that flora to other floras are as well. When taxon relationships are known, conclusions (e.g., biogeographic) about inter-flora relationships can be made. For this purpose we need to know more about a species than its name and distinguishing characters. We will need to know how this particular species is related to species of the same genus, tribe, and family. In this paper three case studies in the angiosperm order Gentianales are discussed to show the utility of a phylogenetic (i.e. cladistic) perspective to floristics: (i) a study of the relationships of the families of Gentianales (Struwe *et al.* 1994), and two species level studies in (ii) *Strychnaceae* (presented here), and (iii) *Potalieae* (*Gentianaceae*; Struwe & Albert, unpublished).

---

[1]The New York Botanical Garden, Bronx, New York 10458-5126, U.S.A.

The concept of diversity and how it should be interpreted and estimated is becoming increasingly important with the destruction, conservation, and reconstruction of biotas around the world. How diversity is best estimated will not be discussed here (cf. Humphries & Williams 1994), but some notions on how taxonomy influences the estimates can be made. There are at least two simple estimates of diversity as viewed through taxonomy: the number of species in an area, and the number of higher taxonomic groups. The first estimate will always be the same regardless of generic, tribal, and familial position of individual species. The latter, however, has a phylogenetic component and could show dramatic changes between different classifications, for example in the number of endemic genera and families. Therefore, phylogenetic hypotheses will influence hierarchy-dependent diversity estimates, so long as classifications reflecting these are provided as well.

Cladistics provides systematists with explicit hypotheses of taxon relationships in the form of a tree. It is also the method for which the result represents the information content in the data with the greatest predictivity (Farris 1979, 1983). Subsequent to the analysis it is possible to identify monophyletic groups. A group is monophyletic if its taxa have synapomorphies (shared derived features) and an exclusive, common origin on a cladogram. However, many of today's recognised genera (and tribes, subfamilies, families, etc.) are para- or polyphyletic, that is, they are linked by symplesiomorphies (shared primitive features) and do not have a cladistic origin that is unique to the group. Usually this is caused by recognition of one well-distinguished monophyletic group (A) that is nested inside another group (B), causing the latter (B) to be paraphyletic. There are two alternative treatments of paraphyletic groups. The first one is a lumper's view, in which the derived group (A) is combined with the paraphyletic one (B), resulting in one large group (A+B) bearing the name with priority. Another option is to split up the paraphyletic genus (B) into smaller monophyletic genera (B, C, D, etc.), thereby retaining the derived genus (A). These two ways of classifying will affect diversity differently; lumping will cause a decrease and splitting an increase in hierarchy-dependent estimates of diversity, and combinations of lumping and splitting in a group will have a complex effect. In practical terms, all such divisions should preferably coincide with the distribution of macromorphological characters that can be used in keys. There are several well-known examples in which derived genera have been included in other genera to preserve monophyly, for example the inclusion of *Lycopersicon* in *Solanum* (Spooner *et al.* 1993) and *Heterogaura* in *Clarkia* (Lewis & Raven 1992; Sytsma & Smith 1992). In both examples, the majority of shared characteristics permit an easy diagnosis; only uniquely derived features differ, but no longer the names. The Gentianales, a major component of the Malesian flora, are discussed below in this light.

GENTIANALES: PHYLOGENETIC DATA

Traditionally, five families have been recognised in Gentianales: *Apocynaceae, Asclepiadaceae, Gentianaceae, Loganiaceae,* and *Rubiaceae.* In Leenhouts' (1963) treatment of *Loganiaceae* for Flora Malesiana a broad circumscription of the family was adopted (cf. Bentham 1857; Solereder 1892 – 1895). The classification of the heterogeneous *Loganiaceae* has been a controversial issue ever since von Martius' (1827) description of the family. Recent classifications have included anywhere from one genus (Takhtajan 1997) to 29 genera in ten tribes (Leeuwenberg & Leenhouts 1980).

Cladistic studies based on morphological, anatomical, and phytochemical data have shown that *Loganiaceae* in the broad sense are paraphyletic (Fig. 1; Struwe *et al.* 1994). Some genera showed closer relationships to Dipsacales (*Desfontainia*) and Scrophulariales (*Retzia, Buddleja, Polypremum,* and *Plocosperma*; Table 1). The

TABLE 1. Classification to order, family, and tribe of genera formerly included in *Loganiaceae* sensu Leeuwenberg & Leenhouts (cf. Struwe & Albert in Struwe *et al.* 1994). Genera distributed in the Flora Malesiana area are printed in bold.

Genus	Order	Family	Tribe
Desfontainia	Dipsacales[1]	*Desfontainiaceae*[2]	
Gelsemium	Gentianales	*Gelsemiaceae*	
Mostuea	Gentianales	*Gelsemiaceae*	
Geniostoma	Gentianales	*Geniostomaceae*	
Labordia	Gentianales	*Geniostomaceae*	
Fagraea	Gentianales	*Gentianaceae*	*Potalieae*
Anthocleista	Gentianales	*Gentianaceae*	*Potalieae*
Potalia	Gentianales	*Gentianaceae*	*Potalieae*
Logania	Gentianales	*Loganiaceae*	
Mitreola	Gentianales	*Loganiaceae*	
Mitrasacme[3]	Gentianales	*Loganiaceae*	
Antonia	Gentianales	*Strychnaceae*	*Antonieae*
Bonyunia	Gentianales	*Strychnaceae*	*Antonieae*
Norrisia	Gentianales	*Strychnaceae*	*Antonieae*
Usteria	Gentianales	*Strychnaceae*	*Antonieae*
Spigelia	Gentianales	*Strychnaceae*	*Spigelieae*
Gardneria	Gentianales	*Strychnaceae*	*Strychneae*
Neuburgia	Gentianales	*Strychnaceae*	*Strychneae*
Strychnos .	Gentianales	*Strychnaceae*	*Strychneae*
Androya	Scrophulariales	*Buddlejaceae*	
Buddleja	Scrophulariales[4]	*Buddlejaceae*	
Emorya	Scrophulariales	*Buddlejaceae*	
Peltanthera	Scrophulariales	*Buddlejaceae*	
Nuxia	Scrophulariales	*Buddlejaceae*	
Gomphostigma	Scrophulariales	*Buddlejaceae*	
Sanango	Scrophulariales	*Gesneriaceae*[5]	
Plocosperma	Scrophulariales	*Plocospermataceae*	
Polypremum	Scrophulariales	*Scrophulariaceae*	
Retzia	Scrophulariales	*Stilbaceae*[6]	

[1] cf. Bremer *et al.* (1994), and Backlund (1996)

[2] *Desfontainiaceae* was recently included in *Columelliaceae* (Backlund 1996). Most characters that support this relationship are anatomical, embryological, cytological, and molecular; the only synapomorphies that could possibly be used as field characters are evergreen leaves and conduplicate leaf vernation, but these are also present in related taxa (Backlund 1996). *Columellia* and *Desfontainia* are macromorphologically very different (cf. Lindley 1853; Leeuwenberg & Leenhouts 1980), and they have never before been included in the same family. To preserve taxonomic stability, and because of the lack of diagnostic field characters, we have chosen to maintain *Desfontainiaceae* separately.

[3] incl. segregate genera *Phyllangium* and *Schizacme* (Dunlop 1996)

[4] cf. Bremer *et al.* (1994)

[5] Dickison (1994), Jensen (1994), Norman (1994), Wiehler (1994)

[6] cf. Bremer *et al.* (1994)

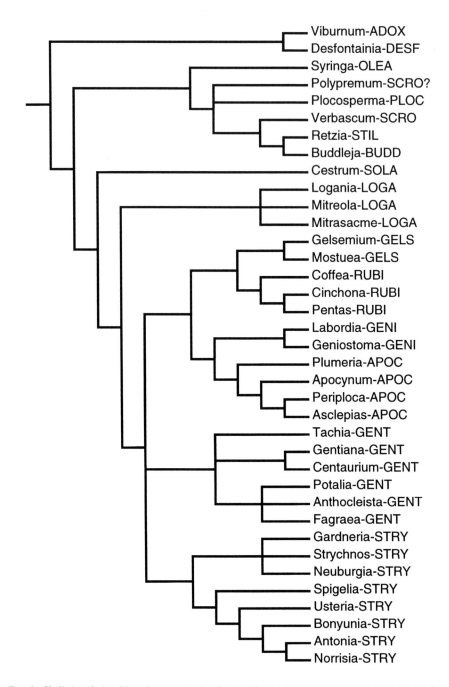

FIG. 1. Cladistic relationships of genera in the Gentianales (strict consensus tree adapted from Struwe *et al.* 1994). Abbreviations in capitals refer to families as follows: ADOX = *Adoxaceae*, APOC = *Apocynaceae*, BUDD = *Buddlejaceae*, DESF = *Desfontainiaceae*, GELS = *Gelsemiaceae*, GENI = *Geniostomaceae*, GENT = *Gentianaceae*, LOGA = *Loganiaceae*, OLEA = *Oleaceae*, PLOC = *Plocospermataceae*, RUBI = *Rubiaceae*, SCRO = *Scrophulariaceae*, SOLA = *Solanaceae*, STIL = *Stilbaceae*, STRY = *Strychnaceae*.

remaining genera of the *Loganiaceae* were distributed among several clades, some of which included genera from *Apocynaceae, Gentianaceae,* and *Rubiaceae* (respectively), thereby making *Loganiaceae* paraphyletic (Fig. 1). To maintain the monophyly of all families in Gentianales a "splitter" reclassification of the *Loganiaceae* was necessary (Struwe & Albert in Struwe *et al.* 1994). The four segregate families (*Loganiaceae, Gelsemiaceae, Geniostomaceae,* and *Strychnaceae*) are presented in a new worldwide key to all families of Gentianales (Appendix 1).

A. *Loganiaceae*

Eleven Malesian genera were included in *Loganiaceae* by Leenhouts (1963). Following cladistic studies, these genera are distributed among six families, reducing the heterogeneity of the *Loganiaceae* drastically (Table 1; Struwe & Albert in Struwe *et al.* 1994). Of these, *Geniostomaceae, Gelsemiaceae, Buddlejaceae,* and *Strychnaceae* are new for the Flora Malesiana project, although *Buddlejaceae* had been recognised earlier in floristic works of adjacent areas (Leeuwenberg & Vidal 1972; Tirel-Roudet 1972).

Of the formerly quite large family *Loganiaceae,* the herbaceous genera *Mitreola* and *Mitrasacme* are the only recognised Malesian members (Struwe & Albert in Struwe *et al.* 1994). The third genus *Logania* comprises herbs and subshrubs from Australia, New Zealand and New Caledonia (Conn *et al.* 1996). *Mitrasacme* is also mainly distributed in the Australian region but a few widespread species reach Malesia, India, and Japan. *Phyllangium* and *Schizacme,* two Australian endemic genera segregated from *Mitrasacme,* were recently described (Dunlop 1996). *Mitreola* has one widespread species in the tropics as well as endemic species in America, Madagascar, Southeast Asia, and Malesia (Leeuwenberg & Leenhouts 1980). *Mitreola* and *Mitrasacme* are cladistically placed as sister taxa (Struwe *et al.* 1994), sharing features such as a herbaceous habit, 2-horned capsules opening by apical slits, (usually) semi-inferior ovaries, and two separate (or nearly) separate styles, whereas differing in the number of corolla lobes and secondary leaf nerves (Leenhouts 1963; Leeuwenberg & Leenhouts 1980). In the cladistic study by Struwe *et al.* (1994), there was some suggestion that *Logania* could be either paraphyletic to *Mitreola* and *Mitrasacme* or their sister taxon. Finally, it should be noted that bootstrap support[1] for this familial circumscription is ambiguous, which could indicate altered relationships as new data (e.g. molecular) are added.

---

[1] Bootstrap support estimates are provided as measures of the data-dependent strength of particular taxon groupings and classifications. Bootstrap resampling analyses of group support were performed using PAUP 3.1.1 (Swofford 1993), following the same protocol used to compare PAUP bootstrapping with parsimony jackknifing (Farris *et al.* 1996): MULPARS off, COLLAPSE on, no branch swapping, random data addition sequence (one each per 1000 bootstrap replicates). Bootstrap values of 50% or below indicate ambiguous support, values between 50% and c. 63% indicate some robustness to extra steps ("Bremer support"), and values above 63% suggest the possibility of support by at least one uncontradicted synapomorphy. Note that bootstrapping, unlike jackknifing, is sensitive to constant columns and autapomorphies; also, results may be affected by PAUP's retention of certain zero-length branches (see Farris *et al.* 1996). Note also that bootstrapping is most frequently applied to molecular data, which are less constrained than morphological data in terms of numbers of characters that could show uncontradicted group support.

B. Geniostomaceae

Another familial segregate of the *Loganiaceae* is the recently described *Geniostomaceae*, distributed in the Mascarenes, Southeast and eastern Asia, Malesia, Australia, and the Pacific. The family consists only of two genera, *Geniostoma* and *Labordia* (the latter is endemic to Hawaii). An unique characteristic for *Geniostomaceae* is a woody capsule with abscising valves that expose the seeds embedded in persistent, pulpy, and red placentas. Whether or not *Labordia* should be included in *Geniostoma* has been a subject of discussion (Leeuwenberg & Leenhouts 1980; Sherff 1939; cf. Conn 1980). In the analysis of Struwe *et al.* (1994) in which polymorphic taxa were recoded into monomorphic units (i.e. "subtaxa"; cf. Nixon & Davis 1991), *Labordia* grouped with the unisexual units of *Geniostoma*, making *Geniostoma* paraphyletic. *Geniostomaceae* were positioned as the sister group to genera from *Apocynaceae* (incl. *Asclepiadaceae*; Struwe *et al.* 1994). One alternative to describing a new family would have been to include *Geniostoma* and *Labordia* in *Apocynaceae*. However, these taxa lack latex, which is a principal key characteristic for *Apocynaceae*, present in all genera. Thus, recognizing *Geniostomaceae* preserved not only monophyly but also the utility of diagnostic characters and the traditional circumscription of a large, well-known family. *Geniostomaceae* are shown to be a well-supported group by bootstrap analysis of the Struwe *et al.* (1994) data matrix, appearing in 61% of replicates; close relationship to *Apocynaceae*, however, is only ambiguously supported.

C. Gelsemiaceae

Gelsemiaceae is the other recently described familial segregate from *Loganiaceae*, also resulting from the cladistic studies of Struwe *et al.* (1994). Distinctive characteristics for the family are white or yellow funnelform flowers with twice dichotomously branched stigmas, capsular fruits, and the presence of complex indole alkaloids (only of the C-17-type; Kisakürek & Hesse 1980). The family includes two genera, *Gelsemium* and *Mostuea*, both of which show disjunct distributions. *Gelsemium* (3 spp.) is a North American/Southeast Asian disjunct with one species in the Malesia area (Leenhouts 1963; Wyatt *et al.* 1993). *Mostuea* is a larger African genus with one species also present in northern South America (Surinam and Brazil; Leeuwenberg 1961c). In studies based on morphological and phytochemical data *Gelsemiaceae* is placed as the sister group to *Rubiaceae* (Struwe *et al.* 1994); however, recent analyses based on molecular data place it as sister group to *Apocynaceae* (Bremer 1996). *Gelsemiaceae* are strongly supported (88%) by bootstrap analysis of the Struwe *et al.* (1994) data matrix, although relationships to other families remain ambiguous.

D. Strychnaceae

Strychnaceae is represented by five genera in Malesia: *Norrisia* of tribe *Antonieae*, *Spigelia* of tribe *Spigelieae*, and *Gardneria*, *Neuburgia*, and *Strychnos* of tribe *Strychneae*. The family has been recognised by several botanists over the years (e.g. Don 1837 – 38; Link 1829 – 1833; and more recently, Hutchinson 1959, and Takhtajan (1997) and now incorporates eight tropical genera (Table 1; Struwe & Albert in Struwe *et al.* 1994). All of these genera were included in *Loganiaceae* by Leeuwenberg &

TABLE 2. Matrix of taxa and characters used in the phylogenetic analysis of *Strychnaceae* (see Appendix 2). The letter "a" in the matrix indicates taxa that are polymorphic (0&1) for specific binary characters, and missing data are indicated by question marks.

```
                               1         2         3         4         5
                      12345678901234567890123456789012345678901234567890 1
```

Antonia ovata	10110001200001?0100010101000000010111000101000001001
Bonyunia antoniifolia	1000000110001110101100020000001111111002101000000?11
Bonyunia aquatica	10000000?00011101011100?000000111111100210100000?11
Bonyunia minor	1010000a?000111a1011100?00000011111110021010000000??
Bonyunia superba	10100001?00011101011100?0000001111111002101000000???
Gardneria ovata	00?1000010010001110010010000011000000200010000002??
Gardneria multiflora	00?10000100000011100000100100 1?0000002003100000 02??
Neuburgia corynocarpa	00?10000100000a?100010010001001?1000120021000000?00
Neuburgia moluccana	00?100001000000?100010010000001?10001200010000??? 00
Norrisia maior	11?000012000101110011010000010011011100?100000 01001
Norrisia malaccensis	11?000012000101110011010000010011011100010000001?01
Spigelia anthelmia	001100001100?1a110100010000000001111111110001010000
Spigelia humboldtiana	a01?0000?100?a001010000?0000000?111111101000 10??? 00
Strychnos nux-vomica	00??1?11?00000111001100?0?00000000000020001010000211
Strychnos axillaris	10??1111?00000111001100?0001001000000200?10000??211
Strychnos ovata	00??0010???0000111101000?0a01000101000200?10100?????
Usteria guineensis	00100000100112111011111000000?00001111001010000 1?10
Labordia waiolani	10?1000110100110200110????0??0?1??001122?00001?0?10
Geniostoma rupestre	0???00001010000??0001001?10a001a?000112031000100?10

Leenhouts (1980). The monophyly of some tribes and genera of *Strychnaceae* remain ambiguous. *Antonieae* were paraphyletic to *Spigelieae* in some of Struwe *et al.*'s (1994) results, but *Spigelieae* were represented by only one taxon in their study (i.e. a result diagnosing either monophyly or paraphyly was not possible). Furthermore, in the analysis in which monomorphic subunits replaced polymorphic taxa, *Neuburgia* and *Gardneria* in tribe *Strychneae* were nested inside the large genus *Strychnos*, rendering that genus paraphyletic (Struwe *et al.* 1994).

To resolve further relationships in *Strychnaceae* (and especially in tribe *Antonieae*), a new matrix of 51 morphological, anatomical, and phytochemical characters for 17 species of *Strychnaceae* (ten Malesian) was compiled for this paper (Appendix 2, Table 2), and analysed with parsimony programs[1]. All species of the four genera in *Antonieae* were included. *Gardneria*, *Neuburgia*, and *Spigelia* were represented by two species each, and three Malesian *Strychnos* species were also included. The three species of *Strychnos* were chosen as representatives from different sections: *Strychnos nux-vomica* from *Strychnos* sect. *Strychnos*, *Strychnos axillaris* from *Strychnos* sect. *Penicillatae*, and *Strychnos ovata* from *Strychnos* sect. *Lanigerae* (following sectional circumscriptions as presented in Leeuwenberg & Leenhouts 1980) *Geniostoma rupestre* and *Labordia waiolani* (*Geniostomaceae*) were used as outgroups to orient the tree (Farris 1982).

Two[2] most-parsimonious trees were found with a length of 103 steps, a consistency index of 0.60 (including autapomorphies), and a retention index of 0.75. In the outgroup-oriented tree two major clades can be distinguished (Fig. 2). One clade contains all species from tribe *Strychneae*, with *Neuburgia* sister to *Gardneria* plus *Strychnos*. The other clade incorporates *Spigelieae* and *Antonieae*. The relationship between *Antonieae* and *Spigelieae* is supported by several unambiguous character changes, e.g., red or purple corollas (ch. 19, with a later reversal in *Antonia* and *Norrisia*) and stalked, axile placentas that are dry in mature fruits (ch. 35, 36). Within *Antonieae*, *Bonyunia* is sister to *Antonia* plus *Norrisia*, with *Usteria* placed basalmost. This arrangement is identical to Struwe *et al.*'s (1994) outgroup-oriented result (Fig. 1).

The two species of *Spigelia* (*Spigelieae*) are the sister group to the *Antonieae* clade, thereby making all tribes within *Strychnaceae* monophyletic in this study. Another classificatory alternative would therefore be to raise all strychnaceous tribes to familial rank, that is, *Antoniaceae*, *Spigeliaceae*, and *Strychnaceae* s.s. This would reduce the heterogeneity of *Strychnaceae*, which was very poorly supported in the bootstrap analysis of the Struwe *et al.* (1994) data matrix. Indeed, within *Strychnaceae*, only core *Antonieae* (*Antonia*, *Bonyunia*, and *Norrisia*) form a group with substantial bootstrap support (67%). *Antoniaceae*, *Spigeliaceae*, and *Strychnaceae* have been used recently with similar circumscriptions (cf. Hutchinson 1959; Takhtajan 1987, 1997), and this classification may prove to be preferable as more data (e.g. molecular) are analysed.

---

[1] The matrix was analysed using the computer program PAUP 3.1.1 (Swofford 1993) with the following options: MULPARS on, COLLAPSE on, TBR branch swapping, and random data addition sequence (100 replicates). Tree length and consistency and retention indices (Farris 1989) were computed using PAUP. Character-state changes unambiguously versus ambiguously mapped to particular tree branches were calculated using MacClade 3.0 (Maddison & Maddison 1992).

[2] PAUP reported four most-parsimonious trees. However, two of the resolutions are supported by a zero-length branches within *Bonyunia*, and therefore the actual number of most-parsimonious topologies are two.

D.1. *Strychneae*

The tribe *Strychneae* contains the large pantropical genus *Strychnos* (c. 190 spp.) together with *Gardneria* (5 spp.) and *Neuburgia* (10 – 12 spp.), both of which have more restricted distributions in the Indo-Pacific area (Leeuwenberg & Leenhouts 1980). *Strychnos* is well-known as the source of several poisonous indole alkaloids (derived from seco-iridoids), e.g. strychnine and brucine, which are used locally as fish and arrow poisons and also have many pharmacological uses (Bisset 1974; Samuelsson 1992).

In the species-level study presented here all genera of *Strychneae* are monophyletic, with *Strychnos* plus *Neuburgia* sister to *Gardneria* (Fig. 2). Indehiscent fruits (ch. 38) are an unambiguous feature supporting the monophyly of tribe *Strychneae*. Synapomorphies (all unambiguous changes) for the *Strychnos* species under study are acrodromous venation (ch. 7), pubescent sepals (ch. 15), and corollas with hairs on the outside (ch. 20). Monophyly of *Strychnos* is narrowly supported by bootstrap analysis, appearing in 51% of replicates. *Gardneria* and *Strychnos* are united by unambiguous characteristics such as persistent styles (ch. 33) and fleshy or leathery mesocarp (ch. 37). An unambiguous synapomorphy for *Neuburgia* is the presence of uniseriate-only rays in the wood (ch. 50), a character that shows homoplastic occurrence in *Spigelieae* and some taxa of *Antonieae* as well. *Neuburgia* is also distinguished from *Gardneria* and *Strychnos* by having drupaceous fruits with only one developing seed in each locule (Leeuwenberg & Leenhouts 1980). Bootstrap support for the monophyly of *Neuburgia* is, however, ambiguous. The monophyly of *Gardneria* is unambiguously supported in this analysis by rotate corollas (ch. 18, also in *Strychnos axillaris*), and coherent or connivent anthers (ch. 30, although the anthers are free in some other *Gardneria* spp.). Bootstrap support for the monophyly of *Gardneria* is substantial at 69%.

Another resolution of *Strychneae* appeared in Struwe *et al.*'s (1994) subtaxa analysis, where *Gardneria* grouped with subtaxa of *Strychnos* characterised by glabrous anthers and cupshaped seeds (e.g. *Strychnos curtisii*, which also has terminal inflorescences with rotate corollas like *Gardneria*; Leenhouts 1963; Struwe *et al.* 1994). *Neuburgia* was also placed inside *Strychnos*, positioned together with *Strychnos* subtaxa having an indumentum on the inside of the corolla as well as flattened seeds (e.g. *Strychnos colubrina*, *Strychnos nux-vomica* and *Strychnos villosa*, which also have terminal inflorescences and salverform corollas like *Neuburgia*; Leenhouts 1963). The positions of *Gardneria* and *Neuburgia* therefore made *Strychnos* paraphyletic in that study. Indeed, *Gardneria* species show no features that do not occur in some species of *Strychnos*, and the unique characters of *Neuburgia* (deciduous style, drupaceous fruits, and different wood anatomy; see Mennega 1980) might simply be derived traits within *Strychnos*. In any case, bootstrap support from both the Struwe *et al.* (1994) and present data matrices is ambiguous for tribe *Strychneae* as well as for any particular arrangement of its three genera. Further studies, perhaps including molecular data and with extended taxon sampling from polymorphic *Strychnos*, may help to clarify generic relationships.

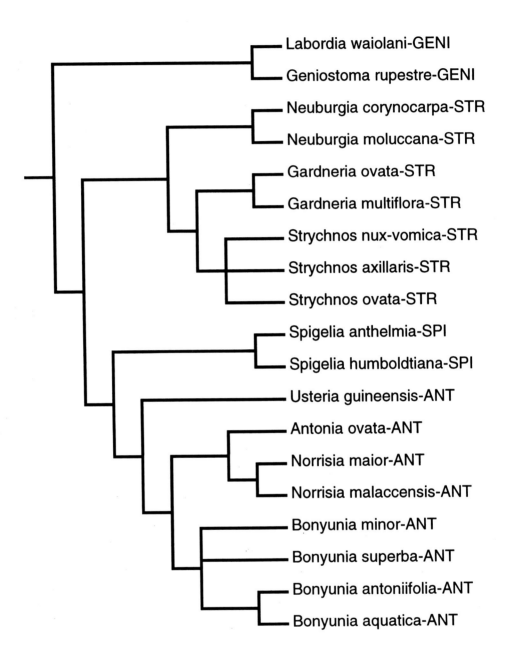

Fig. 2. Cladistic relationships of species of *Strychnaceae*, particularly within tribe *Antonieae*, oriented with *Geniostoma rupestre* and *Labordia waiolani* (both *Geniostomaceae*) as outgroups. Shown is the strict consensus tree of two equally most-parsimonious trees. Familial and tribal groupings are indicated with capital letters according to the following: ANT = tribe *Antonieae*, GENI = *Geniostomaceae*, SPI = tribe *Spigelieae*, and STR = tribe *Strychneae*.

D.2. *Spigelieae*

Spigelia is the only genus in tribe *Spigelieae* (Tab. 1) and is native to the Neotropics, but one species, *Spigelia anthelmia*, is a naturalised weed in western Africa and Malesia (Leeuwenberg 1961b; Leenhouts 1963). Unambiguous character changes for *Spigelia* are cincinnous inflorescences (ch. 10), glabrous corollas (ch. 21), pubescent ovaries (ch. 34), capsules with deciduous valves and a persistent cupular base (ch. 39), ruminate endosperm (ch. 45), and uniseriate-only rays in the wood (ch. 50). Takhtajan (1997) included *Mitreola* and *Mitrasacme* in *Spigeliaceae*, but this is not supported by cladistic studies (Struwe *et al.* 1994). Bootstrap support from the present data matrix is strong for generic monophyly (92%) but ambiguous with respect to relationships with other *Strychnaceae*.

D.3. *Antonieae*

The *Antonieae* clade (Fig. 2) is unambiguously supported by the following characters: calyx without colleters (ch. 4; colleters are present in *Antonia ovata*), hairs on the outside of the calyx and corolla (ch. 15, 20; corollas are glabrous in *Antonia ovata*), and winged seeds (ch. 43; wingless in *Norrisia* spp.). The monotypic genus *Usteria* from West and Central Africa is placed basally in the clade. *Usteria* is a shrub or liana having small flowers with calycophylls, which are single, highly enlarged calyx lobes (Leeuwenberg 1963). Calycophylls are also present in several genera of *Rubiaceae* (Robbrecht 1988; Delprete 1996). The next clade comprises all species of *Bonyunia* (a northern South American genus), which are unambiguously supported by having apiculate anthers (ch. 31), and pubescent styles, stigmas and capsules (ch. 34, 40). The sister group to *Bonyunia* is *Norrisia* (2 spp. in Malesia) plus *Antonia* (one species in northern South America). The relationship between *Antonia* and *Norrisia* is unambiguously supported by 2 – 3 pairs of bracteoles (ch. 9), white or greenish corollas (ch. 19), and wood with uniseriate-only rays (ch. 50). In *Antonia ovata* the bracteoles are enlarged and cover the calyx and the lower part of the corolla tube, whereas *Norrisia* has two small pairs of bracteoles below each flower (all other genera in Strychnaceae have not more than one pair). In fact, *Antonia griffithii* was originally described from Malesia but was later synonymised under *Norrisia malaccensis* (Leenhouts 1963). The presence of lenticels (ch. 2), calyces with equal sized lobes (ch. 14), latrorse anthers (ch. 29), and wingless seeds (ch. 43) are all unambiguous character changes supporting *Norrisia*. Bootstrap support from the present matrix is significant only for the monophyly of *Bonyunia* and *Norrisia* (81% and 97%, respectively), and for one pair of species within *Bonyunia* (*B. antoniifolia* and *B. aquatica*; 56%); all intergeneric groupings are only ambiguously supported. However, *Bonyunia* sister to *Antonia* plus *Norrisia* are supported at the 67% and 53% levels (respectively) by Struwe *et al.*'s (1994) data matrix.

E. *Potalieae* (*Gentianaceae*)

Fagraea is the most species-rich genus in Leenhouts' (1963) treatment of *Loganiaceae* for Flora Malesiana. Leenhouts included 31 species in the genus, which ranges from Ceylon in the west to southern China, through Malesia to northern Australia, and into the Pacific. Recently, 20 new species of *Fagraea* were described from Borneo (Wong & Sugau 1996). *Fagraea* belongs to tribe *Potalieae* (now in *Gentianaceae*) together with the genera *Potalia* and *Anthocleista*. *Potalia* has at least two species and as many as eight in South and Central America (Struwe & Albert, unpublished), and *Anthocleista* contains 14 species from Africa, Madagascar, and the Comores (Leeuwenberg 1961a). The *Potalieae* are all woody plants with a liana, shrub or tree-like habit. The flowers are conspicuous in several species. For example, *Fagraea auriculata* has funnel-shaped flowers up to 30 cm in diameter (Leenhouts 1963) and the flowers of *Fagraea berteroana*[1], widely cultivated in the Pacific, are used for crowns and leis (Leenhouts 1963). *Potalia* and *Anthocleista* are characterised by having 8 – 16 corolla lobes and stamens, which is anomalous in Asteridae (4 – 5 is the common number).

A species-level cladistic study based on morphological data showed that the genera *Anthocleista* and *Fagraea* were paraphyletic, with *Potalia* nested inside *Anthocleista* (Fig. 3; Struwe & Albert, unpublished). *Fagraea* forms five distinct clades below *Anthocleista* and *Potalia*. If this result is upheld by further (e.g. molecular) data, it is obvious that new generic circumscriptions will eventually be required for *Potalieae*.

The phylogenetic results obtained by Struwe & Albert (unpublished) and the character support for different groups are discussed below. For convenience we have used the species circumscriptions accepted by Leenhouts (1963), with the exception of *Fagraea havilandii* (Wong & Sugau 1996). It should be noted that some authors prefer more narrow species concepts for several complex species (e.g. *Fagraea fragrans, F. elliptica, F. racemosa*, and *F. ceilanica*; cf. Wong & Sugau 1996), and future cladistic studies should explore the utility of more restricted terminals.

Placed basalmost in our first approximation of *Potalieae* phylogeny is a clade comprising *Fagraea fragrans* and *F. elliptica*, and in isolated position above it, the widespread and variable species *Fagraea racemosa* (Fig. 3). These two clades could be identified as two separate genera with circumscriptions identical to *Fagraea* sect. *Cyrtophyllum* and the monotypic *Fagraea* sect. *Racemosae* (Leenhouts 1963; cf. Wong & Sugau 1996). Indeed, generic names for both lineages are available: *Cyrtophyllum* and *Utania*. The first clade (*Fagraea fragrans* plus *F. elliptica*) is unambigously supported by having small pollen (less than 20 µm) and exserted stamens. Unambiguous characteristics for the next clade, *Fagraea racemosa*, include blue fruits and a racemose inflorescence, usually hanging and with condensed inflorescence branches along the main axis.

The next clade in the tree (Fig. 3), the *Fagraea ceilanica* group, is comparable to *Fagraea* sect. *Fagraea* in a strict sense (incorporating the type species *Fagraea ceilanica*). This group is unambiguously supported by entire stigmas and white to greyish-white fruits. The relationships of *Fagraea ceilanica* and *F. tubulosa* are poorly resolved within this group, but two groups are well supported. The first of these, *Fagraea auriculata* and *F. involucrata*, shows trends in inflorescence compaction, flower number reduction, and enlarging of floral bracts, and is unambigously supported by having fruits larger than 4 cm in diameter and

[1]Also known as *Fagraea berteriana*, an orthographic variant (see Smith 1988).

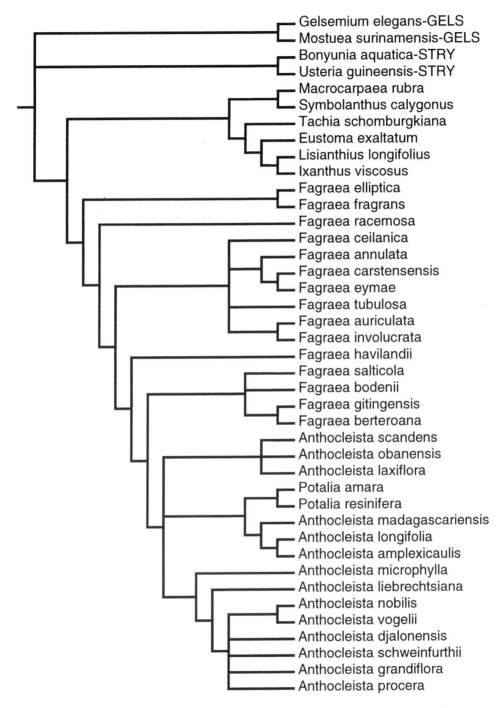

FIG. 3. Cladistic relationships of *Potalieae*, other *Gentianaceae*, *Strychnaceae-Antonieae*, and *Gelsemiaceae*, shown as a strict consensus tree (Struwe & Albert, unpublished). Familial abbreviations for the outgroups are: GELS = *Gelsemiaceae* and STRY = *Strychnaceae*.

reflexed auricles. The second well supported group includes three New Guinean taxa, *Fagraea eymae*, *F. carstensensis*, and *F. annulata* (Fig. 3). These poorly collected species (Leenhouts 1963) are supported as a clade only by pollen characteristics (colpi short and luminae with rounded corners; Punt 1978). Apart from their position on the tree the three species show similarities to other groups. *Fagraea eymae* shows similarity to *Fagraea salticola* of the *Fagraea berteroana* group (leaves with rounded to subcordate leaf-bases and 5 – 6 pairs of secondary nerves are present in both species) and *Fagraea carstensensis* has features in common with *Fagraea involucrata* (glomerulous inflorescences surrounded by an involucrum of reduced leaves). Possibly the only character (except from pollen evidence) that distinguishes *Fagraea annulata* from *Fagraea ceilanica* is the presence of a thickened ring in the corolla tube on which the stamens are inserted (cf. Cammerloher 1922; Leenhouts 1963), which is a feature characterizing the *Fagraea berteroana* group.

Fagraea havilandii and the *Fagraea berteroana* group are the next two clades in the tree, leading up sequentially to *Anthocleista* and *Potalia* (Fig. 3). The monophyly of *Fagraea havilandii* plus the *Fagraea berteroana* group, *Potalia*, and *Anthocleista* is unambiguously supported by linear rather than oblong anthers as well as pollen with perforate tectum (with a change to fine reticulum in the *Fagraea berteroana* group) and distinct endoporus costae (Punt 1978). *Fagraea havilandii*, identified as *F. gardenioides* ssp. *borneensis* by Leenhouts (1963), was raised to species rank by Wong & Sugau (1996).

The sister group to *Anthocleista* and *Potalia* is the *Fagraea berteroana* group (Fig. 3). The monophyly of these three clades is unambiguously supported by having a thickened ring in the corolla tube at the base of the filaments. The annulus is probably a fusion of the lower part of the filaments and shows a transformation series from the thickened ring (annulus) in the corolla tubes of the *Fagraea berteroana* group to nearly completely or completely fused filaments in most *Anthocleista* and *Potalia* species. We suggest that the *Fagraea berteroana* group, *Anthocleista*, *Potalia*, and *Symbolanthus* (which is characterised by a ring-like staminal corona) share ontogenetic homology in their staminal fusions. In the case of the *Potalieae* taxa, they may share historical homology as well (as shown by the phylogenetic analysis). In all other *Fagraea* species (except *Fagraea annulata*) the stamen attachment is simpler, with the filaments diverging separately from the corolla tube. *Fagraea havilandii* lacks the staminal ring, which helps keep it out of the *Fagraea berteroana* group and in an intermediate position.

The species of the *Fagraea berteroana* group, *F. berteroana*, *F. bodenii*, *F. gitingensis*, and *F. salticola*, are easily characterised in the field by having a thickened staminal ring, distinctly bilobed stigmas, and linear anthers. Some of these characters occur in species of the *Fagraea ceilanica* group as well (i.e. as homoplastic characters on the tree), but the combination of characters occurs only in this group. Another synapomorphy for the *Fagraea berteroana* group is pollen with finely reticulate exine (Punt 1978). *Fagraea berteroana* is a widespread and variable taxon, ranging from New Guinea eastwards into the Pacific, whereas *F. salticola* and *F. bodenii* are restricted to New Guinea, and *F. gitingensis* is distributed in the Philippines and Moluccas (Leenhouts 1963).

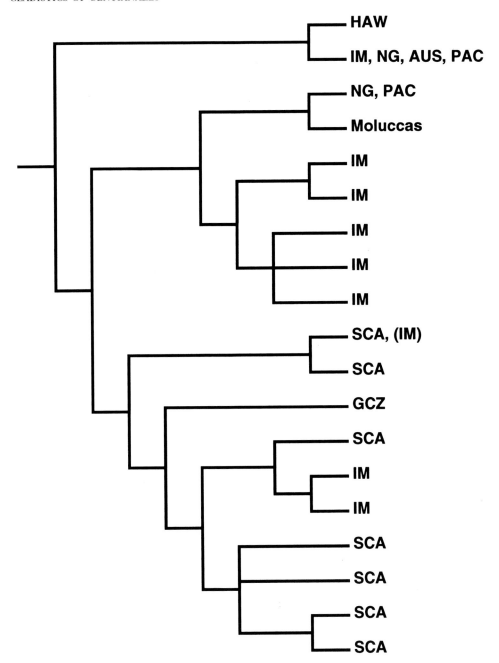

FIG. 4. Unreduced area cladogram developed from cladistic relationships of *Strychnaceae*. Widespread distributions in the Indo-Malesian region are indicated by IM, and narrow endemic distributions are indicated by particular area names (e.g., Moluccas). Other areas are indicated with the following abbreviations: AUS = tropical Australia, GCZ = Guineo-Congolean-Zambesian area in tropical Africa, HAW = Hawaii, NA = North America, NG = New Guinea, PAC = Pacific, SCA = South and Central America. The IM distribution of *Spigelia anthelmia* is hypothesised to represent a later dispersal (Leenhouts 1963).

Anthocleista and *Potalia* are well supported as a monophyletic group by having four free sepals, 8 to 16 corolla lobes and stamens (not five) and pollen with protruding pores (Leeuwenberg 1961a, Punt 1978). *Anthocleista* (from Africa and Madagascar) and *Potalia* (from South and Central America) have been maintained as separate genera for over 175 years, but they are nearly identical morphologically. The principal reason for keeping the genera separate appears to have been their disjunct distribution, but since the type species of the oldest generic name, *Potalia amara*, is nested inside *Anthocleista*, distribution does not correspond to the phylogenetic pattern. This group has three major clades, one including the arborescent species of *Anthocleista* from continental Africa, another including the lianous/shrub-like species of *Anthocleista* also from continental Africa, and the third comprising the Malagasy *Anthocleista* species together with *Potalia*. The similarities among these three clades are much more obvious than their differences, so separation into segregate genera does not seem warranted.

In summary, to classify the *Potalieae* following these phylogenetic results will include problems with paraphyly, priority, support for groups, and presence of field characters. Several monophyletic groups are identified, although some of them have weak support. Of the clades discussed above, bootstrap analysis of our preliminary matrix identifies only *Fagraea elliptica* plus *Fagraea fragrans* and *Anthocleista* plus *Potalia* as a well-supported groups (unpublished results). The genus name with first priority in the tribe, *Potalia*, includes only two described species, and these are nested far inside two other, larger genera. If only one genus were to be recognised in *Potalieae*, all species of *Anthocleista* and *Fagraea* (c. 70 in total) would have to been combined with *Potalia*. Another option is to divide *Potalieae* into several monophyletic genera. *Potalia* (including *Anthocleista*) would stand for the most derived clade. The species of *Fagraea* that belong to the *Fagraea berteroana* group are a well-defined clade that could easily be distinguished from *Fagraea*, both with field characters and synapomorphies. The previously described segregate genera *Cyrtophyllum* and *Utania* stand for easily recognised groups that can be resurrected, further reducing the paraphyly of *Fagraea*. A remaining problem is the position and generic affinities of *Fagraea havilandii*, a question that requires further attention. An additional issue will be incorporation of the many other species of *Fagraea* that are excluded from our current phylogenetic analysis. These points are addressed in our cladistic reclassification of the tribe (in progress).

BIOGEOGRAPHY

1. Malesia: temporal heterogeneity

The study of biogeographic patterns involves taxa: their relationships, and their distributions. Distribution can be explained by vicariance and dispersal events. Biogeographic hypotheses depend on the assumed age of the taxa, that is, a distribution pattern cannot be explained with vicariance or dispersal events that occurred before the origin of the group. Areas can have several biogeographic hypotheses associated with them. The Malesia area, for example, includes elements of Pangaea, which broke up into southern (Gondwana) and northern supercontinents (Laurasia) c. 160 million years (My) ago.

Living and fossil cycads (Cycadales) show a basically Pangaean distribution (cf. Jones 1993; Moretti *et al.* 1993). For example, the Indomalesian, Australian, and Malagasy/African distribution of *Cycas* (cf. Jones 1993: 127) could reflect Pangaean continuity, with Malesia forming the Laurasia/Gondwana contact zone. Further distinction among Australian and Malagasy/African *Cycas* lineages may have occurred when Gondwana split into Africa, Australia, Madagascar, South America, and Antarctica over a c. 30 million-year period beginning by 130 million years ago (cf. Wendel & Albert 1992).

Other plant groups with Malesian representatives show clear evidence of the austral Gondwanan split without any suggestion of Pangaean continuity. *Nothofagus* (*Fagaceae*) shows transpacific relationships between Australia, New Zealand, and southern South America (Humphries 1981; Humphries *et al.* 1986). However, the distribution of Malesian species of *Nothofagus*, which occur in both New Guinea and New Caledonia, could relate to different geologic events such as the formation and accretion of inner Malesian arc elements with other land masses (Duffels & de Boer 1990).

Still other Malesian plant groups show transpacific relationships that are decidely non-austral. The conduplicate-leaved slipper orchids (*Cypripedioideae*) show a disjunct distribution between Indomalesia (incl. the Solomon Islands) and tropical Central-South America (Albert 1994). Here, the boreotropics hypothesis presents a plausible vicariance scenario involving the Eocene/Oligocene breakup of a previously continuous, North Atlantic tropical flora (Lavin & Luckow 1993; Tiffney 1985a, b; Wolfe 1975). This connection, lasting until c. 38 million years ago, extended from Malesia through Asia and northern Africa, across the Atlantic to North and Central America. Temperate climates occurring later in the Tertiary could then have limited tropical taxa to Indomalayan, Caribbean-basin, and South American "refugia". Later dispersal and extinction could have also occurred which could explain the presence or absence of taxa in certain areas. Returning to cycads in this regard, the patchy North-Central American distribution of *Dioon* may be boreotropical when Alaskan and Japanese fossils are considered, whereas post-Eocene climatic events may have shaped the present-day patterns of narrow endemicity (cf. Moretti *et al.* 1993).

In summary, it is clear that the study of biogeography must often include several temporally distinct hypotheses relevant to a single modern-day flora. Each hypothesis involves a specific scenario and explanation for distributional patterns. The Malesia area has a very complex history, incorporating parts of both the Asian and Australian plates as well as climatic changes that caused variations in sea level and humidity (e.g. Duffels & de Boer 1990; Johns 1995).

2. Biogeography of Malesian Gentianales

Several large-scale biogeographical patterns in the Gentianales involving Malesia have been revealed by cladistic studies. Groups showing a pantropical pattern include *Strychnos, Antonieae, Mitreola*, and *Gelsemiaceae*. Disjunct distributions between Malesia and South America are exemplified by the relationship between *Norrisia* and *Antonia*. Another example is *Gelsemium* which is present in Southeast Asia, Malesia (1 sp.), Central America, and southeastern North America (2 spp.; Wyatt *et al.* 1993).

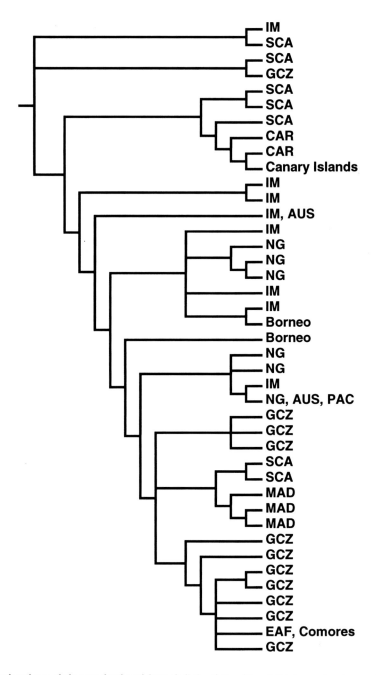

FIG. 5. Unreduced area cladogram developed from cladistic relationships of *Potalieae* and outgroups. Widespread distributions in the Indo-Malesian region are indicated by IM, and narrow endemic distributions are indicated by particular area names (e.g., Borneo). Other areas are indicated with the following abbreviations: AUS = tropical Australia, CAR = Carribean, EAF = East Africa, GCZ = Guineo-Congolean-Zambesian area in tropical Africa, MAD = Madagascar, NA = North America, NG = New Guinea, PAC = Pacific, SCA = South and Central America.

Area relationships on a smaller scale (between geological or vegetational islands, etc.) in Southeast Asia, Malesia, Australia, and the Pacific may become more clear through further studies of species relationships in *Geniostomaceae, Mitrasacme,* and *Strychnaceae* (cf. Axelius 1989, for an example from *Rubiaceae*).

In *Strychnaceae* links among several tropical areas are demonstrated (Fig. 4). The relationships within *Strychneae* are still uncertain since the Indo-Pacific genera *Gardneria* and *Neuburgia* are either nested inside pantropical *Strychnos* (Struwe *et al.* 1994) or *Neuburgia* may be the sister group to *Gardneria* plus *Strychnos* (as reported here). In the *Spigelieae-Antonieae* clade the basal branch (*Spigelia*) represents tropical and subtropical America, which is then sister to tropical areas in South America, Malesia, and Africa (*Antonieae*). Within the *Antonieae* clade Malesia (represented by *Norrisia*) shows a closer relationship to South America (*Antonia* and *Bonyunia*) than to Africa (*Usteria*). This pattern, like that of transpacific slipper orchids (Albert 1994), could suggest a relictual boreotropical distribution for the *Spigelieae-Antonieae* clade.

Biogeographic patterns within *Potalieae* involve nearly all continents (Fig. 5). The neotropical woody gentian outgroup taxa are long-disjunct from Indomalesian *Fagraea* lineages. Within *Fagraea*, with the exception of *Fagraea havilandii*, New Guinea is included within all major components (i.e., most components labelled "Indomalesia" represent widespread taxa also distributed in New Guinea, where other clades so labelled are endemic). New Guinea is therefore directly related to all Malesian areas (including Australian and Pacific distributions) as well as to Africa (at the branch leading to the derived *Anthocleista* and *Potalia* group) and South-Central America (basally, to woody gentian outgroup taxa). In turn, South-Central America (*Potalia*) is directly related to Madagascar (*Anthocleista* pro parte).

New Guinean connections to both Africa and South-Central America could be explained either by the breakup of Gondwana or by the boreotropics hypothesis. The Gondwana hypothesis requires *Gentianaceae* to be c. 130 My old, which has no precedent from the fossil record (the earliest known sympetalous angiosperm, an ericalean, is c. 90 My old; Nixon & Crepet 1993). The alternative is that *Potalieae* (and other woody gentians) show a basically boreotropical distribution, with relictual species distributed in the tropical areas of America, Africa, and Malesia. However, cladistic support for the boreotropics hypothesis — South American lineages derived within diverse North American clades (Lavin & Luckow 1993) — is not provided for *Potalieae* or *Antonieae*, which have no extant or fossil North American clades.

Instead, area relationships implied by the *Strychnaceae* and *Potalieae* cladograms are congruent in their support for the Gondwana hypothesis. The unrooted and reduced area cladogram for *Strychnaceae* (Fig. 6a) shows Indomalesian relationships to South-Central America in three places: between (1) *Antonia* and *Norrisia*, (2) *Strychneae* and *Spigelieae*, and (3) *Gardneria* and *Strychnos* (which is pantropical). Such redundant area relationships are common in cladistic biogeographic studies, and their presence can be confusing when comparing cladograms based on different taxa. However, this is not the case when comparing *Strychnaceae* with *Potalieae*. The reduced area cladogram for *Potalieae* (Fig. 6b) shows a single

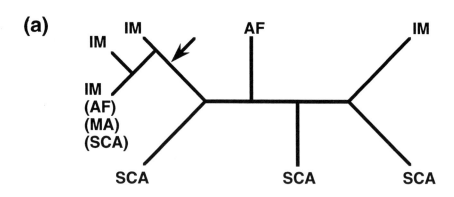

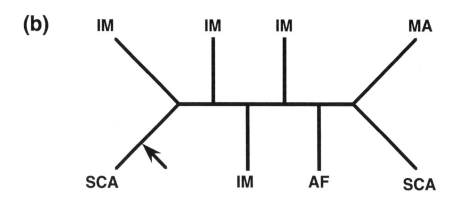

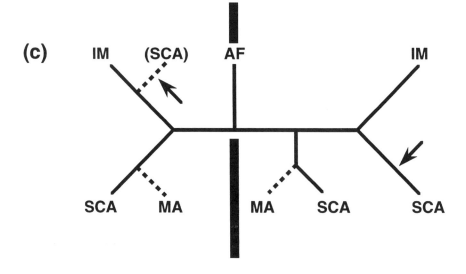

Indomalesia/South-Central America connection, at the root between *Potalieae* and other *Gentianaceae*. With the exception of Madagascar, the unrooted area cladograms of *Strychnaceae* and *Potalieae* are highly compatible. Figure 6c shows both trees reconciled on the area cladogram of *Strychnaceae*. The diverse Indomalesian clades of *Fagraea* can be represented by a single component since relationships within that area are not being considered. For *Potalieae*, Madagascar (*Anthocleista* pro parte) is sister to South-Central America (*Potalia*) and proximal to Africa (*Anthocleista* pro parte), and two placements of Madagascar on the *Strychnaceae* tree fit that condition. Root placement at South-Central America and proximal to Indomalesia, as implied by the *Potalieae* tree and not contradicted by the *Strychnaceae* tree, can also be accommodated at two positions. What results are mirror area relationships on either side of Africa; that is, *Potalieae* relationships can be derived from either half of a reconciled *Strychnaceae* tree.

Potalieae and *Strychnaceae* appear to share a common biogeographic history that is internally replicated in *Strychnaceae*. A simple explanation for such internal replication is the foundation of major clades in a larger area that split and subsequently isolated representatives of each clade. Given the areas involved — Indomalesia, Africa, South-Central America, and Madagascar — that larger area would be Gondwana. New Guinea was connected, via Australia and Antarctica, to the other gondwanic continents listed above until c. 130 million years before present.

Molecular data corroborate a gondwanic explanation for amphioceanic disjunctions involving Malesian Gentianales. The number of nucleotide changes between species on a tree can serve as the basis for estimated time since divergence, that is, if a molecular clock is assumed (Albert *et al.* 1994; Wendel & Albert 1992). Using a tree based on asterid and gentianalean *rbc*L gene sequences (Struwe & Albert, unpublished)[1] divergence times for three species pairs in the Gentianales were calculated (Table 3). Time ranges were derived from three

---

[1] One most parsimonious tree for the data set used in Struwe and Albert (unpublished) was found using PAUP and the options listed in footnote 1, p. 8. All sequences were obtained from GenBank (accession numbers L01902, L01929, L01959, L11678, L11684, L11688, L13931, L14007, L14389, L14390, L14392, L14394, L14396, L14397, L14398, L14402, L14403, L14404, L14410, M16867, X04976, X81095, Z29669, and Z29670).

FIG. 6. Reconciliation of area relationships in *Strychnaceae* and *Potalieae*. Reduced area cladograms obtained from the cladistic relationships of (**a**) *Strychnaceae* and (**b**) *Potalieae*. In *Strychnaceae*, Indomalesia (IM) is related to South-Central America (SCA) in three places: between (1) *Antonia* and *Norrisia*, (2) *Strychneae* and *Spigelieae* (the range of which includes North America), and (3) *Gardneria* and *Strychnos* (which is pantropical). In *Potalieae*, the same major areas are represented with the exception of Madagascar (MA). Arrows indicate root positions inferred by outgroup analysis. (**c**) Area relationships for *Potalieae* reconciled on the area cladogram for *Strychnaceae*. MA, which is sister to SCA in (b) can be added in two places (dotted lines) that preserve its proximity to Africa (AF). A root at SCA and proximal to IM is also possible at two places (one of which requires an added SCA component; dotted line). The result is symmetry in area relationships on either side of Africa (the vertical line shows the axis). The *Potalieae* tree, excluding the redundant IM areas, can be derived by pruning off subtrees on either side of AF. Therefore, area relationships in *Strychnaceae* are no more complex than for *Potalieae*. It is possible that both groups may have diversified before the breakup of Gondwana, by which time *Strychnaceae* had already split into two major clades.

different divergence rate estimates (Albert *et al.* 1994; Wendel & Albert,1992). The pairs *Anthocleista grandiflora/Fagraea berteroana* (Africa/New Guinea-Pacific), *Spigelia anthelmia/Strychnos nux-vomica* (North America/Malesia), and *Gelsemium sempervirens/Mostuea brunonis* (North America/Africa) appear to have diverged contemporaneously. Mesozoic divergence dates are obtained from the woody taxon rate of Albert *et al.* (1994); these are placed well before to precisely at the initial breakup of Gondwana, circa 130 million years before present. Herbaceous taxon rates (Wendel & Albert 1992), which may be faster because of shorter generation times in non-woody plants, yield divergence estimates no earlier than the Cretaceous-Tertiary boundary and no later than Eocene/Oligocene, just prior to (or at) the breakup of the boreotropical interchange. Because nearly all *Potalieae* and *Strychnaceae* are woody plants, we prefer the somewhat heretical Mesozoic estimates that fit the area relationships best and imply an extremely early derivation of the major lineages of *Gentianaceae* and other *Gentianales*.

TABLE 3. Estimates of time since divergence in millions of years for three taxon pairs in Gentianales, based on a cladistic analysis of the *rbc*L gene (Struwe & Albert, unpublished). Percent sequence divergence was calculated by dividing the additive length along a path (e.g. *Anthocleista grandiflora* <—> *Fagraea berteroana* = 46 changes) by the number of nucleotides compared (1,428; 46/1,428 × 100 = 3.22%); these values were divided by independent estimates of rates of *rbc*L divergence (0.02 – 0.07% per million years) to yield approximate divergence times for the taxa compared (see Albert *et al.*, 1994; Wendel & Albert, 1992).

	# nucleotide changes pairwise	divergence rate		
		0.02[1]	0.05[2]	0.07[2]
		time since divergence		
Anthocleista grandiflora/ Fagraea berteroana	46	161.06	64.43	46.02
Gelsemium sempervirens/ Mostuea brunonis	37	129.55	51.82	37.01
Strychnos nux-vomica/ Spigelia anthelmia	42	147.06	58.82	42.02

[1] Albert *et al.* (1994), for woody taxon pairs
[2] Wendel & Albert (1992) for herbaceous taxon pairs
[3] C. F. Bremer (1994)

CONCLUSIONS AND PROSPECTS

A phylogenetic classification will always be the more stable than those erected on other criteria. The principle of parsimony as used in the hierarchical ordering of taxic attributes implies nothing *per se* about evolution or the reality of phylogenetic hypotheses produced. Rather, parsimony analysis simply supplies particular nestings of character states that will have greater explanatory and predictive power. For example, the cladistic classification of Gentianales (Struwe & Albert in Struwe *et al.* 1994) links *Geniostomaceae* with *Apocynaceae* based on several shared morphological character states. If a new organism is described with just these states in the same characters while differing uniquely in several others, its placement within the *Geniostomaceae/Apocynaceae* clade will be unambiguous. Attempting to place the same taxon using a dichotomous key of field characters may prove ambiguous or even impossible; any unique states present in the new taxon might require a fundamentally revised key. However, this is not to say that key characters need interfere with a cladistic classification; indeed, *Geniostoma* and *Labordia* were placed in their own family rather than be included in *Apocynaceae* as laticifer-less members.

A phylogenetic approach to floristics treats areas such as Malesia as non-static. The temporal stratification and spatial dispersion within or between floras (resulting from asymmetrical divergence times and inter-floristic relationships, respectively) can only be explained through a phylogenetic approach. A revision of Malesian *Strychnos*, for example, might not be maximally explanatory (stable) without consideration of other worldwide congeners that could show differential clade relatedness to particular Malesian species. At the same time, relationships within or among floristic regions cannot be assessed solely by species shared (see, e.g. Laumonier 1990). Information on stratification and dispersion is lost when the attributes (characters) of organisms are discarded. After all, it is upon characters that species and higher taxa are erected and identified.

The Gentianales exemplify the utility of a phylogenetic approach to systematics and floristics (cf. Stevens 1990). With respect to Gentianales, Malesia, as an area, shows amphioceanic sister group relationships to South America (*Norrisia/Antonia*) and Africa/Central-South America (*Potalieae*). A gondwanic origin of major clades provides the most plausible biogeographic history for these taxa, which show relictual distributions in tropical areas. With our recent cladistic classification of the Gentianales, two new families have been introduced to Flora Malesiana: *Gelsemiaceae* and *Geniostomaceae*. Furthermore, *Buddlejaceae* and *Strychnaceae* have been resurrected, decreasing the heterogeneity of the *Loganicaeae* while increasing familial diversity in the Malesia area even more. On the generic level, the status of *Gardneria* and *Neuburgia* remain uncertain, whereas the paraphyly of *Anthocleista* and *Fagraea* could be reduced by taxonomic combination or by accepting several segregate genera (respectively). It is expected that future studies of these and other gentianalean taxa, perhaps including molecular data, will contribute to the phylogenetic knowledge necessary for stable classifications and meaningful estimates of diversity within Malesia and related areas.

ACKNOWLEDGMENTS

This research was supported in part by a research grant to LS from the Uddenberg-Nordingska Foundation (Kalmar, Sweden) for the project "Blommorfologi och systematik hos *Fagraea* (*Gentianaceae*)." Additional support was provided by the Lewis B. and Dorothy Cullman Foundation and by Uppsala University, Sweden.

REFERENCES

Albert, V. A. (1994). Cladistic relationships of the slipper orchids (*Cypripedioideae: Orchidaceae*) from congruent morphological and molecular data. Lindleyana 9: 115 – 132.

——, Backlund, A., Bremer, K., Chase, M. V., Manhart, J. R., Mishler, B. D., & Nixon, K. C. (1994). Functional constraints and *rbcL* evidence for land plant phylogeny. Ann. Missouri Bot. Gard. 81: 534 – 567.

Axelius, B. (1989). The genera *Lerchea* and *Xanthophytum* (*Rubiaceae*) from Southeast Asia — taxonomy, phylogeny, biogeography. 119pp. Doctoral dissertation, Department of Botany, Stockholm.

Backer, C. A. & Bakhuisen van den Brink, R. C., Jr. (1965). Flora of Java. Vol. 2. 641pp. Noordhoff, Groningen.

Backlund, A. (1996). Phylogeny of Dipsacales. 238pp. Doctoral dissertation, Department of Systematic Botany, Uppsala.

Bentham, G. (1857). Notes on *Loganiaceae*. J. Linn. Soc. Lond., Bot. 1: 52 – 115.

Bisset, N. G. (1974). The Asian species of *Strychnos*. Part III. The ethnobotany. Lloydia 37: 62 – 107.

Blackwell, W. H. Jr. (1968). Family 159. *Loganiaceae*. In: Woodson, R.E., Schery, R.W., & collaborators. Flora of Panama, part VIII. Ann. Missouri Bot. Gard. 54: 393 – 413.

Bremer, B. (1996). Phylogenetic studies within *Rubiaceae* and relationships to other families based on molecular data. Opera Bot. Belg. 7: 33 – 50.

——, Olmstead, R. G., Struwe, L., & Sweere, J. A. (1994). *rbcL* sequences support exclusion of *Retzia*, *Desfontainia*, and *Nicodemia* from the Gentianales. Pl. Syst. Evol. 190: 213 – 230.

Cammerloher, H. (1922). Die Loganiaceen und Buddleiaceen Niederländisch-Indiens. Bull. Jard. Bot. (Buitenzorg), ser. 3, 5: 294 – 338.

Conn, B. J. (1980). A taxonomic revision of *Geniostoma* subg. *Geniostoma* (*Loganiaceae*). Blumea 26: 245 – 364.

——, Brown, E. A., & Dunlop, C. R. (1996). *Loganiaceae*. Flora of Australia, vol. 28, Gentianales. CSIRO Australia, Melbourne.

Delprete, P. G. (1996). Notes on calycophyllous *Rubiaceae*. Part I. Morphological comparisons of the genera *Chimarrhis*, *Bathysa*, and *Calycophyllum*, with new combinations and a new species, *Chimarrhis gentryana*. Brittonia 48: 35 – 44.

Dickison, W. C. (1994). A reexamination of *Sanango racemosum*. 2. Vegetative and floral anatomy. Taxon 43: 601 – 618.

Don, G. (1837 – 38). A general history of dichlamydeous plants, vol. 4. 908pp. Rivington, London.

Dunlop, C. R. (1996). *Mitrasacme, Schizacme,* and *Phyllangium.* In: Conn, B. J., Brown, E. A., & Dunlop, C. R. *Loganiaceae.* Flora of Australia, vol. 28, Gentianales. CSIRO Australia, Melbourne, pp. 29 – 62.

Duffels, J. P. & de Boer, A. J. (1990). Areas of endemism and composite areas in East Malesia. In: Baas, P., Kalkman, K. & Geesink, R. (eds), The plant diversity of Malesia. Kluwer Academic Publishers, Dordrecht, pp. 249 – 272.

Farris, J. S. (1979). The information content of the phylogenetic system. Syst. Zool. 28: 486 – 519.

—— (1982). Outgroups and parsimony. Syst. Zool. 31: 328 – 334.

—— (1983). The logical basis of phylogenetic analysis. In: Platnick, N. I. & Funk, V. A. (eds), Advances in cladistics, vol. 2. Columbia University Press, New York, pp. 8 – 36.

—— (1989). The retention index and rescaled consistency index. Cladistics 5: 417 – 419.

——, Albert, V.A., Källersjö, M., Lipscomb, D. & Kluge, A. G. (1996). Parsimony jackknifing outperforms neighbor-joining. Cladistics 12: 99 – 124.

Forkmann, G. (1991). Flavonoids as flower pigments: the formation of the natural spectrum and its extension by genetic engineering. Plant Breed. 106:1 – 26.

Hooker, W. J. (1837). *Antonia pilosa.* Ic. Plant. 1: 64.

Humphries, C. J. (1981). Biogeographical methods and the southern beeches (*Fagaceae: Nothofagus*). In: V. A. Funk & D. R. Brooks, (eds), Advances in Cladistics. The New York Botanical Garden, New York, pp. 177 – 207.

——, Cox, J. M., & Nielsen, E. S. (1986). *Nothofagus* and its parasites: A cladistic approach to coevolution. In: A. R. Stone & D. L. Hawksworth (eds), Coevolution and Systematics. Clarendon Press, Oxford, pp. 55 – 76.

—— & Williams, P. H. (1994). Cladograms and trees in biodiversity. In: R. W. Scotland, D. J. Siebert & D. M. Williams (eds), Models in phylogeny reconstruction, Systematics Association special volume 52, Clarendon Press, Oxford, pp. 335 – 352.

Hutchinson, J. (1959). The families of flowering plants — dicotyledons, vol. 1., ed. 2. 510 pp. Clarendon Press, Oxford.

Jensen, S. R. (1992). Systematic implications of the distributions of iridoids and other chemical compounds in the *Loganiaceae* and other families of the Asteridae. Ann. Missouri Bot. Gard. 79: 284 – 302.

—— (1994). A reexamination of *Sanango racemosum.* 3. Chemotaxonomy. Taxon 43: 619 – 623.

Johns, R. J. (1995). Malesia — an introduction. Kew Mag. 12: 52 – 62.

Jones, D. L. (1993). Cycads of the world. 312 pp. The New York Botanical Garden, New York.

Kisakürek, M. V. & Hesse, M. (1980). Chemotaxonomic studies of the *Apocynaceae, Loganiaceae,* and *Rubiaceae* with reference to indole alkaloids. In: J. D. Phillipson & M. H. Zenk (eds), Indole and biogenetically related alkaloids. Academic Press, London, pp. 11 – 26.

Laumonier, Y. (1990). Search for phytogeographic provinces in Sumatra. In: P. Baas, K. Kalkman & R. Geesink (eds), The plant diversity of Malesia. Kluwer Academic Publishers, Dordrecht, pp. 193 – 211.

Lavin, M. & Luckow, M. (1993). Origins and relationships of tropical North America in the context of the boreotropics hypothesis. Amer. J. Bot. 80: 1 – 14.

Leenhouts, P. W. (1963). *Loganiaceae.* In: C. G. G. J. van Steenis (ed.), Flora Malesiana, ser. 1, vol. 6, part 2. Wolters-Noordhoff, Groningen, pp. 293 – 387.

Leeuwenberg, A. J. M. (1961a). The *Loganiaceae* of Africa I. *Anthocleista.* Acta Bot. Neerl. 10: 1 – 53.

—— (1961b). The *Loganiaceae* of Africa. III. *Spigelia* L. Acta Bot. Neerl. 10: 460 – 465.

—— (1961c). The *Loganiaceae* of Africa II. A revision of *Mostuea* Didr. Meded. Landbouwhogeschool 61: 1 – 31.

—— (1963). The *Loganiaceae* of Africa. V. *Usteria* Willd. Acta Bot. Neerl. 12: 112 – 118.

—— (1969a). Notes on American *Loganiaceae.* III. Revision of *Bonyunia* Rich. Schomb. Acta Bot. Neerl. 18: 152 – 158.

—— (1969b). The *Loganiaceae* of Africa VIII. *Strychnos* III. Revision of the African species with notes on the extra-African. Meded. Landbouwhogeschool 69 : 1 – 316.

—— & Leenhouts, P. W. (1980). Taxonomy. In: A. J. M. Leeuwenberg (ed.), Engler and Prantl's Die natürlichen Pflanzenfamilien, ANGIOSPERMAE: Ordnung Gentianales, Fam. *Loganiaceae*, vol. 28b (1). Duncker and Humblot, Berlin, pp. 8 – 96.

—— & Vidal, J. E. (1972). *Buddlejaceae.* In: A. Aubréville & J. -F. Leroy (eds), Flore du Cambodge, du Laos et du Vietnam, vol. 13. Muséum national d'histoire naturelle, Paris, pp. 90 – 98.

Lewis, H. & Raven, P. H. (1992). New combinations in the genus *Clarkia* (*Onagraceae*). Madroño 39: 163 – 169.

Lindley, J. (1853). The vegetable kingdom, ed. 3. 908 pp. Bradbury & Evans, London.

Link, H. F. (1829 – 1833). Handbuch zur Erkennung der nutzbarsten und am häufigsten vorkommenden Gewächse, vol. 1. 864 pp. S. J. Joseephy, Berlin.

Maddison, W. P., & Maddison, D. R. (1992). MacClade: analysis of phylogeny and character evolution, ver. 3.0. Computer program and documentation. Sinauer Associates, Sunderland.

Martius, C. F. P. von. (1827). Nova genera et species plantarum quas in itinere per Brasiliam, vol. 2. 148 pp. V. Wolf, München.

Mennega, A. M. W. (1980). Anatomy of the secondary xylem. In: A. J. Leeuwenberg, (ed.), Engler and Prantl's Die natürlichen Pflanzenfamilien, ANGIOSPERMAE: Ordnung Gentianales, Fam. *Loganiaceae*, vol. 28b (1). Duncker and Humblot, Berlin, pp. 112 – 161.

Moretti, A., Caputo, P., Cozzolino, S., De Luca, P., Gaudio, L., Siniscalco Gigliano, G., & Stevenson, D. W. (1993). A phylogenetic analysis of *Dioon* (*Zamiaceae*). Amer. J. Bot. 80: 204 – 214.

Nixon, K. C. & Crepet, W. L. (1993). Late Cretaceous fossils of ericalean affinity. Amer. J. Bot. 80: 616 – 623.

—— & Davis, J. I. (1991). Polymorphic taxa, missing values and cladistic analysis. Cladistics 7: 233 – 241.

Norman, E. M. (1994). A reexamination of *Sanango racemosum.* 1. Morphology and distribution. Taxon 43: 591 – 600.

Ohwi, J. (1965). Flora of Japan. 1067pp. Smithsonian Institution, Washington D. C.

Punt, W. (1978). Evolutionary trends in the *Potalieae* (*Loganiaceae*). Rev. Palaeobot. Palynol. 26: 313 – 335.

—— (1980). Pollen morphology. In: A. J. M. Leeuwenberg (ed.), Engler and Prantl's Die natürlichen Pflanzenfamilien, ANGIOSPERMAE: Ordnung Gentianales, Fam. *Loganiaceae*, vol. 28b (1). Duncker and Humblot, Berlin, pp. 162 – 191.

—— & Leenhouts, P. W. (1967). Pollen morphology and taxonomy in the *Loganiaceae*. Grana Palynologica 7: 469 – 516.

Robbrecht, E. (1988). Tropical woody *Rubiaceae*. Opera Bot. Belg. 1: 1 – 271.

Rogers, G. K. (1986). The genera of *Loganiaceae* in the southeastern United States. J. Arnold Arb. 67: 143 – 185.

Samuelsson, G. (1992). Drugs of natural origin. 320pp. Swedish Pharmaceutical Press, Stockholm.

Sherff, E. E. (1939). Genus *Labordia*. Field Mus. Nat. Hist., Bot. Ser. 17: 449 – 546.

Smith, A. C. (1988). Flora vitensis nova, vol. 4. 377pp. Pacific Tropical Botanical Garden, Lawai.

—— & Stone, B. C. (1962). Studies of Pacific island plants, XVII. The genus *Geniostoma* (*Loganiaceae*) in the New Hebrides, Fiji, Samoa, and Tonga. Contr. U.S. Natl. Herb. 37: 1 – 41.

Solereder, H. (1892 – 1895). *Loganiaceae*. In: A. Engler & K. Prantl (eds),. Die natürlichen Pflanzenfamilien, vol. 4 (2). Verlag von Vilhelm Engelmann, Leipzig, pp. 19 – 50.

Spooner, D. M., Anderson, G. J., & Jansen, R. K. (1993). Chloroplast DNA evidence for the interrelationships of tomatoes, potatoes, and pepinos (*Solanaceae*). Amer. J. Bot. 80: 676 – 688.

Stevens, P. F. (1990). Nomenclatural stability, taxonomic instinct, and flora writing — a recipe for disaster? In: P. Baas, K. Kalkman & R. Geesink (eds), The plant diversity of Malesia. Kluwer Academic Publishers, Dordrecht, pp. 387 – 410.

Struwe, L., Albert, V. A. & Bremer, B. (1994 [1995]). Cladistics and family level classification of the Gentianales. Cladistics 10: 175 – 206.

Swofford, D. L. (1993). PAUP — Phylogenetic Analysis Using Parsimony, vers. 3.1. Computer program and documentation. Illinois Natural History Survey, Champaign, Illinois.

Sytsma, K. J. & Smith, J. F. (1992). Molecular systematics of *Onagraceae*: examples from *Clarkia* and *Fuchsia*. In: P. S. Soltis, D. E. Soltis & J. J. Doyle (eds), Molecular systematics of plants. Chapman and Hall, New York, pp. 295 – 323.

Takhtajan, A. (1987). Systema Magnoliophytorum. 438pp. Nauka, Leningrad.

—— (1997). Diversity and classification of flowering plants. 643 pp. Columbia University Press, New York.

Tiffney, B. H. (1985a). Perspectives on the origin of the floristic similarity between eastern Asia and eastern North America. J. Arnold. Arb. 66: 73 – 94.

—— (1985b). The Eocene North Atlantic land bridge: its importance in Tertiary and modern phytogeography of the northern hemisphere. J. Arnold. Arb. 66: 243 – 273.

Tirel-Roudet, C. (1972). *Loganiaceae*. In: Aubréville, A. & Leroy, J.-F. (eds), Flore du Cambodge, du Laos et du Vietnam, vol. 13. Muséum national d'histoire naturelle, Paris, pp. 3 – 89.

Wagner, W. L., Herbst, D. R., & Sohmer, S. H. (1990). Manual of the flowering plants of Hawai'i, vol. 1. 988pp. University of Hawaii Press and Bishop Museum Press, Honolulu.

Wendel, J. F. & Albert, V. A. (1992). Phylogenetics of the cotton genus (*Gossypium*): Character-state weighted parsimony analysis of chloroplast-DNA restriction site data and its systematic and biogeographic implications. Syst. Bot. 17: 115 – 143.

Wiehler, H. (1994). A reexamination of *Sanango racemosum*. 4. Its new systematic position in *Gesneriaceae*. Taxon 43: 625 – 632.

Wolfe, J. A. (1975). Some aspects of plant geography of the northern hemisphere during late Cretaceous and Tertiary. Ann. Missouri Bot. Gard. 62: 264 – 279.

Wong, K. M. & Sugau, J. B. (1996). A revision of *Fagraea* (*Loganiaceae*) in Borneo, with notes on related Malesian species and 21 new species. Sandakania 8: 1 – 93.

Wyatt, R., Broyles, S. B., Hamrick, J. L., & Stoneburner, A. (1993). Systematic relationships within *Gelsemium* (*Loganiaceae*): Evidence from isozymes and cladistics. Syst. Bot. 18: 345 – 355.

Yang, S.-Z. & Peng, C.-I. (1994). *Gardneria* (*Loganiaceae*) in Taiwan. Bot. Bull. Acad. Sin. 35: 223 – 227.

APPENDIX 1.

KEY TO THE FAMILIES OF GENTIANALES

1. Laticifers present in vegetative parts; stamens and style often adnate or united to form a gynostegium; cardenolides often present · **Apocynaceae** (incl. *Asclepiadaceae*)

1. Laticifers absent in vegetative parts; stamens not united with the style or stigma; cardenolides absent · 2

2. Ovary inferior (rarely semi-inferior, e.g. in *Gaertnera* and *Pagamea*); interpetiolar stipules usually well-developed; internal phloem absent. · **Rubiaceae** (incl. *Theligonaceae*)

2. Ovary superior (semi-inferior in some *Loganiaceae*, then stipules absent or inconspicuous); leaves amplexicaul or interpetiolar stipules distinct and small (rarely conspicuous) or as interpetiolar lines or ocreas; internal phloem present. · 3

3. Stigma twice dichotomously divided · **Gelsemiaceae**

3. Stigma simple or with one dichotomous division · · · · · · · · · · · · · · · · · · 4

4. Woody dehiscent capsules with seeds embedded in fleshy placentas · **Geniostomaceae**

4. Berries, drupes, or nuts; if capsules, then with dry placentas · · · · · · · · · · · · 5

5. Corolla aestivation contort (rarely imbricate, e.g. in *Bartonia* and *Obolaria*); placentation parietal (or axile, then aestivation contort); inside of corolla often with glands and scales; xanthones present · · · · · · · · · · · · · **Gentianaceae**

5. Corolla aestivation imbricate or valvate; placentation axile; inside of corolla not with glands or scales, but often pubescent; xanthones absent · · · · · · · · · · · 6

6. Berries or drupes with fleshy, leathery, or fibrous fruitwalls $\cdots\cdots\cdots\cdots\cdots\cdots$
    $\cdots\cdots\cdots\cdots\cdots\cdots\cdots\cdots\cdots\cdots\cdots\cdots\cdots\cdots\cdots\cdots\cdots$ ***Strychnaceae****-Strychneae*
6. Capsules with dry fruit walls $\cdots\cdots\cdots\cdots\cdots\cdots\cdots\cdots\cdots\cdots\cdots\cdots\cdots$ 7
7. Capsules with four deciduous valves and a persistent cupular base; inflorescence
    cincinnous $\cdots\cdots\cdots\cdots\cdots\cdots\cdots\cdots\cdots\cdots\cdots\cdots\cdots$ ***Strychnaceae****-Spigelieae*
7. Capsules with two persistent valves; inflorescence cymose, thyrsoid, or flowers
    solitary (cincinnous in *Mitreola*) $\cdots\cdots\cdots\cdots\cdots\cdots\cdots\cdots\cdots\cdots\cdots\cdots$ 8
8. Corolla aestivation valvate; seeds flattened, winged, or slenderly spindle-shaped;
    included phloem present (absent in *Usteria*) $\cdots\cdots\cdots\cdots$ ***Strychnaceae****-Antonieae*
8. Corolla aestivation imbricate (valvate in *Mitrasacme*); seeds rounded, ellipsoid, or
    angular, without wing; included phloem absent $\cdots\cdots\cdots\cdots\cdots\cdots$ ***Loganiaceae***

APPENDIX 2.

CHARACTERS

Characters used in the phylogenetic analysis of *Strychnaceae*. All characters were
treated as unordered except for ch. 49, which reflects a biosynthetic pathway.
Literature consulted for morphology included Backer & Bakhuisen van der Brink
(1965), Blackwell (1968), Conn *et al.* (1996), Hooker (1837), Leenhouts (1963),
Leeuwenberg & Leenhouts (1980), Leeuwenberg (1961b, 1963, 1969a, 1969b),
Ohwi (1965), Rogers (1986), Smith (1988), Smith & Stone (1962), Wagner *et al.*
(1990), and Yang & Peng (1994). For anatomical, palynological, and phytochemical
characters Jensen (1992), Mennega (1980), Punt (1980), and Punt & Leenhouts
(1967) were used as reference works.

1. **Indumentum on branches**: absent (0), present (1). Taxa that have indumentum
    on branches, even if only in very young stages, have been coded with state 1.
2. **Lenticels on branches**: absent (0), present (1). This is a synapomorphy for
    Norrisia. In *Norrisia maior* the lenticels are not very conspicuous, however.
3. **Colleters in leaf axils**: absent (0), present (1). Colleters are specialised
    glandular multicellular hairs that occur in the leaf axils and at the inner bases
    of sepals in nearly all taxa of the Gentianales investigated (e.g. Robbrecht
    1988). Within *Antonieae*, *Bonyunia antoniifolia* and *Bonyunia aquatica* lack
    colleters in the leaf axils.
4. **Colleters inside calyx**: absent (0), present (1). All species of *Antonieae* (except
    Antonia ovata) lack colleters at the inner base of the calyx.
5. **Axillary thorns**: absent (0), present (1). Several species of *Strychnos* have axillary
    thorns (e.g. *Strychnos axillaris*, *Strychnos nux-vomica*, and *Strychnos spinosa*).
6. **Tendrils**: absent (0), present (1). In many species of *Strychnos*, e.g. *Strychnos
    axillaris*, some branches are developed as woody tendrils, positioned either
    solitary or pairwise.
7. **Leaf venation**: eucamptodromous (0), acrodromous (1). Acrodromous leaf
    venation is a synapomorphy for *Strychnos* in this study, although there are other
    Strychnos species not included in this study that have eucamptodromous
    venation (Leeuwenberg & Leenhouts 1980).

8. **Indumentum on leaf, abaxial**: absent (0), present (1).
9. **Number of bracteoles**: absent (0), 1 pair (1), 2-3 pairs (2). Possession of more than one pair of bracteoles forms a synapomorphy for *Antonia* and *Norrisia*.
10. **Inflorescence**: cymose, often thyrsoid (0), cincinnous (1). *Spigelia* is characterised by having cincinnous inflorescences with short-pedicelled flowers.
11. **Flower sexuality**: bisexual (0), gynodioceous or unisexual (1). Unisexual or gynodioceous flowers are only found in *Geniostoma* and *Labordia*.
12. **Perianth**: pentamerous (0), tetramerous (1). A tetramerous perianth is found only in *Usteria* and *Gardneria ovata*. Some *Strychnos* species not included here are also tetramerous (e.g., *Strychnos asterantha*, *Strychnos madagascariensis*, and *Strychnos canthioides*; Leeuwenberg & Leenhouts 1980).
13. **Calyx aestivation:** imbricate (0), valvate or open (1).
14. **Calyx lobes:** equal (0), unequal (1), calycophyllous (2). Within *Labordia*, *Spigelia*, and *Antonieae* there is a tendency toward unequal calyx lobes, with its most prominent expression (a single, highly expanded calycophyll) present only in *Usteria*.
15. **Indumentum on calyx, external:** absent (0), present (1).
16. **Calyx lobe margin:** glabrous (0), ciliate (1).
17. **Corolla aestivation:** contort (0), valvate (1), imbricate (2).
18. **Corolla shape:** salverform (0), rotate (1). Rotate corollas occur in *Gardneria* and some species of *Strychnos* (e.g. *Strychnos ovata*).
19. **Corolla colour at anthesis:** yellow, white, green (0), pink, red, purple (1). The coding of this character reflects hypothetical biosynthetic transformations in the phenylpropanoid pathway to anthocyanins. Yellow or cream colours may reflect accumulation of tetrahydroxychalcone, the product of chalcone synthase activity, whereas green or white colours may indicate the inactivity of this enzyme. The present coding is designed to distinguish a more distinctive step, the conversion of tetrahydroxychalcone derivatives into red-blue anthocyanins by the enzyme dihydroflavonol reductase (cf. Forkmann 1991).
20. **Indumentum on corolla, abaxial**: absent (0), present (1). A thin indumentum on the outside of the corolla is present in *Labordia*, *Strychnos*, and *Antonieae* (except *Antonia ovata*).
21. **Indumentum on corolla, adaxial**: absent (0), present (1). Hairs of various types and positions are common on the inside of corollas in *Strychnaceae*. *Antonieae* are usually irregularly hairy inside the corolla tube, and in *Geniostomaceae* and *Strychneae*, a ring of hairs in the corolla mouth is often present.
22. **Stamen number**: same as corolla lobes (0), one (1). *Usteria guineensis* has only one stamen, in contrast to all other species in *Strychnaceae* and *Geniostomaceae*.
23. **Filaments, length**: short (0), long (1). Most taxa have very short filaments (up to 1 mm), but longer filaments are present in *Spigelia anthelmia*, *Usteria*, *Antonia*, and *Norrisia*.
24. **Filaments in cross-section**: rounded (0), flat (1). Flattened filaments are characteristic for *Gardneria* and *Neuburgia*.
25. **Filaments with hairs at the base**: absent (0), present (1). *Antonia ovata* has filaments that are bearded at the insertion point in the corolla tube.

26. **Indumentum on filaments**: absent (0), present (1). Hairy filaments are present in *Geniostoma* and are sometimes present in *Strychnos ovata*.

27. **Anthers, number of cells**: 2-celled (0), 4-celled (1). Anthers with four cells are present in *Gardneria multiflora* only.

28. **Indumentum on anthers**: absent (0), present (1). Anthers with hairs (i.e., bearded anthers) are scattered among species in *Strychneae* (e.g. *Strychnos axillaris, Strychnos ovata, Neuburgia corynocarpa*) and are sometimes present in *Geniostoma rupestre*.

29. **Anther dehiscence**: introrse (0), latrorse (1). Latrorse anthers are a synapomophy for *Norrisia*. All other taxa have introrse anthers.

30. **Anther fusion**: free, not connivent (0), coherent or connivent (1). Fused (or adherent) anthers are a characteristic feature for most species of *Gardneria*, and forms a synapomorphy for the *Gardneria* species included in this study.

31. **Shape of anther apex:** rounded (0), apiculate, sterile (1). Anthers with apiculate, sterile apices are present in *Geniostoma, Bonyunia, Neuburgia, Gardneria*, and *Strychnos axillaris*.

32. **Indumentum on ovary:** absent (0), present (1). Pubescent ovaries are found in *Labordia, Bonyunia, Norrisia*, and *Strychnos ovata*.

33. **Style:** persistent (0), deciduous (1). Deciduous styles are a synapomorphy for *Spigelieae* and *Antonieae* (*Usteria* has persistent styles), and occur in *Neuburgia* as well.

34. **Indumentum on style and stigma:** absent (0), present (1). A pubescent style and stigma is present in *Bonyunia, Spigelia*, and *Strychnos ovata*.

35. **Position and shape of placenta:** axile, not stalked (0), axile, stalked (1). *Antonieae* and *Spigelieae* have axile, peltate placentas, whereas *Strychneae* and *Geniostomaceae* have inconspicuous axile placentas.

36. **Placenta**: fleshy (0), dry (1). Fleshy placentas are characteristic of all species of *Geniostoma, Labordia*, and *Strychneae*.

37. **Mesocarp:** fleshy or leathery (0), dry (1). *Neuburgia* shows similarity to *Geniostomaceae, Antonieae*, and *Spigelieae* in having fruits with dry mesocarp. In *Neuburgia* the mesocarp is very fibrous. *Gardneria* and *Strychnos* have a fleshy or leathery mesocarp.

38. **Fruit valves:** with torn apex (0), apex not torn (1), indehiscent (2). *Strychneae* have indehiscent fruits. The dehiscent fruits of *Antonieae* and *Spigelieae* have valves, either with torn apices (*Antonia, Bonyunia*, and *Norrisia*) or without (*Spigelia* and *Usteria*). *Labordia* and *Geniostoma* also have capsules without torn apices.

39. **Valves:** persistent (0), deciduous valves with a persistent cupular base (1), completely deciduous valves (2). The capsules of *Spigelia* are 4-valved and circumscissile at the base; the valves fall off, leaving only the base of the capsule, while in *Labordia* and *Geniostoma* the entire valves are deciduous. All other taxa either have indehiscent fruits or fruits with persistent valves.

40. **Fruit skin:** smooth and glabrous (0), warty (1), smooth and pubescent (2). Pubescent fruits are a synapomorphy for *Bonyunia* and are also present in *Labordia waiolani*. Warty capsules are characteristic for *Spigelia anthelmia*.

41. **Fruit colour at maturity:** red, orange (0), brown, beige (1), white (2), black (3). All (capsular) fruits in *Spigelieae* and *Antonieae* are beige or brown. A majority of species in *Strychneae* have red berries or drupes. However, *Neuburgia corynocarpa* has white fruits and *Gardneria multiflora* has black fruits, which are also found in *Geniostoma rupestre*.

42. **Seed shape:** polyhedral or round (0), convex on one side, flat or concave on the other (1). All seeds of *Strychneae* are flattened or nearly so, with one convex side and the other side flat or concave. Flattened, concave seeds also occur in *Geniostoma*.

43. **Seed wing:** absent (0), present (1). A seed wing is present in all genera of *Antonieae*, except in *Norrisia*, which has minute, spindle-shaped seeds.

44. **Seed surface**: not pubescent (0), pubescent (1). Some species of *Strychnos* have seeds with a pubescent (felt-like) testa (e.g. *Strychnos nux-vomica* and *Strychnos ovata*).

45. **Endosperm**: not ruminate (0), ruminate (1). Ruminate endosperm is a synapomorphy for *Spigelia*.

46. **Pollen aperture**: colporate or colpate (0), porate (1). Porate pollen is characteristic for *Labordia* and *Geniostoma* (Punt 1980), whereas all other taxa included in this study have colporate or colpate pollen.

47. **Pollen ornamentation, tectum**: smooth and/or perforate (0), reticulate (1). Pollen with reticulate tecta are present in *Spigelia anthelmia* only (Punt 1980). *Spigelia humboldtiana* has not been investigated.

48. **Scabrae on pollen**: absent (0), present (1). Scabrae, small bumps on the exine, are present in *Usteria*, *Antonia*, and *Norrisia*, but are absent from *Bonyunia*, *Strychneae*, and *Geniostomaceae* (Punt 1980).

49. **Seco-iridoids and complex indole alkaloids**: absent (0), seco-iridoids only (1), seco-iridoids and complex indole alkaloids present (2). This character is ordered in accordance to a biosynthetic pathway for indole alkaloids derived from seco-iridoids (Kisakürek & Hesse 1980; Jensen 1992; see also Struwe *et al.* 1994, for discussion). Seco-iridoids and alkaloids are absent from *Spigelieae* and *Antonieae* (*Usteria* is not investigated), but occur in *Strychnos* and *Gardneria* of tribe *Strychneae*. *Geniostomaceae* have not been investigated.

50. **Rays**: exclusively uniseriate or locally biseriate (0), uniseriate and multiseriate (1). The wood of *Antonia*, *Neuburgia*, *Norrisia*, and *Spigelia* is characteristically uniseriate or rarely biseriate; Mennega 1980).

51. **Included phloem**: absent (0), present (1). Included phloem of the foraminate type is present in *Strychnos*, *Antonia*, *Norrisia*, and *Bonyunia* (Mennega 1980).

Fruits from the tropical rain forest in East Kalimantan and their market potential

JOHAN L. C. H. VAN VALKENBURG[1]

Summary. In a study of the economic and ecological potential of Non-Timber Forest Products of East Kalimantan, fruit trees turned out to be an important commodity group. Fruits were traded in large amounts and fruit trees were planted by all ethnic groups present in East Kalimantan. The diversity of fruit tree species found in a primary forest in East Kalimantan will be compared with the diversity found in the markets of Samarinda, the capital of the province. First selection of species takes place at village level where preferred fruits from the forest are planted and cared for in the vicinity of the village or along tracks. Management systems of these home gardens and orchards differ in levels of maintenance and manipulation. Selection takes place on the basis of personal preferences, ethnic background and possibilities to market surplus amounts. Transportation problems are the prime cause of a further reduction of the diversity of fruits on sale in a major urban centre. Examples will be given of villages of various ethnic groups, with different levels of access to urban centres.

INTRODUCTION

In a study carried out between 1992 and 1994 of the economic and ecological potential of Non-Timber Forest Products of East Kalimantan, fruit trees turned out to be an important commodity group. Descriptions of the species discussed may be found in Valkenburg (1997) and Map 1 gives the locations.

RESULTS AND DISCUSSION

A. *Long Sungai Barang*

Due to its remoteness, the village of Long Sungai Barang, upstream on the Kayan river (Map 1), is unable to market any of its surplus fruits. The gathering of fruit species from the forest and planting in home gardens is considered to be influenced by personal preferences and ecological constraints only.

The edibility of certain wild fruits is open to some question and controversy. Therefore the diversity and abundance of fruit trees will be given by using the list of Jansen *et al.*(1991) and based on the information of local Kenyah informants. This approach in using local informants is comparable with the method of Saw *et al.*(1991) and resulted in a reduction in the number of species bearing edible fruits.

In 1.1 ha of primary forest a total of 26 species with edible fruits were recorded, representing 9% of a total flora of 264 species (trees > 10 cm dbh). These species were present with 92 trees ha[-1] on a total of 670 trees ha[-1], or 13% of all trees.

---

[1]Rijksherbarium/Hortus Botanicus Leiden, P.O. Box 9514, NL-2300 RA Leiden, The Netherlands.

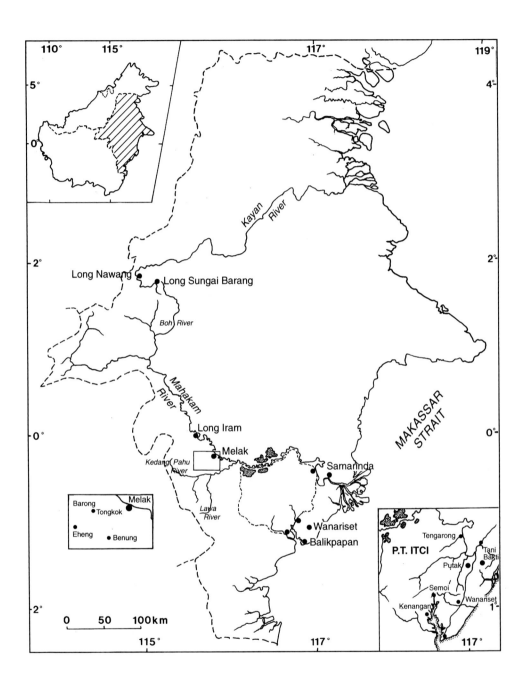

Map of East Kalimantan with the various research sites.

Local informants classified 18 species as edible (3 *Mangifera* species were considered not edible) with a density of 69 trees ha⁻¹, representing 6% of tree species and 10% of trees.

None of the species with edible fruits in the plots is actively gathered; they are collected only as the occasion presents itself. *Baccaurea, Garcinia* and *Nephelium* species are collected in this way. *Durio spp.* and *Parkia speciosa* are actively gathered but were not present in the plots.

Although many fruit species are present in the home gardens, the number of indigenous fruit species planted in the orchards and along tracks is limited to eleven with a clear predominance of five species: *Baccaurea macrocarpa, Dimocarpus longan* subsp. *malesianus, Durio zibethinus, Mangifera pajang* and *Nephelium ramboutan-ake.*

Artocarpus integer, although present in the primary forest surrounding Long Sungai Barang, is not planted as the people do not like the fruit.

B. Barong Tongkok area

The Barong Tongkok area, traditionally inhabited by Benuaq and Tunjung Dayak, is a well known source of fruits in the Samarinda market. The fruits are bought in the local markets by middlemen and shipped downstream from Melak to the market at Samarinda. Only when a sufficiently large volume can be bought, and an adequate profit can be made to defray costs of transportation and handling, is the fruit shipped to Samarinda.

In addition to planting fruit trees in home gardens in the vicinity of villages, planting of fruit trees is incorporated in the swidden cycle. This management system is generally referred to as Lembo-culture. An important difference with the swidden cycle as practised in Long Sungai Barang, is that individual households will stay for up to four years in their swidden, working on alternate pieces of land, thereby allowing for maintenance of the planted fruit trees. In Long Sungai Barang a group of households will clear a large swidden and move to another area the next year. The great diversity of fruit species found in these home gardens (Lembo) and the multitude of species traded in local markets has been described by Sardjono (1990). The present study concentrated on the indigenous fruit species traded in local markets, with special attention to the village of Benung. A total of 31 fruit species were observed or reported on sale in local markets (Table 1). No distinction was made in 'cultivars' or selections of e.g. *Nephelium lappaceum* and *N. ramboutan-ake,* but an exception was made for two 'cultivars' of *Dimocarpus longan* subsp. *malesianus.* The cultivation status of the various fruits is based on interviews with local informants and observations in home gardens. Ten species were shipped downstream to Samarinda market. None of the *Mangifera* species was shipped, as the fruits are apparently too delicate to be shipped in bulk, or are simply not favoured by buyers in Samarinda.

C. Vicinity of Samarinda

The indigenous fruit trees present in home gardens in two villages at less than one hour by road from the Samarinda market were compared. Tani Bakti, a transmigration village with Javanese farmers established in 1973 and Putak, a

TABLE 1. Important indigenous fruit trees in the village of Benung, Barong Tongkok area. Trade in local markets and in Samarinda markets downstream

	Local market	Samarinda	Status
Archidendron jiringa	*	*	cultivated
Artocarpus anisophyllus	*		wild & cultivated
Artocarpus integer	*	*	primarily cultivated
Artocarpus lanceifolius	*		wild
Baccaurea macrocarpa	*	*	primarily cultivated
Baccaurea pyriformis	*	*	wild & cultivated
Baccaurea spp. (3)	*		primarily wild
Dacryodes rostrata	*		wild & cultivated
Dimocarpus longan 'green'	*		primarily cultivated
Dimocarpus longan 'yellow'	*	*	primarily cultivated
Durio dulcis	*		primarily wild
Durio kutejensis	*	*	primarily cultivated
Durio oxleyanus	*		primarily wild
Durio zibethinus	*	*	primarily cultivated
Lansium domesticum	*	*	wild & cultivated
Mangifera spp. (8)	*		wild & cultivated
Nephelium lappaceum	*	*	cultivated
Nephelium spp. (4)	*		wild & cultivated
Parkia speciosa s.l.	*	*	wild & cultivated

village inhabited by Tunjung Dayak. The diversity of indigenous fruit species planted in the Dayak village far outnumbers the indigenous species planted in the transmigration village, 22 species versus 9 (Table 2). Noteworthy is that *Mangifera caesia* and *M. similis* are planted in the transmigration village, whereas these species are collected from the surrounding forest in the dayak village. These two *Mangifera* species are reported in cultivation in Java (Verheij & Coronel 1991). Therefore the transmigrant farmers are familiar with the species and have incorporated them. The transmigrant farmers have up to now restricted themselves to planting fruit trees they were familiar with in Java, with the exception of *Durio kutejensis*.

D. Samarinda market

The major indigenous fruits in the Samarinda markets originate from home gardens (Table 3). The supply may come from the vicinity of Samarinda, or areas upstream as much as two days travel by boat, depending on the fruit species. Also, fruits transported by road from Banjarmasin are occasionally sold in the market.

TABLE 2. Indigenous fruit trees present, cultivated and traded in two villages in the vicinity of Samarinda

	Putak	Tani Bakti	Trade	Status
Archidendron jiringa	*	*	P, T	cultivated
Artocarpus anisophyllus	*			primarily wild
Artocarpus elasticus				wild
Artocarpus integer	*	*	P, T	primarily cultivated
Artocarpus lanceifolius		+		wild
Artocarpus odoratissimus	*			primarily cultivated
Baccaurea macrocarpa	*		P	primarily cultivated
Dacryodes rostrata	*			wild & cultivated
Dimocarpus longan 'green'	*		P	primarily cultivated
Dimocarpus longan 'yellow'	*		P	primarily cultivated
Durio dulcis	*		+#	primarily wild
Durio kutejensis	*	*	P	primarily cultivated
Durio oxleyanus		*	+#	primarily wild
Durio zibethinus	*	*	P	primarily cultivated
Lansium domesticum	*	*	P	wild & cultivated
Litsea garciae	*			primarily wild
Mangifera caesia	*	+		primarily wild
Mangifera decandra	*		P	wild & cultivated
Mangifera laurina	*		P	primarily cultivated
Mangifera pajang	*			primarily cultivated
Mangifera similis	*	+		wild & cultivated
Nephelium cuspidatum	*			wild
Nephelium lappaceum	*	*	P, T	cultivated
Nephelium ramboutan-ake	*			primarily cultivated
Parkia speciosa s.l.	*	*	P	wild & cultivated
Sandoricum koetjape	*		+	primarily cultivated

P   traded in Putak
T   traded in Tani Bakti
+   sold in Samarinda market, but originating elsewhere
+#   sold in stalls along the Balikpapan-Samarinda road

TABLE 3. Indigenous fruits on sale in two major markets in Samarinda, East Kalimantan, 1992-1994

	Importance	1992	1993	1994	Status	Vernacular name	Price (Rp.)
Archidendron jiringa	++	*	*	*	cultivated	jengkol	1500 – 2000/kg
Artocarpus integer	++++	*	*	*	primarily cultivated	cempedak	500 – 1000/fruit
Artocarpus lanceifolius	+	*		*	wild	keledang	150/fruit
Baccaurea macrocarpa	+++	*		*	primarily cultivated	kapul	500 – 1000/kg
Baccaurea motleyana	++++	*	*	*	cultivated	rambai	300 – 500/kg
Baccaurea pyriformis	+++	*		*	primarily cultivated	cantik manis, jentikan	500 – 1000/kg
Baccaurea macrophylla	+	*		*	cultivated	ramania	50 – 75/fruit
Dialium indum	+			*	wild	asam keranji	2000/kg
Dimocarpus longan subsp. *malesianus*	+++	*		*	cultivated	ihau, mata kuching	1350 – 2000/kg
Durio kutejensis	++++	*	*	*	cultivated	lai	500 – 1000/kg
Durio zibethinus	++++	*	*	*	cultivated	durian	500 – 5000/fruit
Garcinia ?dulcis	++	*	*	*	?wild	asam kandis	5000 – 10000/kg
Garcinia mangostana	++	*	*	*	cultivated	manggis	30 – 100/fruit
Lansium domesticum	++++	*	*	*	cultivated & wild	langsat	500 – 1000(–1500)/kg
Mangifera caesia	+			*	primarily wild	wanyi	800/kg
Mangifera similis	+			*	primarily wild	asam putaran	50 – 150/fruit
Nephelium lappaceum	++++	*	*	*	cultivated	rambutan	500 – 650/kg
Pangium edule	++	*	*	*	wild	kluwak	2000/kg
Parkia speciosa s.l.	++	*	*	*	cultivated & wild	peteh	500 – 1500/3 pods
Sandoricum koetjape	++++	*	*	*	primarily cultivated	kecapi	40 – 50/fruit

+   occasionally in the market in small amounts   +++   occasionally in the market in large amounts

++   regularly in the market in small amounts   ++++   regularly in the market in large amounts

Fragile fruits like *Bouea macrophylla*, *Mangifera caesia* and *M. similis*, originate from the vicinity of Samarinda. The same applies for the forest-collected *Artocarpus lanceifolius*.

In the market a distinction is made between forest-collected and cultivated *Lansium domesticum*. The forest-collected fruits have a somewhat thicker skin. Likewise a distinction is made between 'peteh hutan' and cultivated peteh commonly called 'peteh jawa'. Both types are considered to belong to *Parkia speciosa*. The peteh hutan has narrower pods, the seeds are smaller and have a more bitter taste.

Irregular fruiting (McClure 1966, Medway 1972), with indigenous fruit trees fruiting only once in 3 – 5 years, was observed in several areas in East Kalimantan. Also fruiting seasons differ considerably between areas. These two phenomena are obscured in the Samarinda markets because the area that the Mahakam river basin covers, with its multitude of tributaries, is enough to allow fruits to be harvested from somewhere within it at any one time.

General comments

Regional differences within East Kalimantan, with respect to natural distribution of fruit trees, are considerable. *Dialium indum* and other *Dialium* species with edible fruits, though omnipresent in the Balikpapan-Samarinda area, were not encountered in the vicinity of Long Sungai Barang where *Lansium domesticum* and *Durio kutejensis* were also not found.

Fruit species reported to be planted and traded in Sabah (William *et al.* 1992) or Sarawak (Verheij & Coronel 1991), e.g. *Artocarpus odoratissimus* are of no importance in East Kalimantan. However, *Durio kutejensis* often considered as a minor species (Verheij & Coronel 1991), is of major economic importance in East Kalimantan.

Although access to urban markets may result in a shift towards fruit species that can be sold in these markets (e.g. *Durio zibethinus*, *D. kutejensis*, *Artocarpus integer*, *Lansium domesticum*), many species are still planted because people prefer them for home consumption with any surplus sold in local markets. A good example are *Nephelium ramboutan-ake* and *Mangifera pajang*, equally favoured by Benuaq, Tunjung and Kenyah Dayak.

Potential economic species

The genus *Mangifera* with its multitude of species that are found both wild and (semi-)cultivated, especially along the Mahakam river, yield many good-quality fruits (Bompard & Kostermans 1992, Verheij & Coronel 1991), but trade is limited to local markets. The thick rind of *Mangifera pajang* might overcome the problems of rough handling during transportation, an essential problem in marketing. However, the fact that non-indigenous people are unfamiliar with the fruit limits its market potential at present.

The fruits of *Durio kutejensis* have good potential for markets outside East Kalimantan. Not only is the fruit a common sight in urban markets, the tree also starts fruiting at an early age when only 4 – 5 m tall. The fruits are medium-sized and have soft flexible spines, which makes handling easier. The yellow colour of the fruit, and especially the dark yellow to orange colour of the aril, are attractive. In flavour and texture of the aril the species somewhat resembles the ordinary durian.

Another example of a fruit species with potential for markets outside East Kalimantan, is *Baccaurea macrocarpa*. The thick woody rind makes it a sturdy fruit, resistant to rough handling in transportation.

The thick woody rind also extends the shelf life of this species, as compared to the more commonly sold rambai (*Baccaurea motleyana*). The variation in sweetness and size of the fruits points to the great potential for selection.

Finally a species that is considered to have potential by the author, but is rarely planted in East Kalimantan and was never found in the market. The fruits of *Artocarpus rigidus* have a delicious citrus-like refreshing sweet flavour and an attractive orange coloured perianth.

TRANSPORT AND HANDLING

Before any of the above mentioned species with economic potential can really be marketed, two major problems need to be solved.

Transportation time needs to be reduced and handling/packaging needs to be improved, both at regional level to increase market potential of fragile fruits and especially for inter-island trade. Port facilities for export are absent at present.

The problem of often irregular fruiting, resulting in an unpredictable supply, may partly be solved through selection, and possibly also by applying fertilizer.

ACKNOWLEDGEMENTS

Fieldwork for the present study was carried out within the framework of the international MOF-Tropenbos Kalimantan Project.

REFERENCES

Bompard, J. M. & Kostermans, A. J. G. H. (1992). The genus *Mangifera* in Borneo: results of a IUCN-WWF/IBPGR Project. In: Ghazally Ismail, Murtedza Mohamed and Siraj Omar. Proceedings of the International Conference on Forest Biology and Conservation in Borneo, July 30 – August 3, 1990 Kota Kinabalu, Sabah, Malaysia. Yayasan Sabah, Sabah.

Jansen, P. C. M., Lemmens, R. H. M. J., Oyen, L. P. A., Siemonsma, J. S., Stavast, F. M. & Valkenburg, J. L. C. H. van (1991). Plant Resources of South-East Asia. Basic list of species and commodity grouping (final version). 372 pp. Pudoc, Wageningen, The Netherlands.

McClure, H. E. (1966). Flowering, fruiting and animals in the canopy of a tropical forest. Malay. For. 29: 182 – 203.

Medway, L. (1972). Phenology of a tropical rain forest in Malaya. Biol. J. Linn. Soc. 4: 117 – 146.

Sardjono, M. A. (1990). Die Lembo-Kultur in Ost-Kalimantan. PhD thesis, Universität Hamburg. 212 pp.

Saw, L. G., LaFrankie, J. V., Kochummen, K. M., & Yap, S. K. (1991). Fruit trees in a Malaysian rain forest. Econ. Bot. 45(1): 120 – 136.

Valkenburg, J. L. C. H. van (1997). Non-Timber Forest Products of East Kalimantan. Potentials for sustainable forest use. Tropenbos Series 16. The Tropenbos Foundation, Wageningen, The Netherlands. Pp. 202.

Verheij, E.W.M. & Coronel, R. E. editors (1991). Plant Resources of South-East Asia. No. 2 Edible fruits and nuts. Pudoc, Wageningen, The Netherlands. 446 pp.

William, W., Wong, W. & Lamb, A. (1992). Species diversity of wild fruit trees in the forests of Sabah as illustrated by the genera *Artocarpus, Durio* and *Mangifera.* In: Ghazally Ismail, Murtedza Mohamed and Siraj Omar. Proceedings of the International Conference on Forest Biology and Conservation in Borneo, July 30 – August 3, 1990 Kota Kinabalu, Sabah, Malaysia. Yayasan Sabah, Sabah.

Increased speciation in New Guinea: tectonic causes?

PETER C. VAN WELZEN[1]

Summary. New Guinea, in comparison to other areas in the Malay Archipelago, has a relatively high number of endemic species. These species are mainly found in the northern half of the island and the mountain ranges, and seem to correlate with the tectonic history of New Guinea. The South has always been a geologically stable area, while the northern half is composed of more than 30 terranes (± microplates), which accreted in groups during different periods over the last 40 MA. The correlation between geology and plant distributions is examined by recording the distributions and dividing these into patterns, which are used for the correlation. The distribution data of 961 endemic species are compiled from 10192 specimens. 21 patterns are established, ranging from widespread throughout New Guinea to a local distribution on a single mountain top. The patterns show that 876 out of the 961 species are present in the northern half of New Guinea. There is a slight correlation, mainly only the Finisterre region, between individual terranes and distributions. A high correlation is found between species distribution and groups of terranes which accreted during a particular time period; c. two-thirds of the species are restricted to terranes which accreted during one period. Most endemic species are found in the mountains and thus the correlation between terrane accretion and plant distributions is probably not direct, but indirect with associated orogenic activity, and perhaps also climate. (Unfortunately, the mountains have also been most collected so this might prejudice the conclusion!).

INTRODUCTION

During the second Flora Malesiana Symposium the distribution data of all indigenous species published in Flora Malesiana series 1, 4 – 11(2), amounting to one sixth of the estimated flora, were compiled and presented (van Welzen 1992, following an idea in van Welzen 1989). This major project suggested that large islands in the Malay Archipelago (Malesia), such as Borneo and New Guinea, usually show more species complexes and harbour more species than the other island groups in Malesia. However, Fig. 1 (left columns) and Table 1 show this hypothesis to be falsified; smaller areas like Sumatra and especially the Malay Peninsula have almost equal numbers as Borneo and New Guinea.

Endemic species show a completely different picture (Fig. 1, right columns). Borneo and especially New Guinea have significantly more endemic species than the other phytogeographic areas. The numbers of endemic species deviate significantly from an estimated number based on the numbers of species (Table 1), the flora of Borneo and New Guinea comprises more endemic species than expected, the other areas far less except the Philippines.

The endemic species in Borneo and New Guinea are mainly found in the northern halves of both islands; on Borneo they are found in Sarawak, Brunei,

[1]Rijksherbarium/Hortus Botanicus, University of Leiden, P.O. Box 9514, 2300 RA Leiden, The Netherlands.

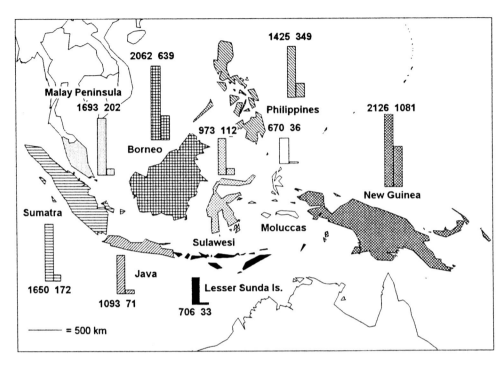

FIG. 1. Phytogeographic areas of Malesia. Numbers of indigenous species are shown in the left columns, numbers of endemic species in the right columns. The distribution data of the species are compiled from Flora Malesiana ser. 1: 4-11(2).

Sabah and N. Kalimantan; in New Guinea in the Vogelkop (the 'Bird's head' peninsula), the central mountain range, and the area north of it. These are the areas with most recent (micro)plate tectonic movements (Daly *et al.* 1991, Michaux 1991, Pigram & Davies 1989), which may indicate that tectonic movements might induce speciation. The speciation can then be traced in the record of endemic species. The central hypothesis in this paper will be the idea that plate tectonic movements indeed cause speciation. This hypothesis will be tested with distributional data of endemic New Guinea plants.

In fact, Lyell's paradigm (1830 – 33) 'The present is the key to the past' is reversed. Nowadays, consensus more or less exists about the fact that every c. 26 million years a mass extinction event occurred, which was followed by a rapid unrelated radiation in other groups of organisms than those which were abundant before the· extinction (Raup 1991). The mass extinctions were worldwide phenomena, but on a more local scale similar processes occurred, e.g. the Deccan traps in India. Extinction in a flora is often difficult to trace, but speciation can be estimated by the numbers of endemic species. Therefore, we let the past be the key to the present, we will use the geological history to see if it correlates with speciation or, more specifically, with distributions of species.

TABLE 1. Indigenous and endemic species in the different floristic areas of Malesia (data from Flora Malesiana ser. 1: 4 – 10(3)). The expected numbers of endemic species are based on the numbers of indigenous species (expected number of endemic species per area = sum of all endemic species/sum of all indigenous species × number of indigenous species per area). The floristic provinces were taken from van Steenis (1950).

	Indigenous spp	Endemic spp	Expected endemics
Sumatra	1650	172	359
Malay Peninsula	1693	202	368
Borneo	2062	639	448
Java	1093	71	238
Philippines	1425	349	310
Sulawesi	973	112	212
Lesser Sunda Islands	706	33	153
Moluccas	670	36	146
New Guinea	2126	1081	462

DATA AND METHODS

The data on endemism in New Guinea were taken from Flora Malesiana series 1, vol. 4 – 11(2) and Orchid Monographs 1 – 7. Only those endemic species (961 in total) were used in the analysis which were represented by specimens in the Rijksherbarium in Leiden, The Netherlands (10,192 specimens in total). The label data were used to produce detailed distribution maps per species with the computer programs COOR (coordinates database of collecting localities, © P.C. van Welzen, Rijksherbarium Leiden) and KORT (dot map drawing program, © B. Hansen, Herbarium, Copenhagen, Denmark). The final analysis and the maps reproduced in this article were prepared with the computer program MapInfo for Windows ®.

The distributions of the endemic species were arranged into patterns, whereby similar distributions which overlapped each other for more than 90% were regarded as the same pattern. If too many intermediate distributions appeared between different patterns then the patterns were united; otherwise, the overlap is indicated in Table 2. Finally, the patterns were compared with the separate terranes and the different phases in the accretion.

Species at different altitudes show different patterns. Therefore, a division was made between high altitude plants (>1500 m) and low altitude plants (< 1500 m). The 1500 m border is arbitrary, but worked quite well when its application was not too strict. Plants occurring between 1000 – 2000 m were recorded to be high altitude plants, while plants occurring between 800 – 1600 m were recorded as low altitude plants. Sometimes, the altitudinal range of the species was too large,

TABLE 2. Main distribution patterns and sub(sub)patterns of endemic plants in New Guinea and their correlation with the geological history of the island in the column Correlating phase. *Italics* = transition patterns with distributions which usually overlap with several geological accretion phases.

No.	Pattern	Number	Subpattern	Number	Subsubpattern	Number	Altitude[1]	Overlap[2]	Correlating phase	Figure
1	Eastwest	15					1 (1 l/h)	2(1)	-	Fig. 4
2	Widespread	46					1 (8 l/h)	3(1)	-	Fig. 5
3	Craton	24					1 (2 l/h)	7b2(2), 7c(5)	Craton	Fig. 6
4	North widespread	54					1 (25 l/h, 4 h)	2(15)	-	Fig. 6
5	25 MA	31					1 (1 l/h)	2(2)	25 MA	Fig. 7
5.a			widespread	2			1		25 MA	Fig. 7
5.b			Irian Jaya	11			1	2(1)	25 MA	Fig. 7
5.c			Papua New Guinea	18			1 (1 l/h)	2(1)	25 MA	Fig. 7
6	*25-10(-2) MA*	*19*					*l*	*4(1)*	-	*Fig. 7*
7	Mountain	419					h (15 l, 23 l/h)		-	Fig. 8-13
7.a			Vogelkop-Peninsula	29			h	2(8)	-	Fig. 8
7.b			central part	58			h (5 l, 8 l/h)		25 MA	Fig. 9
7.b.1					widespread	30	h (3 l/h)		25 MA	Fig. 9
7.b.2					Star mountains	20	h (5 l, 1 l/h)		25 MA	Fig. 9
7.b.3					eastwest	8	h (5 l, 4 l/h)		25 MA	Fig. 9
7.c			Central Irian Jaya	129			h (2 l, 3 l/h)	1(2)	25 MA	Fig. 10
7.c.1					widespread	27	h (1 l/h)		25 MA	Fig. 10
7.c.2					Idenburg	9	h		25 MA	Fig. 10
7.c.3					Carstensz mts	37	h (2 l, 1 l/h)		25 MA	Fig. 10
7.c.4					Doorman mt	15	h		25 MA	Fig. 10
7.c.5					Snow mts.	41	h (1 l/h)		25 MA	Fig. 10
7.d			Vogelkop-Central NG	16			h (3 l/h)	1(1), 7a(2)	-	Fig. 11
7.e			Papua New Guinea	47			h (2 l/h)		25 MA	Fig. 10, 11
7.e.1					single mountain	17	h		25 MA	Fig. 10
7.e.2					central mt group	30	h (2 l/h)		25 MA	Fig. 11
7.f			Finisterre-Central NG	49			h (3 l/h)	7e2(4)	-	Fig. 12
7.g			East New Guinea	91			h (3 l, 4 l/h)		-	Fig. 13
7.g.1					Pen.-central PNG	40	h (1 l/h)	7e2(1)	-	Fig. 13
7.g.2					Pen.-border	18	h (2 l, 1 l/h)		-	
7.g.3					Pen.-Vogelkop	33	h (1 l, 2 l/h)	1(10), 4(2)	-	
8	15 MA Mountain	80					h (2 l/h)		15 MA	Fig. 14, 15
8.a			widespread Peninsula	8			h (1 l/h)		15 MA	Fig. 14
8.b			North + central Penin.	53			h (1 l/h)		15 MA	Fig. 15
8.c			East Peninsula	19			h		15 MA	Fig. 15

Table 2 continued

No.	Pattern	Number	Subpattern	Number	Subsubpattern	Number	Altitude[1]	Overlap[2]	Correlating phase	Figure
9	15 MA Low alt.	58								
9.a			widespread Peninsula	9			1		15 MA	Fig. 16, 17
9.b			central Peninsula	6			1		15 MA	Fig. 16
9.c			North coast Peninsula	14			1		15 MA	Fig. 17
9.d			South coast Peninsula	19			1		15 MA	Fig. 17
9.e			Southeast Peninsula	10			1		15 MA	Fig. 17
10	*15 MA-Finisterre Mt.*	*20*					*h (1 l/h)*		-	
11	*15 MA-Fin.-10 MA low*	*3*					*l*	*4(3)*	-	
12	*15 MA-Finisterre low*	*18*					*l (3 l/h)*			
13	Finisterre	37								Fig. 14, 15
13.a			mountain	25			h (2 l/h)		10 MA	Fig. 14
13.b			low alt.	12			l (1 l/h)		10 MA	Fig. 15
14	*Finisterre - 10 MA*	*3*					*l (1 l/h)*		*10 MA*	*Fig. 16*
15	10 MA	31					1 (2 l/h, 2 h)		10 MA	Fig. 12
15.a			NW shoulder Irian Jaya	4			1		10 MA	Fig. 12
15.b			East Irian Jaya	15			l (1 l/h, 2 h)	15a(1)	10 MA	Fig. 12
15.c			West Papua New Guinea	2			l		10 MA	Fig. 12
15.d			Central Papua New G.	10			l (1 l/h)		10 MA	Fig. 12
16	*10 MA-Vogelkop*	*5*					*l*		*10 MA*	*Fig. 13*
17	Vogelkop	81					h (28 l, 4 l/h)		10 MA	Fig. 14, 15
17.a			mountain	52			h (4 l/h)	7d(1)	10 MA	Fig. 14
17.b			low alt.	29			1		10 MA	Fig. 15
18	*Vogelkop-Waigeo*	*1*					*l*		-	*Fig. 16*
19	Waigeo	3					1		2 MA	Fig. 17
20	Bismarck Archipelago	11							2 MA	Fig. 16, 17
20.a			widespread	2			l (2 l/h)		2 MA	Fig. 16
20.b			New Britain	4			l (1 l/h)		2 MA	Fig. 17
20.c			New Ireland + Manus	5			1		2 MA	Fig. 17
21	Bougainville	2					l (1 l/h)		2 MA	Fig. 16

[1]Altitude of the plants. l = low altitude, < 1500 m; h = high altitude, > 1500 m; l/h = present from low to high altitude. First character applies to all endemic plants in the pattern concerned, between brackets numbers of species deviating from the prevalent altitude.

[2]Distributions are sometimes difficult to refer to one pattern, the numbers before the brackets refer to the number of the alternative pattern with which overlap occurs, between brackets the numbers of species with overlapping distributions.

then the plants were noted as low/high altitude species. Plants in the latter category were usually united with the most common (low or high altitude) category in a certain pattern.

The geological history of New Guinea is based on Pigram & Davies (1987). The southern part of New Guinea was part of the Australian Craton (below, 'the Craton'; craton = geologically stable area). The northern part of New Guinea ('the North') consists of more than 30 different terranes (or 'microplates') of various origin: sea floor, microcontinents, parts of craton margin, volcanic arcs, etc. (Fig. 2). Most terranes are relatively small in area because they are partly covered by later terranes and by sedimentation from post-accretion basins which occurred at plate borders. Later, terranes were often tilted during the uplift of the central mountain ranges, which decreased the surface plane of the terranes even more. The terranes accreted in several phases (Pigram & Davies 1987, their Fig. 4, see also Fig. 3). At 40 MA only the Craton was present, while by 25 MA the terranes had accreted which now form part of the central mountain range together with the northern edge of the Craton. Around 15 MA the East Papuan Peninsula (below, 'the Peninsula') accreted, the terranes making it up having already amalgated before they joined New Guinea. At 10 Ma the northern edge and the Vogelkop were formed, and 2 MA several additional terranes around the Vogelkop docked. Next in line to accrete with New Guinea are the Bismarck Archipelago and the Solomon Islands.

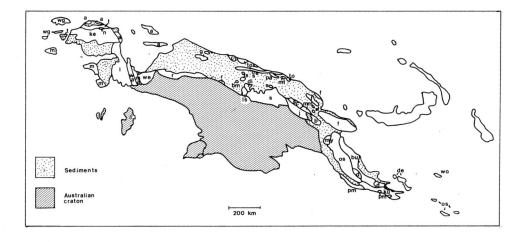

FIG. 2. The different terranes present in New Guinea as described by Pigram & Davies (1987). a = Arfak; b = Bena Bena; bm = Boder mountains; bu = Bowutu; c = Cyclops; d = Dayman; de = D'Entrecasteaux; di = Dimaie; f = Finisterre; g = Gauttier; j = Jimi; ke = Kemum; ku = Kutu; l = Lengguru; ls = Landslip; m = Misool; ma = Mangguar; mm = Marum; mt = Mount Turu; my = Menyamya; n = Netoni; os = Owen Stanley; pa = Prince Alexander; pm = Port Moresby; r = Rouffaer; s = Sepik; sc = Schrader; t = Tamrau; to = Torricelli; wa = Wandamen; we = Weyland; wg = Waigeo; wo = Woodlark. Drawing after Fig. 2 of Pigram & Davies (1987).

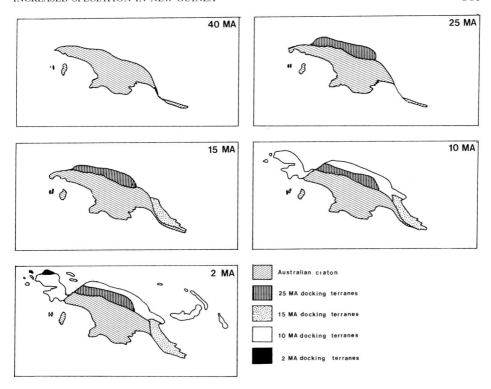

FIG. 3. Phases in the accretion history of New Guinea. Drawing after Fig. 4 of Pigram & Davies (1987).

PATTERNS

The different distributional patterns and their subdivisions are shown in Table 2 in which data about the number of species per (sub)pattern, altitude at which the species are found, overlap with other (sub)patterns, and the correlation with any of the geological phases can also be found.

Several patterns cannot be linked to the accretion history of the terranes:

— East – west distribution (pattern 1; Fig. 4). 15 species have a disjunct distribution in the east and the west of New Guinea. Perhaps the species delimitation has to be re-evaluated.

— Widespread distribution (pattern 2; Fig. 5). 32 species are widespread in New Guinea. Not all of them are found throughout New Guinea, some are found quite locally, but the defining criterion for widespread was presence in both the south (the craton area) and the north (terrane area).

The other patterns can more or less be correlated with the geological history of New Guinea. However, several northern patterns cannot be correlated with any exact geological period. These are:

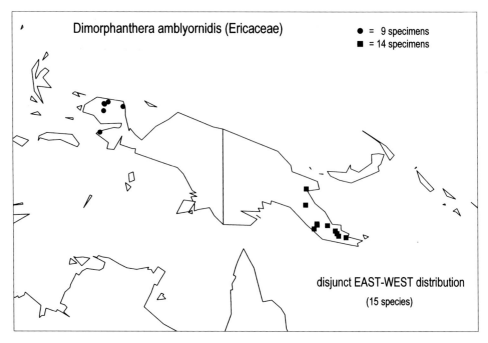

FIG. 4. Pattern 1: Disjunct East-West distribution.

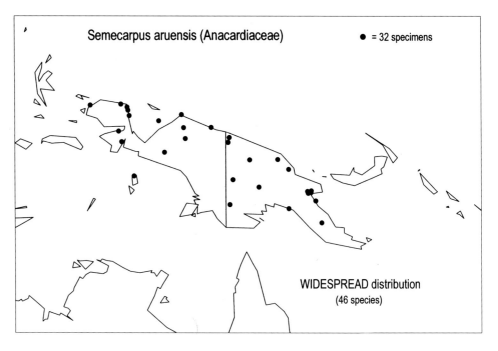

FIG. 5. Pattern 2: Widespread distribution.

— North-Widespread distribution (pattern 4; Fig. 6, stars). 55 species are found throughout the North (including the central mountain range which contains part of the craton edge). This pattern, like the next one, is correlated with the history of terrane accretion, the species are limited to the North, but due to dispersal they are now widespread. The plants are mainly found at lower altitudes.

— Mountain distribution (pattern 7; Fig. 8, 11 – 13). Several subpatterns cannot be linked to a particular geological phase:

    — widespread from Vogelkop to the Peninsula (pattern 7a; Fig. 8). This pattern shown by 29 species is closely linked to the former North-Widespread pattern, only these species are found at higher altitudes throughout the mountain ranges.

    — widespread from Vogelkop to central New Guinea (pattern 7d; Fig. 11, stars). 15 species are widespread in this part of the Mountain ranges.

    — widespread from Finisterre to central New Guinea (pattern 7f; Fig. 12, dots). 49 species occupy only this part of the mountain ranges.

    — widespread in East New Guinea (pattern 7g; Fig. 13, dots). 90 species are found from the Peninsula up to the more central parts of New Guinea. Three subpatterns may be distinguished, 39 species are found from the Peninsula to Central Papua New Guinea, 18 species from the Peninsula to the border region between Irian and Papua, and 33 species occur from the Peninsula to the border of the Vogelkop area.

Several 'transition' patterns exist, distributions which overlap several of the geological phases. All are of minor importance.

— 25 – 10(– 2) MA distribution (pattern 6; Fig. 7, stars). 19 species are found in the lowlands of the 25 MA accretion part as well as the 10 up to even the 2 MA old parts.

— 15 MA – Finisterre Mountains distribution (pattern 10). 20 species are found at higher altitudes in the Peninsula (15 MA) as well as the Finisterre Mountains, which accreted c. 10 MA.

— 15 MA – Finisterre Low altitude distribution (pattern 12 in Table 2). 18 species show the same distribution as the former pattern, only at altitudes below 1500 m.

— 15 MA – Finisterre – 10 MA distribution (pattern 11 in Table 2). In comparison to the former pattern only 3 species are even further distributed along the North coast of New Guinea, from the Peninsula up to Central New Guinea.

— Vogelkop – Waigeo distribution (pattern 18; Fig. 16, squares). One species is found in the 10 MA area of the Vogelkop and the 2 MA old area of Waigeo island.

All other patterns can be correlated to a special accretion phase. These will be discussed under the accretion phase they reflect.

Craton area

— Craton distribution (pattern 3; Fig. 6, squares). 24 species are found in the Craton area. It is, however, uncertain if these species grew in this area throughout the past 40 MA since most of this area is of very low altitude and may have been inundated during one or more interglacial periods.

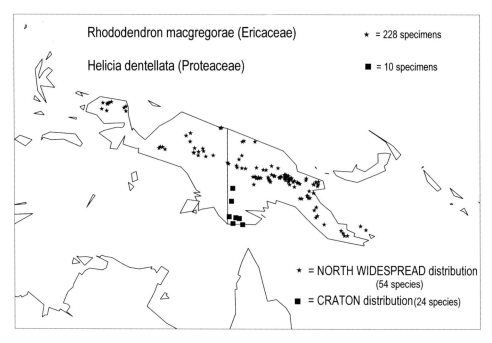

FIG. 6. Pattern 3: Craton distribution; and Pattern 4: North Widespread distribution.

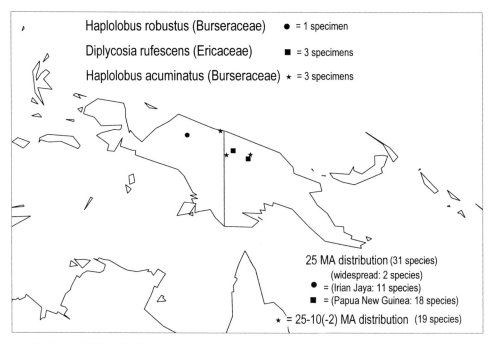

FIG. 7. Pattern 5: 25 MA distribution with pattern 5.b (species only in Irian Jaya), pattern 5.c (species in Papua New Guinea); and 25-10(-2) MA distribution, a transition distribution.

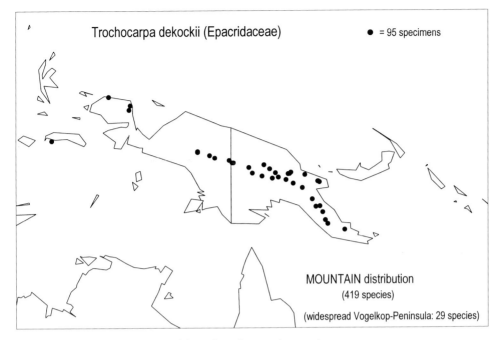

FIG. 8. Pattern 7.a: Species widespread throughout the central mountain ranges.

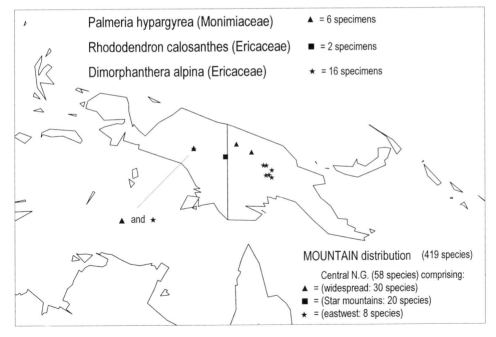

FIG. 9. Pattern 7.b: Species present in central part of mountain ranges with pattern 7.b.1 (widespread in central part), 7.b.2 (Star Mountains), and 7.b.3 (east-west disjunct distribution).

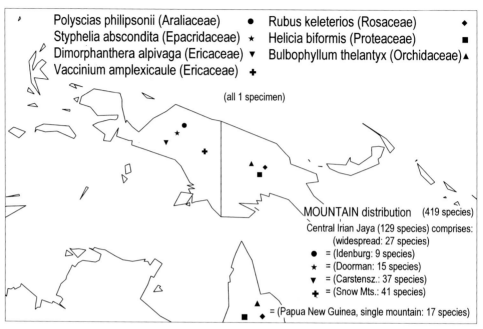

FIG. 10. Pattern 7.c: Species only in the central Irian Jaya part of the central mountain ranges with pattern 7.c.2 (Idenburg River), 7.c.3 (Carstensz. Mountains), 7.c.4 (Doorman Mountain), and 7.c.5 (Snow Mountains); and pattern 7.e.1 (species on single mountains in central Papua New Guinea).

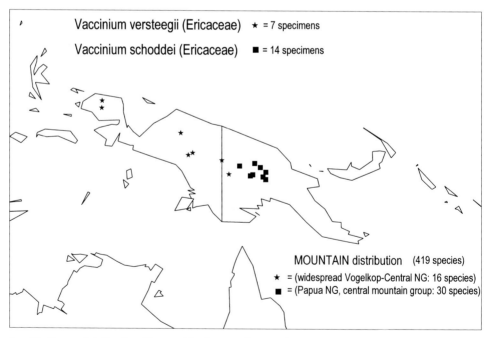

FIG. 11. Pattern 7.d: Species widespread in the central mountain ranges from the Vogelkop to central New Guinea; and pattern 7.e.2 (species present in the central mountains of Papua New Guinea).

25 MA area
— 25 MA lowland distribution (pattern 5; Fig. 7). 31 species are present at the lower altitudes in this accretion region. Of these species, 18 are found only in Papua New Guinea, 11 only in Irian Jaya, and only 2 are 'widespread' in both areas.
— Mountains (pattern 7; Fig. 9 – 11). Several subpatterns can be distinguished:
    — Central part of New Guinea (pattern 7b; Fig. 9). 58 species are present in this region, of which 20 are only found in the Star Mountains (Fig. 9, squares), 8 have a more or less eastwest distribution (Fig. 9, stars) and 30 are widespread in this area (Fig. 9, triangles).
    — Central Irian Jaya (pattern 7c; Fig. 10). Of these 129 species, 41 are present in the Snow Mountains only (Fig. 10, plus), 37 in the Carstensz. Mountains (Fig. 10, downwards pointing triangle), 16 on the Doorman Mountain (Fig. 10, star), 9 in the Idenburg area (Fig. 10, dot) and 27 are widespread in this part of Irian Jaya.
    — In Papua New Guinea (pattern 7e; Fig. 10, 11) the same phenomenon can be found, here too, species are present on single mountains (17 species, Fig. 10 upward pointing triangle, square, and diamond) or they (30 species) are present in the mountains of the Central part of Papua New Guinea (Fig. 11, squares).

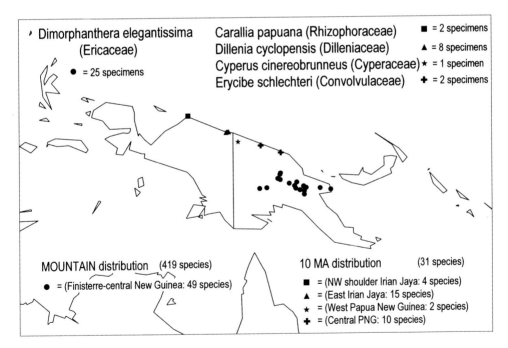

FIG. 12. Pattern 7.f: Plants present in the mountains of the Finisterre Range up to Central New Guinea; and pattern 15: 10 MA distribution with pattern 15.a (N.W. shoulder of Irian Jaya), pattern 15.b (East Irian Jaya); pattern 15.c (West Papua New Guinea), and pattern 15.d (Central Papua New Guinea). (The report of a specimen of *Dimorphanthera elegantissima* from Milne Bay in Sleumar's Flora Malesiana account was based on a doubtful identification and is here discounted.)

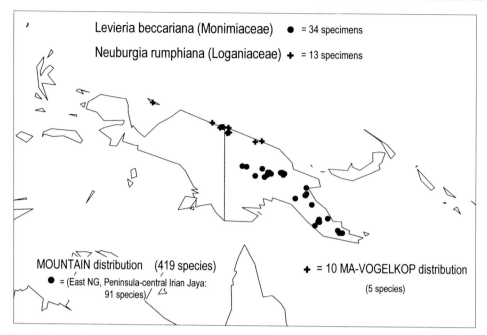

FIG. 13. Pattern 7.g: Plants widespread throughout the mountains from the Peninsula even up to the Vogelkop; and pattern 16: 10 MA-Vogelkop distribution, a type of transition distribution.

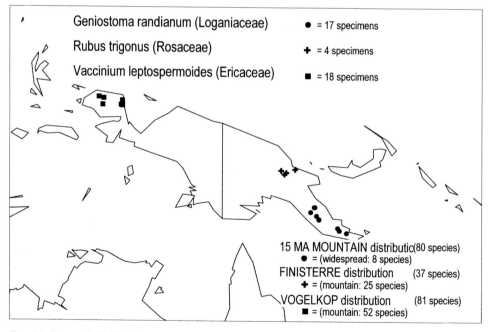

FIG. 14. Pattern 8: 15 MA Mountain distribution with pattern 15.a (widespread in Peninsula); Pattern 13: Finisterre distribution with pattern 13.a (high altitudes of Finisterre Range); and Pattern 17: Vogelkop distribution with pattern 17.a (high altitudes of the Vogelkop).

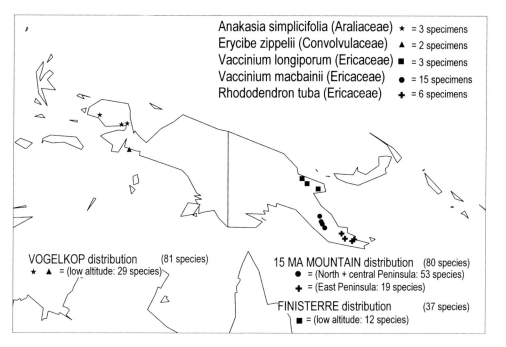

Fig. 15. Pattern 8: 15 MA Mountain distribution with pattern 15.b (North and central Peninsula), pattern 15.c (East Peninsula); Pattern 13: Finisterre distribution with pattern 13.b (low altitudes of Finisterre area); and Pattern 17: Vogelkop distribution with pattern 17.b (low altitudes of the Vogelkop).

15 MA area (Peninsula)

— 15 MA Mountain distribution (pattern 8; 80 species; Fig. 14, 15). At the higher altitudes 8 species are widespread in the Peninsula (Fig. 14, dots), 53 are restricted to the northern and central part of the Peninsula (Fig. 15, dots), and 19 are found in the east of the Peninsula (Fig. 15, plus signs).

— 15 MA Low altitude distribution (pattern 9; 58 species; Fig. 16, 17). At lower altitudes 9 species are widespread throughout the Peninsula (Fig. 16, triangles), or they are found more locally, i.e. the central part (Fig. 17, square), the Southeast (Fig. 17, stars), the South coast (Fig. 17, triangles), and the North coast (Fig. 17, plus signs).

10 MA area (Finisterre, North coast, Vogelkop)

The Finisterre region is treated separately. Many species are limited to this area and the overlap with the rest of the North coast of New Guinea is minimal — only three species (Fig. 16, dots). The Vogelkop is also treated separately for the same reason; furthermore, there is a disjunction between this area and the North coast. Only five species are found in both the Vogelkop and the North coast (pattern 16 in Table 2, Fig. 13, plus signs).

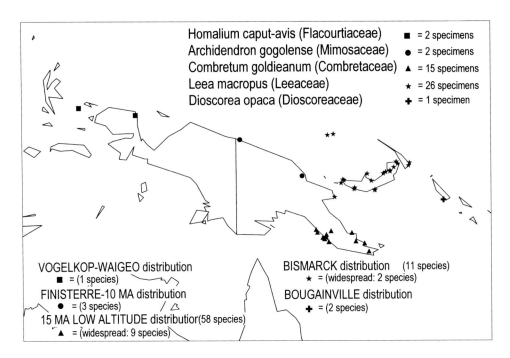

Fɪɢ. 16. Pattern 9: 15 MA Low altitude distribution with pattern 15.a (widespread in Peninsula); Pattern 14: Finisterre-10 MA distribution, a kind of transition distribution; Pattern 18: Vogelkop-Waigeo distribution, also a transition distribution; Pattern 20: Bismarck Archipelago distribution with pattern 20.a (widespread throughout the archipelago); and Pattern 21: Bougainville distribution.

— Finisterre distribution (pattern 13; 37 species; Fig. 14, 15). This distribution is again subdivided into a mountain distribution (25 species, Fig. 14, plus signs) and a lowland distribution (12 species, Fig. 15, squares). The widespread mountain species are often present in the Central mountain range and the Finisterre mountains (Fig. 8).

— North coast (pattern 15; 31 species, Fig. 12). These species are only local, being restricted to the NW shoulder of Irian Jaya (four species, Fig. 12, square), East Irian Jaya (15 species, Fig. 12, triangles), West Papua New Guinea (two species only, Fig. 12, star), or Central Papua New Guinea (ten species, Fig. 12, plus signs).

— Vogelkop (pattern 17; 81 species, Fig.14, 15). Most species are found in the mountains (Fig. 14, squares), far fewer at lower altitudes (Fig. 15, stars and triangle) and then mainly in the southern part of the Vogelkop and the 'neck' of the Vogelkop.

2 MA area (Waigeo, Bismarck Archipelago, Bougainville)
— Waigeo (pattern 19; Fig. 17, downward pointing triangle). Only three species are endemic on this island.

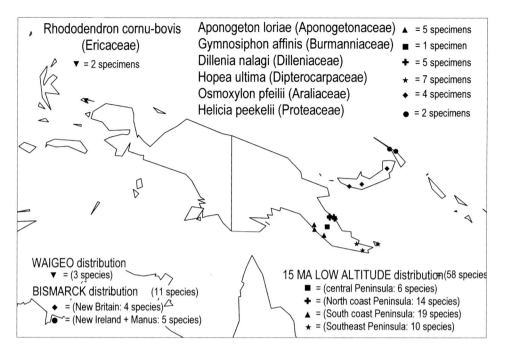

FIG. 17. Pattern 9: 15 MA Low altitude distribution with pattern 15.b (central part Peninsula), pattern 15.c (North coast), pattern 15.d (South coast), pattern 15.e (S.E. Peninsula); Pattern 19: Waigeo distribution; and pattern 20: Bismarck Archipelago distribution with pattern 20.b (New Britain) and pattern 20.c (New Ireland and Manus).

— Bismarck Archipelago (pattern 20; Fig. 16, 17). 11 species are endemic in this archipelago, two are widespread (Fig. 16, stars), four are restricted to New Britain (Fig. 17, diamonds), and five are only found on (Manus and) New Ireland (Fig. 17, dots).

— Bougainville (pattern 21; Fig. 16, plus). Only two species are endemic on this island. There might be more endemic species, but because this island is outside the original circumscription of Malesia (van Steenis 1950) it is often not included in the revisions. Using generic distributions, van Balgooy (1971: 90 – 91) demonstrated that the Solomons, to which Bougainville belongs, should have been included in the Malesian region, because they contain a mainly Malesian flora, although somewhat depauperate.

It is important to note that the patterns and subpatterns found do not represent a hierarchical classification, partly because there is overlap between various (sub)patterns and partly because several subpatterns are not exclusive, e.g. pattern 7.e.1 (single mountains in Papua New Guinea) can easily be included in 7.e.2 (mountain groups in Papua New Guinea).

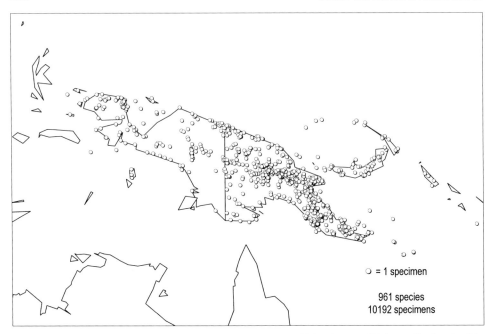

FIG. 18. Collecting sides of the 10,192 specimens used to compile all distributions. Highest collecting densities are found in Papua New Guinea, especially in the central mountain ranges. The lowest collecting density is found in South New Guinea.

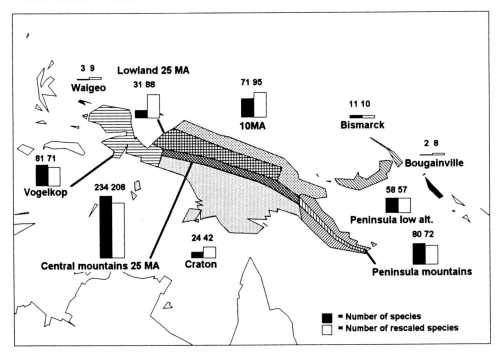

FIG. 19. Different accretion phases of New Guinea with their numbers of endemic species (left columns) and rescaled endemic numbers (right columns; necessary to suppress the influence of non-random collecting, for formula see discussion).

CORRELATIONS AND DISCUSSION

The main purpose of this paper is to see if correlations can be made between the patterns and the geological history. Before this can be done, several remarks have to be made.

It is assumed that small distributions, such as those of endemic species, reflect the place of origin of those species, mainly because the species have evolved in geologically recent times. However, some endemic distributions may be relics; species may have had more widespread distributions that have since become restricted or even fragmented. Disjunct distributions are then to be expected, like the east/west distributions in pattern 1 and 7.b.3. Since such highly disjunct distributions are rare, the initial assumptions seem reasonable.

Collecting has not been random in New Guinea, which probably influenced the present recognition of patterns. Fig. 18 shows that most collecting is concentrated in Papua New Guinea, especially in the mountain ranges. Irian Jaya and the south of New Guinea show a relative undercollecting, just like all other lowland areas. Independently, Conn (1994, his Fig. 3 & 4) described the same phenomenon. Possible results of non-random collecting are:

a. Most endemic species are found at higher altitudes (596 out of 961 examined).

b. Most patterns are found in the mountains.

c. Several patterns, like 7.g.2, are probably artificial due to undercollecting in Irian Jaya; some patterns end at the border.

A completely different factor may also have affected the evaluation of distribution patterns — the nature of the species concepts of the various researchers who revised the families for Flora Malesiana. A narrow species concept will result in numerous endemic species, of which the distribution might easily correlate with terranes, while a broad species concept will usually result in broader distributions which do not correlate with terranes. Within New Guinea it is predominantly the *Ericaceae* which are prominent in most (mountainous) patterns, 377 endemic species are known in New Guinea, slightly more than one third of all endemic species in this sample. Fortunately, though the species concept in the *Ericaceae* is perhaps a bit narrow, it satisfies the needs of most researchers and horticulturists.

Correlations between distributions and individual terranes are barely evident. The best correlation is found between the Finisterre region (**f** in Fig. 2) and pattern 13 (Figs 14 and 15). Other correlations are less impressive:

— Waigeo (**wg** in Fig. 2) and pattern 19 (three species only).

— Rouffaer (**r** in Fig. 2) and several 7c patterns, although the latter are more restricted to just a single mountain.

— the Owen Stanley Range in the Peninsula (**os** in Fig. 2) and pattern 8b.

— Bowutu (**bu**, also Peninsula) and some distributions in pattern 9c.

It is not surprising that the correlation between patterns and terranes is weak. The terranes often do not constitute homogeneous areas, they do not contain a single soil type, for instance. They are also of very diverse origin (Pigram & Davies 1987), several are of oceanic origin and could never have carried any land plants, like Bowutu (see above), Cyclops, Dayman, Kutu, Marum, Menyamya, Port Moresby, Rouffaer (see above), Mt Turu, and perhaps Waigeo, Gauttier and Sepik. The surface of the terranes is usually small, due to different combinations of several causes: they were originally small, they have been partly overthrust by newly accreting terranes, they have been tilted and lifted during orogeny, and they have been partly covered by sediment (dotted in Fig. 2; after Pigram & Davies 1987). Therefore, even though a taxon may once have been an endemic species on a terrane, dispersal has resulted in much wider distributions than the terranes themselves. All these reasons explain why the correlation between patterns and terranes is poor.

On the other hand, the correlation between the different accretion phases (Fig. 3, after Pigram & Davies 1987) and distribution patterns is surprisingly high (Tables 3 & 4 and Fig. 19). About 600 out of 961 species are restricted to one of the different accretion phases (the figure 595 in Table 4 is because the 10 MA area is separated from the Vogelkop and, therefore, pattern 16 (represented by five species) could not be included). Of the remaining 361 species, 300 are also limited to the mountains and northern half of the island; thus the distributions of these species are correlated also with the accretion history of New Guinea.

In Table 4 and Fig. 19 an attempt has been made to mitigate the influence of unequal collecting. Several algorithms exist which, at best, should include a factor for the size of the areas and the levelling off of species numbers in climax vegetation. Unfortunately these factors are not known and therefore a much simpler algorithm has been used:

$$\text{Species}_{\text{rescaled}} = \frac{\sum_{i=1}^{n} \text{Specimens}_i}{\sum_{i=1}^{n} \text{Species}_i} * \frac{\text{Species}_i}{\text{Specimens}_i} * \text{Species}_i$$

(Rescaled number of species i = Mean number of specimens per species i for all areas, multiplied by the inverse of the mean number of specimens per species for a single area, multiplied by the number of species for that area)

The fraction which consists of the first two parts of the algorithm is 1 if the mean number of collections for an area equals the mean value for all areas. If the fraction is less than 1, then the area is relatively overcollected and the rescaled number will be lower, meaning that several species would be expected to be more widespread if surrounding areas were better sampled. If the fraction is higher than 1, then the area is undercollected and the rescaled number will be higher, equals an expectation that more endemic species will be found once the area is better collected.

Table 4 and Fig. 19 show that the mountainous areas in particular are relatively overcollected (see also p. xxx & Fig. 18) and the rescaled numbers are lower than the original numbers. Nevertheless, even the rescaled numbers show that

TABLE 3. Numbers of species (at altitudes below and above 1500 m and their total) which correlate with the different accretion phases (upper half) or with the complete accretion history (bottom half) of North New Guinea.

Distributions correlated with phases	All species			Species represented by more than 6 specimens			Patterns from Fig. 2
	Low altitude species (specimens)	High altitude species (specimens)	Total species (specimens)	Low altitude species (specimens)	High altitude species (specimens)	Total species (specimens)	
Craton	24 (56)	-	24 (56)	2 (19)		2 (19)	3
25 MA accretion phase	31 (44)	234 (1066)	265 (1110)	-	39 (616)	39 (616)	5, 7[b, c, e]
15 MA accretion phase	58 (238)	80 (362)	138 (600)	12 (135)	16 (188)	28 (323)	8, 9
10 MA accretion phase	80 (245)	77 (373)	157 (618)	9 (109)	23 (230)	32 (339)	13-17
2 MA accretion phase	16 (55)	-	16 (55)	1 (26)	-	1 (26)	19-21
Widespread Northern half	95 (1675)	205 (4852)	300 (6527)	58 (1540)	156 (4658)	214 (6198)	4,6,7[a, d, f, g],10-12,18
Widespread in New Guinea	61 (1226)	-	61 (1226)	40 (1149)	-	40 (1149)	1, 2
Total	365 (3539)	596 (6653)	961 (10192)	122 (2978)	234 (5692)	356 (8670)	

TABLE 4. Numbers of species used in Fig. 19 (left) and the same for the species with a well represented distribution (right).

Area	All species			Species represented by more than 6 specimens			
	Species	Spp$_{rescaled}$	Specimens	Species	Spp$_{rescaled}$	Specimens	
Craton	24	42	(56)	2	3	(19)	
Central mountains 25 MA	234	208	(1066)	39	33	(616)	
Lowland 25 MA	31	88	(44)	-	-	-	
Peninsula mountains	80	72	(362)	16	18	(188)	
Peninsula low alt.	58	57	(238)	12	14	(135)	
10 MA	71	95	(216)	2	3	(21)	
Vogelkop	81	71	(372)	21	25	(233)	
Waigeo	3	9	(4)	-	-	-	
Bismarck	11	10	(49)	1	1	(26)	
Bougainville	2	8	(2)	-	-	-	
Total	595		(2409)	93		(1238)	
specimen / species		4.05			13.31		

that most endemic species are found in the mountain ranges. The same can be observed in Table 3 (N.B. note, too, that most species widespread in the north are also found in the mountains). Even though the mountains are by far the best collected areas, the conclusion can be drawn that most endemic species occur at higher altitudes.

Most endemic species are also poorly represented in herbaria, often only by one or a few specimens, which means that knowledge of their distribution is probably (very) incomplete. If the only species to be considered are those which are represented by at least six specimens, then only 356 are left (Table 3). Most patterns also disappear because they are represented by less than five species (Table 3 & 4). The mountain patterns all remain.

The high endemicity of the mountain regions suggests that the focus on terranes as being rafts, each bringing their complement of species and each contributing to the floristic diversity of the New Guinea region, is misplaced. In New Guinea, orogeny is caused by the accretion of terranes. Endemic distributions hardly correspond with the individual terranes and are seemingly not directly linked with plate tectonic movements, but they do correspond with the age of the collisions, though with a lapse of time, because new terranes first have to dock before older terranes are uplifted. This means that most endemic distributions will postdate the collision of the terranes with the Craton. It also means that results of plate tectonic movements, like orogeny, are probably the major causes of speciation. Other results of tectonic movements are often changes in the climate and pattern 13b (Finisterre, low altitude) and 9d (South Coast Peninsula) may be distribution patterns which are a result of the drier climate in the Markham valley and the savannah around Port Moresby respectively. Another correlation is between orogeny, climate, and natural catastrophes (for data, see Johns 1986) — earthquakes with associated landslides, volcanic eruptions, cyclones — occur in the north of New Guinea, just like the majority of endemic species. Earthquakes and volcanoes occur in tectonically active areas and even typhoons or tropical storms will have their worst effects in such areas. Slopes in the actively orogenic areas of New Guinea are young and unstable, and slumps and mudslides induced by heavy rains are frequent. Catastrophes can leave large areas barren and especially in the mountains it may take a long time before the forest regenerates. In the meantime, these barren places act as barriers for genetic exchange between populations of the same species, barriers which might result in the independent development (speciation) of the two separated groups.

The distributions of most endemic species are found in the central and northern part of New Guinea. Very few are present in the Craton area, while this is by far the largest area. The former discussion suggests that this is a direct result of the fact that the Craton is a geologically stable area. However, this area is clearly undercollected. A third suggestion for the relative lack of endemic species might be the that, during glacial periods, species could easily disperse from or to Australia on the land bridges revealed. This is seemingly the case with savannah animals, but is less likely with rain forest animals (Kikkawa et al. 1981). The same perhaps applies to the plants. Savannah species, like several Mimosaceae, are found in New Guinea and North Australia where (for rain-forest plants) there is overlap only in generic distributions, but far less in species distributions; forest species are usually endemic in either New Guinea or Australia. Conclusions concerning a correlation between distribution patterns and the accretion history of New Guinea have to be made with much prudence.

Independent confirmation of a correlation between different tectonic periods and distributions may be found in phylogenetic information in combination with distribution data. The older species should be found in the older accretion areas. Turner (1995) combined all phylogenetic biogeographical information on *Sapindaceae* and his Component Compatibility Analysis (p. 119) provides a reasonable confirmation of the accretion history as sketched by Pigram & Davies (1987). However, far more phylogenetic information of especially New Guinean taxa is needed for a statistically sound conclusion.

CONCLUSIONS

1. A correlation between distributions patterns and single terranes is weak and seems to exist for only a few terranes, mainly the Finisterre terrane.

2. A correlation exists between distributions and groups of terranes which accreted in the same period, which indicates that increased tectonic activity induces speciation.

3. The main correlation is found between altitude and speciation, most endemic species are found in the mountains.

4. Therefore, the following prudent general conclusion can be made, speciation is not directly caused by plate tectonic activity, but by the results of tectonic movements, mainly orogeny.

In addition a correlation between climate and distributions may also exist, whereby the climate may also be one of the results of plate tectonic movements.

ACKNOWLEDGEMENTS

I warmly thank Peter Stevens (Harvard University) for his constructive remarks and for rephrasing some of my English; I also thank Mark Coode (Kew) who provided some other suggestions.

REFERENCES

Balgooy, M. M. J. van (1971). Plant geography of the Pacific. Blumea Suppl. 6: 1 – 222.
Conn, B.J. (1994). Documentation of the flora of New Guinea. In: C.-I. Peng & C. H. Chou (eds), Biodiversity and terrestrial ecosystems: 123 – 156. Academia Sinica Monogr. Ser. 14, Taipei.
Daly, M. C. *et al.* (1991). Cenozoic plate tectonics and basin evolution in Indonesia. Mar. Petr. Geol. 8: 2 – 21.
Johns, R. J. (1986). The instability of the tropical ecosystem in New Guinea. Blumea 31: 341 – 371.

Kikkawa, J., Monteith, G. B. & Ingram, G. (1981). Cape York Peninsula: Major region of faunal interchange. In: A. Keast (ed.), Ecological biogeography of Australia: 1695 – 1742. Dr. W. Junk B. V. Publishers, The Hague, Boston, London.

Lyell, C. (1830 – 33). Principles of Geology. John Murray, London.

Michaux, B. (1991). Distributional patterns and tectonic development in Indonesia: Wallace reinterpreted. Aust. Syst. Bot. 4: 25 – 36.

Pigram, C. J. & Davies, H. L. (1989). Terranes and the accretion history of the New Guinea orogen. BMR J. Austr. Geol. Geoph. 10: 193 – 211.

Raup, D. M. (1991). Extinction. Bad genes or bad luck? 210 pp. W. W. Norton & Cie., New York, London.

Steenis, C. G. G. J. van (1950). The delimitation of Malaysia and its main plant geographical divisions. In: C. G. G. J. van Steenis (ed.), Flora Malesiana ser. 1, 1: lxx – lxxv. Noordhoff-Kolff N. V., Djakarta.

Turner, H. (1995). Cladistic and biogeographic analyses of *Arytera* Blume and *Mischarytera* gen. nov. (*Sapindaceae*) with notes on methodology and a full taxonomic revision. Blumea Suppl. 9: 1 – 230.

Welzen, P. C. van (1989). *Guioa* Cav. (*Sapindaceae*): taxonomy, phylogeny, and historical biogeography. Leiden Bot. Ser. 12: 48 – 51.

— (1992). Species richness and speciation in Malesia. Programme & Summaries of Papers and Posters Second Flora Malesiana Symposium: 43.

The phytogeography of Malesian *Euphorbiaceae*[1]

T. C. WHITMORE[2]

Summary. Euphorbiaceae s.l. are the fourth largest family of vascular plants in Malesia with 91 genera and 1354 species so far recorded. A modern critical taxonomic revision exists for all regions except the Philippines. Analysis of the distribution of the family across Malesia divided into five regions chosen to reflect geological history and present-day climate shows that it is most strongly represented by species and endemics in the rain forest regions of west and east Malesia and the Philippines. It is much poorer in the centre (Sulawesi, Maluku) and south (western Nusa Tenggara), where there are extensive areas of seasonal climates and forests. In the past during Glacial epochs rain forests have been reduced in area.

Malesian Euphorbiaceae are richest today in the parts of Malesia that have had rain forests continuously through past climatic changes and it is deduced that they have speciated mainly in perhumid climates in tropical rain forests.

INTRODUCTION

Soon after I began my studies on the forest floras of the insular Far East I developed an interest in *Macaranga*. This is a big genus of c. 280 species, many of which occur in secondary forest. I studied *Macaranga* in vivo in the Solomon Islands, Malaya (Peninsular Malaysia), and to a lesser extent in Papua New Guinea during 1962 – 72, and came to know them as a conspicuous component of secondary forests along roads and on abandoned farmlands. In any one locality some 3 – 10 species can be found, often as the dominant trees. More recent visits to various parts of Sumatra and Kalimantan confirm this observation. It came as a surprise to discover, on visits to Sulawesi in 1974 and 1984 and to Maluku in 1985, that there *Macaranga* does not fill this niche largely to the exclusion of other trees. In these parts of central Malesia it is much less commonly seen.

These perceptions were the stimulus for the present investigation of the phytogeography of Malesian *Euphorbiaceae*. Does analysis confirm the impression gained in the field? What is the range, richness and degree of endemism in different parts of Malesia of *Macaranga* and of all *Euphorbiaceae*? Are there common patterns, and if so can we explain them?

THE DATA

A. *Euphorbiaceae*

The *Euphorbiaceae* are a good family for phytogeographical enquiry. In the broad sense (i.e. *Euphorbiaceae s.s.* plus close allies) they have 91 genera and 1354 species. They rank fourth in size amongst the five largest families of Malesian vascular plants

[1] in memoriam H.K.Airy Shaw, magistri sapientis Euphorbiacearum, dedicavit

[2] Geography Department, Cambridge University, Cambridge CB2 3EN England

that each have over 1000 species (Table 1). *Euphorbiaceae* are the only one of these families to have a comprehensive recent taxonomic treatment. They have been recently critically revised by a single botanist, the late H. K.Airy Shaw (1975, 1980, 1981, 1982, 1983), who was active until the early 1980s. He was assisted with *Macaranga* (the second largest genus) by myself and with *Euphorbia* by A. Radcliffe-Smith, and augmented for Malaya by Whitmore (1972) and for Java by Backer & Bakhuizen (1963).

TABLE 1. Malesian vascular plant families with 1000 species or more
(Soepadmo 1995; this study).

Orchidaceae	6500
Rubiaceae	2000
Myrtaceae	1600
Euphorbiaceae s.l.	1354
Melastomataceae	1000

Except for the Philippines, the taxonomy of *Euphorbiaceae* is on a sound and consistent basis. The copious collections made in many parts of Malesia during the 1950s to 1970s have been taken into account. Airy Shaw worked at Kew where the Philippines flora is poorly represented and he apologised that his treatment (Airy Shaw 1983) is an uncritical updating of the list of the family made by Merrill (1923). The Philippines have 388 known *Euphorbiaceae* and nearly half were described either by Merrill (33 %) or Elmer (13 %) who are both known to have had a very narrow species concept. Moreover, in their day, before aerial communications had developed, tropical botanists worked very much in isolation within national boundaries with scant attention to the floras of adjacent countries.

The structure of *Euphorbiaceae*, its division into groups and subgroups, according to the recent classification of Webster (1994), is shown in Table 2.

The size within Malesia of the genera found there is summarised in Table 3. Ten genera have over 50 species each; 68 genera have ten species or fewer, and of these 41 are monospecific in Malesia.

Antidesma, Galearia and *Microdesmis* are included in *Euphorbiaceae* following Webster (1994) although Airy Shaw put them in adjacent small families. Two other genera traditionally placed in *Euphorbiaceae* are now believed not to be closely related, namely *Buxus* (*Buxaceae*) and *Daphniphyllum* (*Daphniphyllaceae*). These are excluded from consideration in the present analysis. *Bischofia* is also excluded, even though Webster did include it, because Airy Shaw (1965) demonstrated its close similarities with *Staphyleaceae* rather than *Euphorbiaceae*.

TABLE 2. Synopsis of Malesian *Euphorbiaceae* following the classification of Webster (1994). Number of species occurring in Malesia shown.

I (Subfamily) *Phyllanthoideae*
  1 (Tribe) *Wielandieae*
        6 *Actephila* (6 spp.)
  3 *Briedelieae*
        13 *Cleistanthus* (78)
        14 *Briedelia* (16)
  4 *Phyllantheae*
    4b (Subtribe) *Leptopinae*
        16 *Leptopus* (2)
    4f *Flueggeinae*
        25 *Flueggea* (4)
        26 *Richeriella* (2)
        28 *Margaritaria* (1)
        29 *Phyllanthus* (104)
        31 *Sauropus* (14)
        32 *Breynia* (15)
        33 *Glochidion* (187)
  5 *Drypeteae*
        35 *Drypetes* (64)
  6 *Antidesmeae*
    6c *Scepinae*
        44 *Baccaurea* (40)
        46 *Ashtonia* (2)
        47 *Aporosa* (60)
    6d *Antidesminae*
        52 *Antidesma* (98)
  7 *Hymenocardieae*
        57 *Hymenocardia* (1)

II *Oldfieldioideae*
  11 *Caletieae*
    11b *Dissiliariinae*
        66 *Austrobuxus* (1)
        70 *Choriceras* (1)
    11c *Petalostigmatinae*
        72 *Petalostigma* (1)
    11d *Pseudanthinae*
        73 *Kairothamnus* (1)

III *Acalyphoideae*
  15 *Chaetocarpeae*
        91 *Trigonopleura* (1)
        92 *Chaetocarpus* (1)
  17 *Cheiloseae*
        94 *Cheilosa* (2)
        95 *Neoscortechinia* (4)
  18 *Erismantheae*
        96 *Erismanthus* (1)
        97 *Moultonianthus* (1)
        98 *Syndyophyllum* (1)

  19 *Dicoelieae*
        99 *Dicoelia* (1)
  20 *Galearieae*
        100 *Microdesmis* (1)
        101 *Galearia* (6)
  22 *Agrostistachydeae*
        105 *Agrostistachys* (4)
        108 *Chondrostylis* (2)
  23 *Chrozophoreae*
    23c *Doroxylinae*
        115 *Doroxylon* (1)
        116 *Sumbaviopsis* (1)
        118 *Melanolepis* (1)
    23d *Chrozophorinae*
        119 *Chrozophora* (1)
  26 *Pycnocomeae*
    26b *Blumeodendrinae*
        131 *Blumeodendron* (7)
        133 *Ptychopyxis* (13)
        134 *Botryophora* (1)
  27 *Epiprineae*
    27a *Epiprininae*
        135 *Epiprinus* (1)
        139 *Koilodepas* (11)
        140 *Cladogynos* (1)
    27b *Cephalomappinae*
        143 *Cephalomappa* (5)
  29 *Alchorneae*
    29a *Alchorneinae*
        151 *Alchornea* (9)
  30 *Acalypheae*
    30e *Cleidiinae*
        164 *Wetria* (1)
        165 *Cleidion* (7)
    30f *Macaranginae*
        167 *Macaranga* (168)
    30g *Claoxylinae*
        169 *Claoxylon* (63)
        173 *Micrococca* (3)
    30i *Rottlerinae*
        176 *Mallotus* (60)
        179 *Coccoceras* = 176
        180 *Trewia* (1)
        181 *Neotrewia* (1)
        183 *Octospermum* (1)
    30j *Acalyphinae*
        184 *Acalypha* (30)
    30k *Lasiococcinae*
        185 *Lasiococca* (2)
        186 *Spathiostemon* (2)
        187 *Homonoia* (1)

TABLE 2 continued

III *Acalyphoideae* (contd.)
    31 *Plukenetieae*
        31a *Plukenetiinae*
            193 *Plukenetia* (1)
        31b *Tragiinae*
            195 *Cnesmone* (4)
            196 *Megistostigma* (4)
            202 *Pachystylidium* (1)
        31c *Dalechampiinae*
            203 *Dalechampia* (1)
    32 *Omphaleae*

            204 *Omphalea* (4)

IV *Crotonoideae*
    35 *Adenoclininae*
        35b *Endosperminae*
            216 *Endospermum* (10)
    36 *Gelonieae*
            217 *Suregada* (6)
    37 *Elateriospermeae*
            218 *Elateriospermum* (1)
    38 *Jatropheae*
            222 *Deutzianthus* (1)
            225 *Anneisjoa* (1)
    39 *Codieae*
            228 *Ostodes* (1)
            232 *Dimorphocalyx* (11)
            233 *Fontainea* (1)
            234 *Codiaeum* (15)
            237 *Blachia* (1)
            238 *Strophioblachia* (1)
            240 *Baliospermum* (1)

    40 *Trigonestemoneae*
            241 *Trigonostemon* (34)
    41 *Ricinocarpeae*
        41a *Ricinocarpinae*
            243 *Alphandia* (1)
        41b *Bertyinae*
            248 *Borneodendron* (1)
    42 *Crotoneae*
            250 *Fahrenheitia* (2)
            252 *Croton* (61)
    44 *Aleuritideae*
        44a *Aleuritinae*
            256 *Aleurites* (1)
            257 *Reutealis* (1)
        44c *Grosserinae*
            262 *Tapoides* (1)

V *Euphorbioideae*
    45 *Stomatocalyceae*
        45a *Stomatocalycinae*
            273 *Pimelodendron* (4)
    46 *Hippomaneae*
        46b *Carumbiinae*
            279 *Homalanthus* (26)
        46c *Hippomaninae*
            281 *Sebastiana* (4)
            282 *Stillingia* (1)
            284 *Excoecaria* (12)
            288 *Sapium* (3)
    49 *Euphorbieae*
        49c *Euphorbiinae*
            304 *Euphorbia* (23)

B. Plant collecting intensity in Malesia

Figure 1 shows the status of plant collecting in 1972, as collections per 100 km^2 of different parts of Malesia. Java and Malaya had been extremely heavily collected compared to everywhere else. The Philippines came third but a very long way behind. During the third quarter of this century copious collections were made, mainly by government Forest Departments, in Malaya (Cockburn, Ng, Whitmore), Borneo (mainly Brunei, Sabah and Sarawak by Anderson, Ashton, Cockburn, Meijer and Wood, also east Kalimantan by Kostermans et al.) and New Guinea (Womersley et al. in Papua New Guinea, Dutch forest botanists in the BW series in West New Guinea prior to its fusion with Indonesia in 1963). This burst of collecting then fizzled out. Its effect has been to make collections from Malesia much more uniform than they were earlier (van Steenis 1950),

TABLE 3. Size in Malesia of the genera of *Euphorbiaceae*

Number of species in genus	Number of genera this size	Genera
1 – 10	68 (41 monotypic)	
11 – 20	8	
21 – 30	3	*Euphorbia* (23) *Homalanthus* (26) *Acalypha* (30)
31 – 40	2	*Trigonostemon* (34) *Baccaurea* (40)
41 – 50	–	
51 – 60	2	*Aporosa* (60) *Mallotus* (60)
61 – 70	3	*Croton* (61) *Claoxylon* (63) *Drypetes* (64)
78		*Cleistanthus*
98		*Antidesma*
104		*Phyllanthus*
168		*Macaranga*
187		*Glochidion*

mainly by filling in knowledge of parts of Borneo and of New Guinea. However, both these islands, and Sulawesi and Sumatra also, remain least well collected and the continuing considerable unevenness of collections is bound to colour any analysis unless it is allowed for.

C. The physical setting of Malesia

Malesia is a clearly defined and demarcated phytogeographical region. It encompasses the great archipelago that sweeps eastwards from Sumatra to the Bismarcks beyond New Guinea. To the north it is delimited at a line connecting Kangar to Pattani in the Malay peninsula (Whitmore 1975). To the south, the Torres Strait between Thursday Island off the northern tip of Queensland and the other

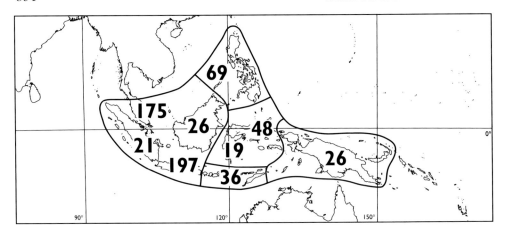

FIG. 1. Status of plant collecting in Malesia, 1972, as collections per 100 km² (van Steenis-Kruseman 1974).

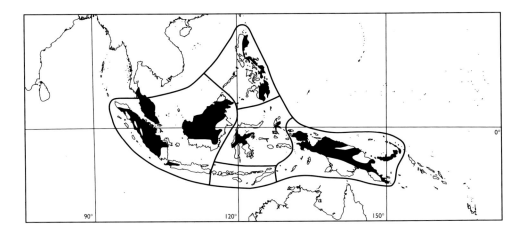

FIG. 2. Perhumid (shaded) and seasonally dry climates of Malesia. (Based on climatic classification of Schmidt & Ferguson (1951) as presented by Whitmore (1975, Fig. 3.1); their A type climates shaded here as perhumid).
   Note the perhumid western and eastern blocks centred respectively on the Sunda and Sahul continental shelves and the region of seasonal climates that runs north to south and also forms a belt in southern Malesia.

islands south of Papua forms the boundary. The island arcs of Melanesia that lie beyond Australia, south and east of the Bismarck archipelago, are not included. Luzon is the northern limit.

   The analysis presented here is of all *Euphorbiaceae* recorded to occur naturally within Malesia. A tally is included of how many extend beyond, either northwards into continental southeast Asia or to the east or south into Melanesia or Australia.

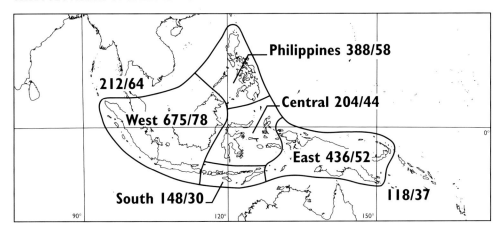

FIG. 3. The different regions of Malesia recognised for this paper showing the numbers of species and genera of *Euphorbiaceae s.l.* within each region and extending beyond Malesia.

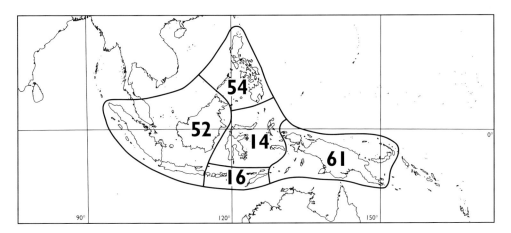

FIG. 4. Percentage of species of Malesian *Euphorbiaceae* endemic to each region.

D. Palaeogeography

Over the past two decades there has been a revolution in knowledge of tropical palaeogeography and palaeoclimates.

It is now known that, in simple terms, Malesia, the Malay archipelago, was created about 15 Ma ago, in the mid-Miocene, by the collision of part of the southern supercontinent Gondwanaland, as it drifted northwestwards, with the southeast extremity of the northern supercontinent Laurasia. The collision was at or slightly east of present-day Sulawesi (Audley-Charles 1981, Audley-Charles et al. 1981, Hall

& Blundell 1996). Thus the lands of the Sunda continental shelf (Sumatra, Malaya, Java, Borneo and Palawan) are Laurasian and those of the Sahul shelf (New Guinea and attendant small islands) are Gondwanan.

Prior to the collision fragments of Gondwanaland had broken off between the Cambrian and late Jurassic and drifted northwards to collide with and become embedded in Laurasia as terranes (Audley-Charles 1987, Burrett et al. 1991).

Wallace's Line, one of the world's strongest zoogeographical boundaries, is the main biological reflection of this geological history (Cranbrook 1981). But this history can also be seen in the range of many plants, some of which have a mainly northern (Laurasian) distribution while others are mainly southern (Gondwanan). In addition, a few plant groups are bicentric. This carries the implication either that they reached Malesia on both the northern and southern supercontinents, or alternatively some could have migrated between the supercontinents on the 'stepping stones' provided by the drifting terranes mentioned above (Dransfield 1987, Whitmore 1981a).

E. Palaeoclimates

We live today at the height of an Interglacial epoch. During Glacial epochs equatorial climates have been slightly drier, slightly cooler, and more seasonal. Sea level has been as much as 180 m lower, which means in Malesia that the Sunda and Sahul shelves have been exposed as dry land (Whitmore 1981b).

Detailed reconstruction of Malesian palaeoclimates is incomplete but perhumid (aseasonal) lowland climates became less extensive and seasonal climates became more extensive during the Glacial maxima.

Figure 2 shows the areas with present day perhumid and seasonal climates in Malesia and adjacent lands. It can be seen that the lands of the Sunda and Sahul shelves are big blocks of perhumid climates. There is a belt of seasonal climates that runs north to south from the Philippines through Sulawesi and Maluku in central Malesia. It is probable that this belt became enlarged during Glacial epochs and that the big blocks of perhumid climate became confined to part of the hearts of the Sunda and Sahul shelf regions (see Morley & Flenley 1987, Fig. 5.5). Perhumid climates are of only limited extent beyond Malesia northwards into continental Asia, and are not found to the south in Australia, but do occur in the island arcs of Melanesia east of Malesia.

There has been a sequence of alternating Glacial and Interglacial epochs during the Pleistocene (the past two million years) and repeated contraction and expansion of the blocks of perhumid climate. Before the Pleistocene, the late Tertiary also had similar alternating Glacials and Interglacials (Morley & Flenley 1987). Evidence gets increasingly fragmentary further back in time but it is clear that since Malesia came into existence c. 15 Ma ago the region has had a continuously fluctuating climate. Forest types and species will have expanded and contracted their geographical range in consort with climatic change. Tropical rain forests occur in places with a perhumid climate. Today, at the height of an Interglacial, rain forests are near their maximum extent, and this has been uncommon during the Pleistocene, which has mostly had bigger areas of seasonal climates.

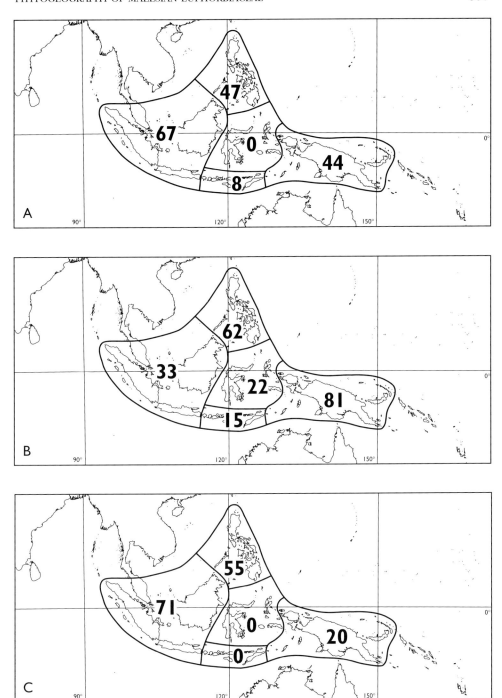

FIG. 5. Regional percentage species endemism **A** *Briedelieae* (2 gen. 94 spp.), **B** *Phyllantheae* (8 gen. 329 spp.), **C** *Drypeteae* (1 gen. 64 spp.).

ANALYSIS

F. General

What is the distribution of *Euphorbiaceae* in Malesia and does this reflect palaeogeography or palaeoclimate?

To address this question, firstly Malesia is subdivided into five regions, Fig. 3.

W: the west — Sumatra, Malaya, Java and Borneo,
P: the Philippines,
C: the centre — Sulawesi (with Sangihe and Talaud) and Maluku,
S: the south — Nusa Tenggara[3] east as far as Timor,
E: the east — New Guinea, small islands south to Thursday, the Bismarcks, Nusa Tenggara east of Timor and Seram Laut, Gorong, Watubela and Kai.

W is the Sunda Shelf[4] (climatic types A & B in the Schmidt & Ferguson (1951) classification, see Whitmore (1975, Fig. 3.1). S is the dry part of Nusa Tenggara. It is the only part of Malesia with strongly seasonal climates today (Fig. 2; Schmidt & Ferguson climatic types E and F). E is the Sahul Shelf plus the islands south of New Guinea that are perhumid, seasonal or only slightly seasonal, including easternmost Nusa Tenggara (Schmidt & Ferguson climatic types A, B, C and D).

The map of present day climates (Fig. 2) shows that the western (W) and eastern (E) regions encompass the big western and eastern blocks of present-day perhumid climates. Today, perhumid climates also occur in the eastern half of the Philippines, region P, and about half the central region C. In the drier, more seasonal, Glacial epochs, perhumid climates are likely to have been mainly confined to areas W and E.

G. Detail

The richness of *Euphorbiaceae* across the region is shown on Fig. 3. The number that range outside northwards up the Malay peninsula, eastwards beyond the Bismarck archipelago, or southwards into Australia are also included in the analysis. In fact, relatively few Malesian *Euphorbiaceae* have ranges extending outside the region (Fig. 3), only 16% of species extend northwards and 9% to the south plus east. In terms of both species (675) and genera (78) region W is richest. Both P and E are about equally rich in genera (58, 52), with E leading in species (388, 436). These results however inevitably reflect differences in intensity of collecting between different parts of Malesia. To allow for this, and also for the different sizes of the different parts of Malesia considered here, fuller analysis is made on the percentage of the known species of each region which are endemic to that region. This gives a measure of the amount of speciation that has occurred, unbiased by either collecting intensity or area.

A map of percentage species endemic to each region for all *Euphorbiaceae* is given in Fig. 4. In Figs. 5 – 7 maps are given for all the tribes of the family with 25 species or more and in Fig. 8 *Macaranga* is shown.

---

[3] viz the Lesser Sunda Islands
[4] except for Palawan because the data do not enable Palawan to be distinguished from others of the Philippine islands

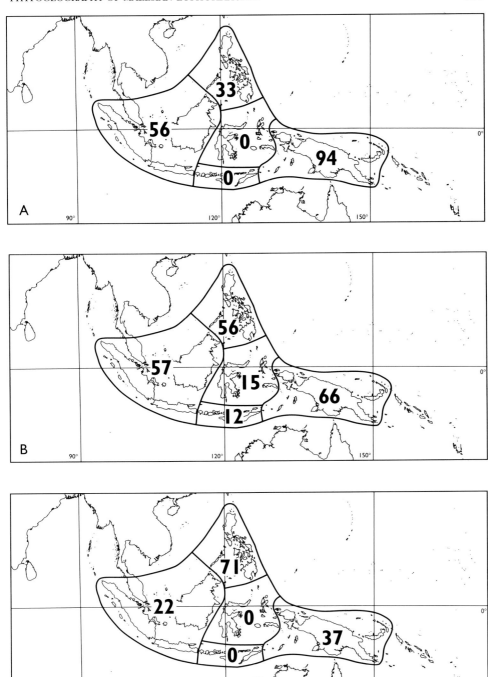

FIG. 6. Regional percentage species endemism **A** *Antidesmeae* (4 gen. 200 spp.), **B** *Acalypheae* (14 gen. 340 spp.), **C** *Codieae* (7 gen. 31 spp.).

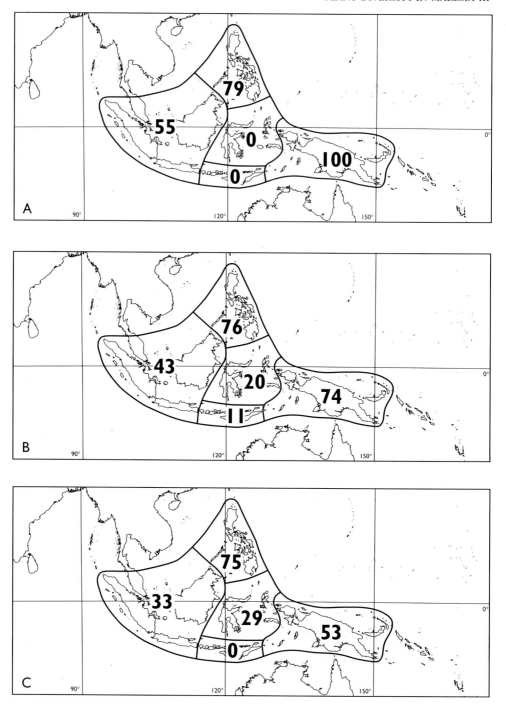

Fig. 7. Regional percentage species endemism **A** *Trigonostemoneae* (1 gen. 34 spp.), **B** *Crotoneae*, (2 gen. 63 spp.), **C** *Carumbiinae* (1 gen. 26 spp.).

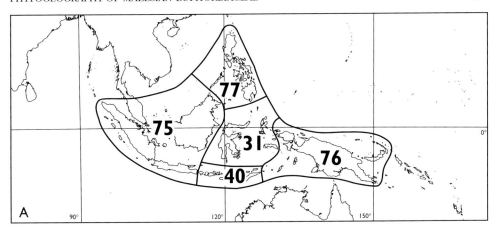

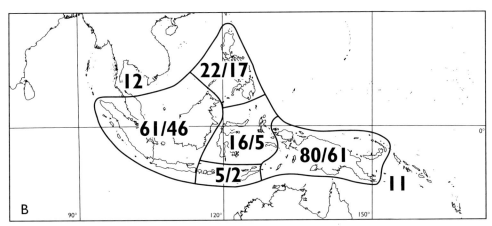

FIG. 8. *Macaranga* **A** percentage endemism, **B** numbers of species and endemics for each region.

CONCLUSIONS

Examination of these maps shows that the family as a whole and all individual groups exhibit the same pattern of low endemism in Centre and South regions and high endemism in most or all of West, Philippines and East regions. That is to say, the distribution of endemics in the *Euphorbiaceae* (*s.l.*) as a whole and its larger component tribes is strongly concentrated on the rain forested parts of Malesia. These, as shown above, are the regions which have had rain forests continuously through past periods of more extensive seasonal climates. The Centre and South regions, with seasonal climates today, and even more seasonality in the past, have few endemic *Euphorbiaceae*.

Some support for this finding that in Malesia *Euphorbiaceae* have speciated largely in the perhumid rain forests is provided by plot enumerations. These have shown *Euphorbiaceae* to be the most species-rich family amongst trees of 10 cm diameter and over in places likely to have remained perhumid, namely Kalimantan (Wanariset, Lempake), Sumatra (Ketambe) and Malaya (Pasoh), but only fourth in rank in north Sulawesi (Toraut) which is likely to have had seasonal climates in the past (Whitmore & Sidiyasa 1986, Table 2, Kochummen et al. 1990).

Euphorbiaceae are well distributed across the region. There is no evidence at the level of the family or its larger tribes for either a Laurasian or a Gondwanan bias.

THE PHILIPPINES

It is possible that modern critical revision of Philippine *Euphorbiaceae* will diminish the number of endemics at present recorded. In the extreme case this could reduce endemism to a level comparable to regions C and S rather than to those of W and E as shown at present. Today about half the Philippines have a rain forest climate, Fig. 3. At past Glacial maxima there would have been less rain forest and it is surprising (and perhaps therefore an artefact) that the Philippines are revealed on present evidence to be about equally as rich in percentage species endemism as W and E which have very probably always had extensive rain forest. If the Philippines really do have a rich *Euphorbiaceae* flora, I predict that *Macaranga* will be a conspicuous and common component of the wayside and secondary forest tree flora, as it is in west and east Malesia. If, however, this richness is an artefact due to over-optimistic description of species then I predict *Macaranga* will be as inconspicuous as it is in central Malesia. I do not have the necessary experience to adjudicate this matter.

DISCUSSION

No apology is made for the presentation here of a narrative account of the phytogeography of Malesian *Euphorbiaceae*. The conclusions have been reached by a process of ratiocination based on weighing up the evidence of the physical setting (palaeogeography, palaeoclimate) and the distribution of the family and its larger component tribes in a manner which allows for incompleteness of these data. It is not considered that a conclusion would be more valid had numerical formulae and mathematical manipulation been utilised. In particular there is nothing magic about processing data through a computer. By whatever method the reckoning is approached subjective decisions must be made and deficiencies in the data allowed for. With the approach used here subjective choices have been made concerning data input and in the interpretation of the output. Numerical techniques would add an extra stratum of subjectivity in the decision on which formulae and parameters to be employed. The two major weaknesses of the data are the unevenness of collecting and the lack of a modern critical revision of Philippines' *Euphorbiaceae*. These colour the results howsoever the analysis is conducted, and have to be allowed for by rational argument. It is submitted that for the kind of data presented here the human brain is a very good mechanism with which to weigh up all the various lines of evidence and reach a conclusion.

ACKNOWLEDGEMENTS

This research was conducted while the author was Professor at Ehime University, Matsuyama, Japan, under sponsorship from the British Council and Monbusho. Professor K. Ogino is thanked for his hospitality and provision of resources.

REFERENCES

Airy Shaw, H. K. (1965). Bischofiaceae. Kew Bulletin 18, 252 – 4.
—— (1975). The *Euphorbiaceae* of Borneo. Kew Bull. Add. Ser. 4.
—— (1980). The *Euphorbiaceae* of New Guinea. Kew Bull. Add. Ser. 8.
—— (1981). The *Euphorbiaceae* of Sumatra. Kew Bull. 36, 239 – 374.
—— (1982). The *Euphorbiaceae* of Central Malesia (Celebes, Moluccas, Lesser Sunda Is.). Kew Bull. 37, 1 – 40.
—— (1983). An alphabetical enumeration of the Euphorbiaceae of the Philippines. Royal Botanic Gardens, Kew.
Audley-Charles, M. G. (1981). Geological history of the region of Wallace's Line. In: Whitmore, T. C. (ed.). Wallace's Line and plate tectonics. Clarendon Press, Oxford.
—— (1987). Dispersal of Gondwanaland: relevance to evolution of the angiosperms. In: Whitmore, T. C. (ed.). Biogeographical evolution of the Malay archipelago. Clarendon Press, Oxford.
Audley-Charles, M. G., Hurley, A. M. & Smith, A. G. (1981). Continental movements in the Mesozoic and Cenozoic. In: Whitmore, T .C. (ed.). Wallace's Line and plate tectonics. Clarendon, Oxford.
Backer, C. A. & Bakhuizen van den Brink Jr., R. C. (1963). Flora of Java vol. 1. Noordhoff, Groningen.
Burrett, C., Duhig, N., Berry, R. & Vane R. (1991). Asian and southwest Pacific continental terranes derived from Gondwana and their biogeographical significance. *Austral. Syst. Bot.* 4, 13 – 24.
Cranbrook, Lord. (1981). The vertebrate faunas. In: Whitmore, T. C. (ed.). Wallace's Line and plate tectonics. Clarendon Press, Oxford.
Dransfield, J. (1987). Bicentric distribution in Malesia as exemplified by palms. In: Whitmore, T.C. (ed.). Biogeographical evolution of the Malay archipelago. Clarendon Press, Oxford.
Hall, R. & Blundell, D.J. (eds.) (1996). Tectonic evolution of southeast Asia. Geological Society Special Publication 106.
Kochummen, K.M., LaFrankie, J. V. Jr., Manokaran, N. (1990). Floristic composition of Pasoh forest reserve, a lowland rain forest in Peninsular Malaysia. *J. of Trop. For. Sci.* 3, 1 – 13.
Merrill, E. D. (1923). An enumeration of Philippine flowering plants 2, 389 – 464. Manila.
Morley, R. J. & Flenley, J. R. (1987). Late Cenozoic vegetational and environmental changes in the Malay archipelago. In: Whitmore, T. C. (ed.). Biogeographical evolution of the Malay archipelago. Clarendon Press, Oxford.

Pflanzenreich IV, 147.
Schmidt, F. H. & Ferguson, J. H. A. (1951). Rainfall types based on wet and dry
period ratios for Indonesia with western New Guinea. Verhandelinge Djawatan
Meteorologi dan Geofisik, Djakarta.
Soepadmo, E. (1995). Floristic diversity of southeast Asian tropical rain forest and
its ecogeographical significance. In: Lovejoy, T. E. & Primack, R. B. (eds.).
Ecology, conservation and management of southeast Asian rain forest. Yale
University Press.
Steenis, C. G. G. J. van (1950). Survey of botanical collecting density in Malaysia.
Flora Malesiana I, 1, cviii.
Steenis-Kruseman, M. J. van (1974). Desiderata for future exploration. Flora
Malesiana I, 8, iii – iv.
Webster, G. L. (1994). Synopsis of the genera and suprageneric taxa of
Euphorbiaceae. Ann. Missouri Bot. Gard. 81, 33 – 144.
Whitmore, T. C. (1972). *Euphorbiaceae.* In: Whitmore, T. C. (ed.). Tree flora of
Malaya 2. Longman, Kuala Lumpur & London.
—— (1975b). Tropical rain forests of the Far East. Clarendon Press, Oxford.
—— (1981a). Wallace's line and some other plants. In: Whitmore, T.C. (ed.).
Wallace's line and plate tectonics. Clarendon Press, Oxford.
—— (1981b). Palaeoclimate and vegetation history. In: Whitmore, T.C. (ed.).
Wallace's line and plate tectonics. Clarendon Press, Oxford.
Whitmore, T. C. & Sidiyasa, K. (1986). Composition and structure of a lowland rain
forest at Toraut, northern Sulawesi. Kew Bull. 41, 747 – 56.

The need for intensive fieldwork in preparing a definitive Flora Malesiana treatment: experience with bamboos with special reference to Indonesia

Elizabeth A. Widjaja[1]

Summary. Bamboos have been neglected by taxonomists for many years. A comparison of bamboo studies carried out in Malesia suggests that more fieldwork is required to make the bamboo treatment more definitive. The experience of a three year project to carry out bamboo collecting in Indonesia shows that intensive collecting can successfully be achieved and will significantly contribute to the Flora account. This work has increased the number of bamboo specimens in Herbarium Bogoriense by 100%. Similarly, the 125 species now recorded for Indonesia represent almost twice the number previously estimated.

INTRODUCTION

Although bamboos represent one of the most economically important plant groups in the Malesian region, they have been neglected from in-depth taxonomic study for many years. In the past, botanists have rarely collected bamboos during their general botanical explorations although in recent years fieldwork concentrating on bamboos has been carried out in several areas of Malesia following increasing recognition of bamboo as an economic resource. Over the past 16 years interest in the taxonomy of bamboos has increased as a result of work carried out by S. Dransfield (Dransfield 1981, 1983, 1992), K.M. Wong (Wong 1995), C. Roxas and E.A. Widjaja (Widjaja 1987).

During the bamboo workshop at the First Flora Malesiana Symposium in 1989, a proposal was put forward to complete the treatment of the Malesian bamboos by 1996. However, specialists were aware that this could only be fulfilled with additional support to carry out intensive fieldwork, particularly in Indonesia and the Philippines. Any increase in the number of bamboo collections will help to complete the Flora account.

BAMBOO STUDIES AND FIELDWORK IN MALESIA

Malaysia

Prior to 1979 bamboo exploration in Malaysia was carried out mainly in the Malay Peninsula by staff of the Forest Research Institute, Malaysia; this work resulted in an extensive published account by Holttum (1958). Since 1979, further collections have been made in the Malay Peninsula, Sabah and Sarawak particularly by Wong and S. Dransfield. From these, revisions of bamboos in the Malay Peninsula (Wong 1995) and Sabah (Dransfield 1992) have been prepared.

---

[1] Herbarium Bogoriense, Puslitbang Biologi – LIPI, Bogor, Indonesia.

Brunei Darussalam

Recent collections have been made by Wong and S. Dransfield. A checklist of the taxa occurring in Brunei has been compiled by Dransfield (1996).

The Philippines

Most of collections were made by E.D. Merrill and other staff of PNH before World War II. Since then the only botanist attempting to revise the bamboos was Santos who made a large number of collections. These were kept in his private herbarium after his death although Dilliman University and the Philippines National Museum have been attempting to transfer them to a public herbarium. More recently, botanists such as J. & S. Dransfield, Roxas and Widjaja have made sporadic explorations, especially in Palawan, Luzon and Mindanao. However the collections are still inadequate to prepare a complete account.

New Guinea

An account of the bamboos of New Guinea was prepared by Holttum (1967), after fieldwork carried out in 1963. Since then little work has been undertaken with the exception of E.E. Henty who attempted to revise the work of Holttum (1967) by making new collections and compiling further data on bamboos in the region. However, his work was never published although a draft revision of the New Guinea bamboos was at an advanced stage of preparation. Bamboo exploration in Irian Jaya has not been attempted with the exception of some collections made by the present author in 1993 and S. Dransfield in 1995. In the absence of extensive collections it will be difficult to revise the bamboos of New Guinea, especially from Irian Jaya.

Indonesia (except Irian Jaya)

Prior to 1990, knowledge of bamboos was based on old collections such as those of Blume, Zollinger, Koorders, and Backer. A grant from the Canadian International Development Research Centre (IDRC) between 1990 – 1993, allowed bamboo exploration in Indonesia to be undertaken more systematically. This grant enabled the author to carry out intensive fieldwork in Sumatra, Kalimantan, Sulawesi, Java, Bali, Lesser Sunda Islands and the Moluccas (as well as Irian Jaya indicated above). There are still several areas, especially in central Kalimantan and the Moluccas which need to be explored. More species are expected to occur in these areas.

INTENSIVE BAMBOO COLLECTIONS
FOR FLORA MALESIANA CAN BE COMPLETED

Some difficulties hamper a timely completion of the bamboo account for Flora Malesiana. Firstly, the authors have to deal with other commitments and tasks from their own institution. Secondly, bamboo collections are still insufficient to present a complete picture of the richness in the Malesian region as some areas are still under-collected. However, the IDRC-funded project has demonstrated that intensive collecting can successfully be achieved which, in turn, will significantly contribute to the Flora account.

Fig. 1 shows areas of Indonesia where collections were made prior to 1990. Fig. 2 shows the extent of these areas on completion of the project in 1993. Collecting activities during the project resulted in a 100% increase in the number of bamboo specimens preserved in Herbarium Bogoriense. The number of known species was doubled (Table 1). It can now be reported that Indonesia has 125 species of bamboo growing naturally or in cultivation, the latter including both native and introduced species. This figure represents about 10% the world's bamboo species. Previously, only c. 60 species were recorded for Indonesia, making very clear the extent to which this project has increased our knowledge of Indonesian bamboo diversity. The taxonomic implications of the project are staggering because it yielded so many new species; about 50% of the species recognized are new records or new taxa. The collections have even led to the establishment of new genera. A taxonomic account of Indonesian bamboo novelties will be published, and a Field Guide to the Indonesian Bamboos is in preparation. Regrettably the development of a database of Malesian bamboos, for which Indonesia agreed to take responsibility at the Second Flora Malesiana Symposium in 1992, has not been completed through lack of funds.

TABLE 1. Number of genera, species and specimens collected in Indonesia

Genera	No. species before 1990	No. species after 1990	Notes
Bambusa	8	14	
Cephalostachyum	0	1	
Dendrocalamus	4	10	
Dinochloa	3	15	
Fimbribambusa	0	2	Widjaja 1997, in press
Gigantochloa	13	30	
Melocanna	1	1	
Nastus	7	9	
Neololeba	0	5	From *Bambusa* (Widjaja 1997, in press)
Phyllostachys	2	2	
Parabambusa	0	1	Widjaja 1997, in press
Pinga	0	1	Widjaja 1997, in press
Racemobambos	5	7	
Schizostachyum	11	24	
Shibatea	1	1	
Sphaerobambos	1	1	
Thyrsostachys	1	1	
Total no. genera	12	17	
Total no. species	57	125	
Total no. specimen	1939	4089	

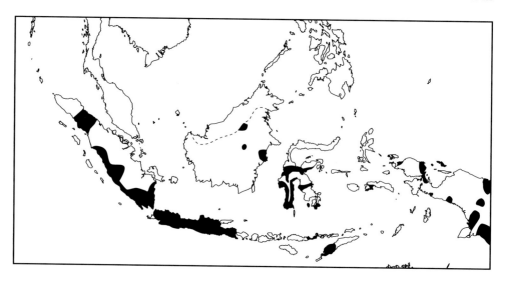

FIG. 1. Bamboo collection in Indonesia before 1990.

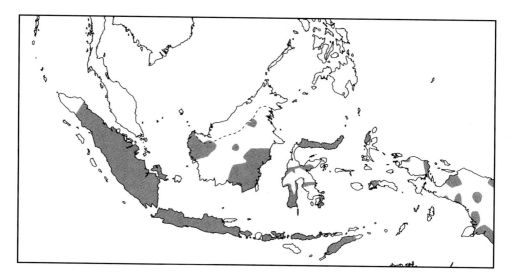

FIG. 2. Bamboo collection in Indonesia after 1990.

Since the Indonesian bamboo flora is now relatively well known, the Philippines is now the main area where intensive exploration is needed. Learning from our experience in Indonesia, it seems that without similar activity in the Philippines, it is impossible to complete the Malesian bamboo flora in the short time. Therefore, additional funding for fieldwork in the Philippines is needed.

CONCLUSION

From the above experience with bamboos it can be concluded that, without good collections, it can be difficult to complete Flora Malesiana accounts. Apart from increased collecting, the number of taxonomists working on Flora Malesiana should be increased. Similarly, facilities should be provided from the Foundation or from the countries concerned with Flora Malesiana. The Flora Malesiana Foundation and the Malesian nations are advised to raise funds for field trips in order to allow for sound revisions of under-collected taxa.

REFERENCES

Dransfield, S. (1981). The genus *Dinochloa* (*Gramineae-Bambusoideae*) in Sabah. Kew Bull. 36(3): 613 – 633.

Dransfield, S. (1983). The genus *Racemobambos* Holttum (*Gramineae-Bambusoideae*). Kew Bull. 37 (4): 661 – 679.

Dransfield, S. (1983) Notes on *Schizostachyum* (*Gramineae-Bambusoideae*) from Borneo and Sumatra. Kew Bull. 38 (2): 321 – 332.

Dransfield, S. (1992). The bamboos of Sabah. Sabah Forest Records No. 14. Forestry Department, Sabah, Malaysia. 94 pp.

Dransfield, S. (1996). *Gramineae-Bambusoideae*. In: M. J. E. Coode, J. Dransfield, L. L. Forman, D. W. Kirkup & I. M. Said (eds), A checklist of the flowering plants and gymnosperms of Brunei Darussalam. pp. 371 – 374. Ministry of Industry and Primary Resources, Brunei Darussalam.

Holttum, R. E. (1958). The bamboos of the Malay Peninsula. Gard. Bull., Singapore 16: 1 – 135.

Holttum, R. E. (1967). The bamboos of New Guinea. Kew Bull. 21: 263 – 292.

Widjaja, E. A. (1987). A revision of Malesian *Gigantochloa* (*Poaceae-Bambusoideae*) Reinwardtia 110 (3): 291 – 380.

—— (1997). New taxa in Indonesian bamboo. Reinwardtia 11(2), (in press).

Wong, K.M. (1995). The bamboos of Peninsular Malaysia. Malayan Forest Records No. 41. Forest Research Institute, Malaysia. 200 pp.

The seafaring people of South Sulawesi and their role in the conservation of mango germ-plasm diversity

Nengha Wirawan[1]

Summary. This study is part of an assessment of the diversity, ecology, value and management of 'native' mangoes in South Sulawesi, Indonesia. While local myths connect mangoes to the early history of the former local kingdom, the gregarious growth of diverse and valuable mangoes in certain villages in South Sulawesi suggests the important role of the seafaring Buginese, Makassarese, and Mandarese people in the enrichment and maintenance of germ-plasm diversity in the Province. Aside from taking mangoes as fresh fruits and trading goods, these people use mature but unripe fresh mangoes for the preparation of simple but appetising fish dishes. For long sailing trips, dried sliced unripe mangoes known as 'pakacci', 'pacukka', 'kaloko', 'kaloe' or 'pamaissang' often form the main spices carried for these journeys. Although such spices can be kept for a long time, fresh replenishments may be made wherever and whenever they come ashore. With limited flying animals that could help the inter-island dispersal of mangoes, the high diversity of non-endemic cultivars growing gregariously in many villages in South Sulawesi is a good indication of what these seamen have brought home. In contrast to the solitary growth that limits hybridisation in natural habitats, the aggregation and maintenance of such cultivars provide good opportunities for hybridisation and the development of new cultivars. In the present study, from 525 trees selected from almost all ecological zones, more than 250 out of the 270 local names are cultivars of *Mangifera indica, M. laurina* and other species that also grow in other places in the Archipelago.

INTRODUCTION

This paper is part of a study of the mangoes of South Sulawesi, home of four major ethnic groups, the Buginese, Makassarese, Mandarese, and the Torajans (Maps 1 & 2). While the Torajans are generally rice and hill farmers, the other three groups are also well known seafaring traders and fishermen. The objective of this paper is to support the contention that these seafaring people play important roles in the enrichment and maintenance of mango germ-plasm diversity in this Province.

We started our mango study through a series of interviews with store keepers and peddlers of mangoes in the capitals of the Province, districts and subdistricts or with shop-keepers along roads when we visited villages in all of the 23 districts in the Province. Typical questions we asked included what were the local names of mangoes that they knew and the villages where those mangoes were known to grow or from where they actually bought the fruits that they were selling. We then visited those villages and asked similar questions of whoever we met in the villages on our way to see the head of each village.

---

[1]Department of Forestry, Hasamuddin University, Ujung Pandang, Indonesia. Present address: The Nature Conservancy, Jalan Radio IV No. 5, Kebayoran Baru, Jakarta, Indonesia.

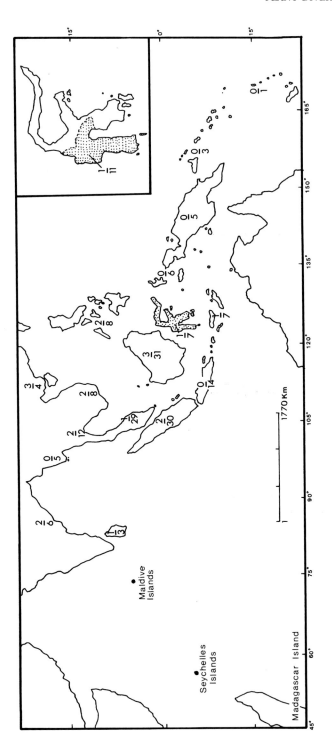

MAP 1. Density of *Mangifera* species for each island, region or island group. Endemic species above and total number of species below the hyphen.

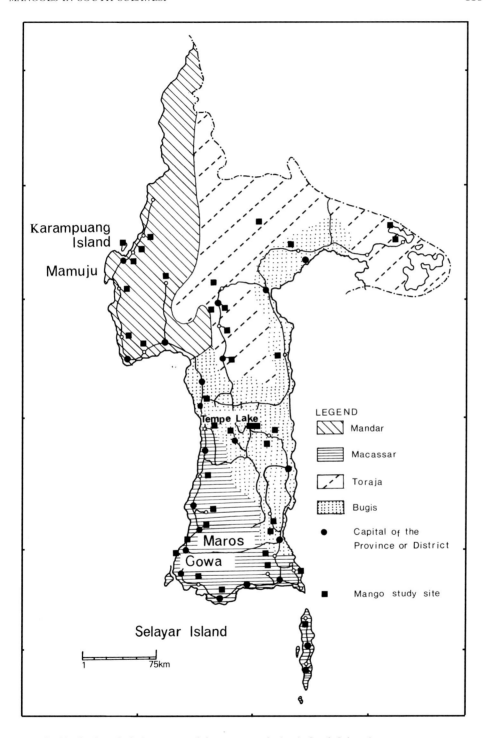

MAP 2. Distribution of ethnic groups and the mango study sites in South Sulawesi.

After explaining the purpose of our visit, we asked the village head to help provide us with knowledgeable persons who could take us around to locate the mango trees that the villagers had been describing or the trees whose names we already knew. In addition, we also visited some logging areas in the northern part of the Province to look for mango species growing in primary forest. The generalised location of the villages/areas we visited are shown in Map 2.

In addition to recording the local names and specific locations of the mango trees, whenever available we collected flowering and fruiting specimens for our taxonomic studies and for that purpose tagged each tree with a unique reference number. We also collected information on economic value, traditional uses, management and other relevant aspects that could give a better understanding of the status of the South Sulawesi mangoes.

Analyses of our data, particularly those that relate to the diversity, distribution, economic value and the traditional uses of mangoes, as well as local myths and aspects of the general livelihood of the people, suggested good evidence for supporting the thesis that the seafaring Buginese, Makasssarese, and Mandarese play an important role in the enrichment and maintenance of mango germplasm diversity.

SUPPORTING EVIDENCE

There are seven major points which we consider as evidence for supporting the thesis. These are:

A. *Almost all mango species occurring in South Sulawesi also occur on other islands outside Sulawesi*

The recent study of Kostermans & Bompard (1993) shows Sulawesi to have a total of seven species of mango, of which one is endemic. With more specimens becoming available for our present study, for South Sulawesi alone we now recognise eleven species of which ten also occur outside Sulawesi and one, the endemic *Mangifera sulavesiana* Kosterm., is still restricted to the island or Province. As shown in Table 1, together with their occurrence in the province, three of the ten non-endemic species also occur on islands west and east of Sulawesi, six species occur only on the western part and one species only on its eastern side.

In addition four of these species *(Mangifera indica, M. laurina, M. pedicellata* and *M. minor)* have 11–90 local names, which together with relatively high numbers of local names found in a particular area (up to 40 names), indicate their potential richness in varieties or cultivars.

B. *The inter-island or inter-continental spread of mangoes is carried out by man*

Except for *Mangifera gedebe* which grows in swamp areas and is dispersed widely from Burma through Malesia to New Guinea by water, all mangoes are dispersed by relatively large animals including hornbills, monkeys, and terrestrial mammals

TABLE 1. The mango species of South Sulawesi, their occurrence in Sulawesi (S) and west (W) or east (E) of it, together with the number of local names belonging to each of the species

No. Mango species	Distribution			Number of local names
	W	S	E	
1. *Mangifera indica* L	x	x	x	62
2. *Mangifera odorata* Griff.	x	x	x	2
3. *Mangifera foetida* Lour.	x	x	x	2
4. *Mangifera laurina* Bl.	x	x	–	90
5. *Mangifera rigida* Bl.	x	x	–	1
6. *Mangifera similis* Bl.	x	x	–	1
7. *Mangifera dongnaiensis* Pierre	x	x	–	2
8. *Mangifera pedicellata* Kosterm.	x	x	–	11
9. *Mangifera lalijiwo* Kosterm.	x	x	–	1
10. *Mangifera minor* Bl.	–	x	x	23
11. *Mangifera sulavesiana* Kosterm.	–	x	–	1

(elephants, porcupines, man, etc.) (Kostermans & Bompard 1993). Since none of these large animals, except man, are able to cross the seas or straits of the Indo-Malayan Archipelago, they cannot be responsible for the wide inter-island distribution of mangoes, including the ten non-endemic species of South Sulawesi.

The role of man in the inter-island or inter-continental dispersal of mangoes is exemplified by the world-wide spread of *Mangifera indica*. This species, which has an enormous number of cultivars and is now cultivated all over the tropics, is believed to have originated in Assam, India and parts of neighbouring Burma. As described by Kostermans & Bompard (1993), the dissemination of mangoes throughout the tropics was rather slow. The Arabs and Phoenicians introduced this species to East Africa during the 14th–16th Centuries and the Portuguese took it to West Africa and Brazil in the 18th Century. While the Tamils brought it at an early stage to the Malay Peninsula, no mention was made of how this species spread eastwards to Sulawesi and other islands in the Indo-Malayan Archipelago.

C. *The distribution of mangoes coincides with the sailing range of the South Sulawesi people*

As shown in Map 1, except for *Mangifera indica*, all the mango species are restricted to tropical Asia, where the highest concentration of species occurs in the Malay Peninsula, Sumatra, Java, and Borneo and decreases rapidly to the east in the Pacific area (Micronesia) with one species and to the west in India with six species. This range of mango distribution is well within the sailing range of the seafaring people of South Sulawesi.

As stated by Lineton (1975), long before the West had made contact with the Malay world and also long before the Bugis states of South Sulawesi had become incorporated in the Dutch colonial empire after the fall of Makassar in 1669, the Bugis people had already developed a well-established pattern of sailing routes. She also stated "for at least two centuries the Bugis people have been spreading out from their homeland .... to other parts of the Indonesian archipelago, to the Malay Peninsula, and even as far as Siam".

Historical records and traditional stories also suggest that the Bugis and Makassar people had already visited northern Australia and nearby islands (Spillett 1988) as well as Madagascar (KOMPAS 1991) through the use of their traditional *phinisi* boats, more than 300 years ago. Although there were other ethnic groups that were also roaming the archipelago at that time, such as the Bajo or Bajao people, from whom the South Sulawesi people perhaps learned the skills of constructing such a traditional boat and sailing it without any modern navigating instruments (Dr. Mukhlis Paeni, pers. comm.), the South Sulawesi people proudly showed their sailing skills when Makassarese people with their traditional boats successfully retraced their ancestor's routes to northern Australia in 1988 (Spillet 1988) and to Madagascar in 1991 (KOMPAS 1991). Earlier, the ruggedness of such a boat was also successfully tested against the Pacific Ocean when the Government of Indonesia brought such a boat for exhibition to the 1986 EXPO in Vancouver, Canada (PEDOMAN RAKYAT 1986).

That they may also have visited India a long time ago, could be suggested by two local legends. In a Bugis-Makassar legend, the first Ruler of Gowa (the first Kingdom in the area) was a beautiful lady who was found meditating under a prolific *taipa jombe-jombea* mango tree (Dr. Mukhlis Paeni, pers. comm.). In a Torajan legend, mango was the tree that gave Lakipadada (the grandson of the first man on earth, *Tomanurung*)plenty of fruit to eat and to carry home when he was stranded on an island on his return from the 'world's end' where God granted him a five generation length of time to live. It was also on that mango tree that he was able to cling to the leg of the mythical *Garuda* bird (who also fed on the mangoes) who then brought him to the Gowa Kingdom where he married the King's daughter. This marriage gave birth to three sons and a daughter who ultimately became the first Kings and Queen in the four Kingdoms (Gowa, Luwu, Sangalla and Bone) developed from the original Gowa Kingdom. As symbols of their authorities, each one of them was given a flag, a sword, and shells of the mango seeds (Puang W. P. Sombolinggi, pers. comm.).

One shell of this 'mango' seed is still kept by the descendant of the Sangalla King (Puang W. P. Sombolinggi) in the Museum Buntu Kalando in Tana Toraja. Others are kept by descendants of the Wajo King at Sengkang, and in the Museum La Galigo in Ujung Pandang. Unfortunately, through comparison with specimens in Herbarium Bogoriense, these shells are not mango but rather the Coco de Mer palm, *Lodoicea maldivica* (Gmel.) Pers. This palm grows naturally in the Seychelles Islands, in the Indian Ocean. Regardless of this confusion, while the first legend indicates a very close resemblance with the Indian myth, the second legend and the existence of the shells of those 'mango' seeds strongly suggest that the seafaring people of South Sulawesi must also have visited India and its surrounding islands.

D. *Because of its importance, collecting mango fruits wherever they came ashore during their sailing trips and then discharging the seeds elsewhere else helped extend the mango's inter-island dispersal*

Mango as part of their heritage

The two legends mentioned above also show the importance of the mango as part of the cultural background of the people in this Province. While the mango is highly regarded spiritually in the first legend, in the second it is highly valued because of the nourishment that the fruits provide. If mango fruits form an important food for the *Garuda* bird and saved and nourished Lakipadada during his journey, such fruits must be important to keep.

Mango as part of their favourite delicacies

Mango fruits are generally eaten by all ethnic groups in the Archipelago, either as table or fresh fruits when ripe, pickled when slightly young, or as 'rujak' in which green fruits are eaten with a paste made from a mixture of palm sugar, the pulp of tamarind, salt, chillies, and with or without shrimp paste. A specific way of consuming mango, which may not exist anywhere else, is to eat very young fruits of a special cultivar (locally known as *lana'bu*) by dipping them in pickled shrimps (Dr. Mukhlis Paeni, pers. comm.).

Another traditional way of eating mango, specifically the well-ripened fruits of *Mangifera odorata*, is with steamed sticky rice. While this seems to be the favourite way villagers enjoy this fruit, the dish is also considered to be one of the favourite traditional delicacies in Thailand (Pinsuvana 1976). This similarity is not only limited to how the fruit is eaten but also its name; in South Sulawesi the fruit is known as *pao macang* (Bugis) or *taipa macang* (Makassar) while in northern Malaya and Peninsular Thailand its vernacular name is *maa-chang* (Kostermans & Bompard 1993). As people liked the dish very much and therefore considered it very important, the travelling Buginese not only enjoyed the dish during their journeys but also apparently learned how to serve the fruit, taking the fruit and its Thai name home. *Mangifera odorata* is an everwet tropical rain forest species that occurs mostly on the western side of Sulawesi (Kostermans & Bompard 1993).

Mango as key ingredient of their fish dishes

As inter-island traders and fishermen, the Buginese, Makassarese and Mandarese consider fish as their main source of protein. They have various ways of preparing fish (dried or fresh), but in the majority of cases they prefer to have unripe mango as one of the ingredients in their spices. Though I consider the sour taste of the young unripe fresh mango is still rather strong — mixed with garlic, shallots, candle-nut, chillies, and fried salted fish or steamed fresh fish — the chopped unripe fresh mango is considered highly appetising in fish dishes.

Since mangoes are not available throughout the year, young unripe mangoes are usually sliced, soaked in salted water and then dried before being kept in a dry, well aerated basket, to meet the need for cooking fish. In this form, which is known as *pacukka* or *kaloko* (Bugis), *pakacci* or *kaloe* (Makassar) or *pamaissang* (Mandar), mango fruit can last for up to a few years. For preparing simple but tasty fish dishes, the dried mango is usually used, together with turmeric powder, shallot and chillies, for cooking fresh tuna or other fish. Since fish can be caught during the journey and the dried mango can last for a long time, such a dried mango usually forms the main ingredient of spices that are taken on sea journeys. Even now, this dried mango is still commonly sold in traditional markets outside Ujung Pandang, and very often also brought to other islands as trading goods where it is sold to South Sulawesi people who have already settled there.

While sweet mangoes with the least number of fibres are preferred for table fruits and may be brought for the journey as a source of fresh food that can last a few weeks, any kind of mango can be used in the preparation of the fish dishes. Although such spices can be kept for a long time, fresh mango replenishments of any kind may be made whenever or wherever the seafarers come ashore, particularly during long sailing trips.

Mango as trading goods that they brought around from one island to another

Because of their high value, mangoes have apparently been an important trading item at least since Rumphius' time. As cited by Kostermans & Bompard (1993), in 1741 Rumphius already mentioned that the *rawa-rawa, pali-pali* or *lumis* mango was sold in the markets of Makassar (now Ujung Pandang) at a high price. Based on their taxonomic interpretation, Kostermans & Bompard considered that these mangoes were not local cultivars or species but may have been imported from Borneo. This suggests that inter-island trading of mangoes, which is still done by the South Sulawesi people, was already known at least since that time. In other words, mangoes have been brought around from one island to another at least since that time.

E. Seeds of the mango fruits brought home from their journeys were thrown away around their houses which then grow into dense stands of mango trees

As stated by Kostermans & Bompard (1993), in their natural habitats, mango species rarely grow gregariously. They usually occur as .... "single individuals, widely scattered, sometimes only one per square kilometre. Groups of trees originating from forest fruits brought home and thrown-away seeds are found near dwellings of the forest population".... In principle, these statements are quite true for the South Sulawesi mangoes. So far, the endemic mango species of South Sulawesi (*M. sulavesiana*) occurs only in primary forest where it grows solitarily and sparsely while all the other species and cultivars occur in or near villages or settlements either solitary, in groups of small numbers of individuals, or gregariously, forming dense mango stands.

The slight difference, however, relates to the origins of trees that grow together in groups or patches. In Kostermans & Bompard's case, the reason for the groups of mango trees occuring by dwellings near the forest is because the seeds come from the nearby forest. In South Sulawesi, the groups of mango trees, which in several cases grow gregariously forming dense stands, do not occur in the hinterland near the forest but in villages along the coast or at the margins of an inland lake (Lake Tempe) that used to be accessible by sea-going boats up the Cenrana River.

Since the villages where these dense stands of mangoes occur are inhabited by inter-island traders and fishermen or are former centres of trading and fishing, such mango stands are likely to be derived from seeds of mango fruits brought in from other islands in the Archipelago by the seafaring traders or fishermen. Remaining stands of such mangoes can still be observed, for example, in villages near Sengkang (on the shores of Lake Tempe), Ujung Pandang, Maros, Gowa and in Karampuang Island on the northwestern coast of the Province near Mamuju. For example, in a 25 × 30 m plot in a home garden at Limbung near Ujung Pandang, 45% of the 114 trees larger than 10 cm diameter are mango trees. Although the number of trees in the 35 × 50 m plot of home garden in Karampuang Island is less than at Limbung, 58% of the 80 trees are mangoes. To visualise the densities of trees (particularly the mangoes) in these two plots, projections of their crowns are shown in Figs. 1 & 2.

F. *The dense aggregations of various kinds of mangos around the dwellings of coastal people increase the potential for hybridisation and thus the creation of new cultivars*

Based on studies in *M. indica,* pollen of the mango is scant and pollination, mostly by flies and thrips, is compulsory for fruit set (Kostermans & Bompard 1993). Since thrips are weak fliers and most effective only over short distances (Whitmore 1988), they are unlikely to effect cross pollination if the trees are far apart. The aggregation of different mango cultivars within a home-yard or dwelling, therefore, increases the potential for hybridisation and also the creation of new cultivars, thus enriching the mango germ-plasm diversity of the area.

Although we have not yet finished our taxonomic studies — particularly at the level of cultivar — the size, shape, aroma, fibre content, etc. of the fruits indicate peculiarities and perhaps also richness that does not seem to occur in other places in Java, Kalimantan, Bali and the Lesser Sunda Islands. Although certain mango trees in more than ten districts in the province are locally called *bali* or *balis* this is not because it was brought from the island of Bali but because the word literally means 'turned'. The villagers told us that the name relates to the fact that the fruits of the trees differ from the fruits originally planted.

G. *In spite of continuing strong pressure for change of land use, people still try to maintain parts of their mango stands*

Change of land-use practice is often a continuous process that goes along with an increase in development activities that try to meet the needs of the increasing population. Aside from the mango trees growing in home gardens of the seafaring villagers of South Sulawesi (Figs. 1 & 2) mango trees can also be seen surviving where the original land-use has been converted for rice cultivation.

In villages around Ujung Pandang, Maros and Gowa, mango trees can still be seen scattered in the middle of rice fields growing solitarily, forming walls or lines along the rice-field bunds, or growing together to form dense stands of trees. Aside from the trees along the bunds (Fig. 3), an analysis of a 20 × 25 m plot within one of the islands of such mango stands (Fig. 4) shows 88% of the 87 trees growing in the plot are actually mango trees. The preponderance of such a landscape could be interpreted as the insistence of the traditional coastal people of South Sulawesi to maintain mango trees to provide them with some of their daily needs.

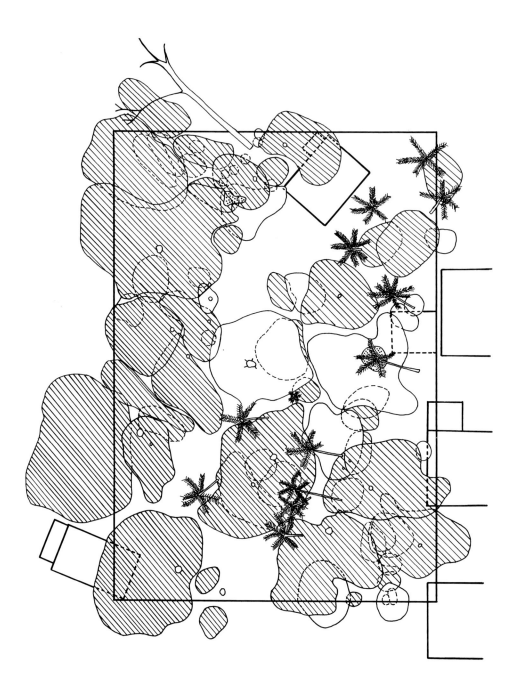

FIG. 1. Home gardens of villagers at Karampuang Island.

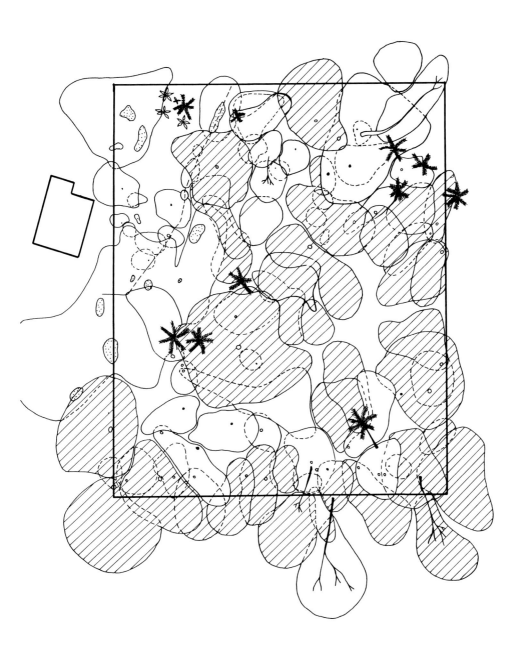

Fɪɢ. 2. Home garden of a villager at Limbung near Ujung Pandang.

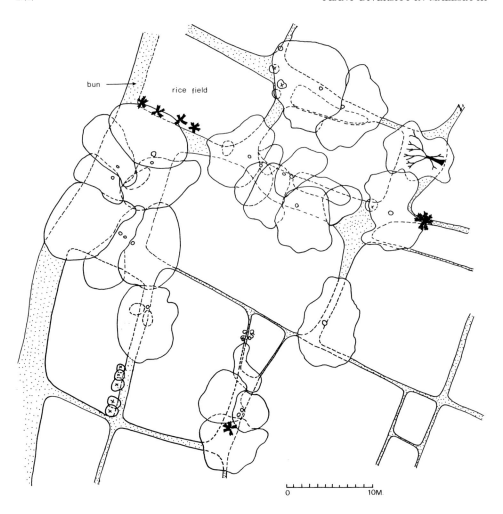

FIG. 3. Mango trees accommodated along rice-field bunds, near Ujung Pandang.

CONCLUSIONS AND RECOMMENDATIONS

The above evidence suggests that, at least during the last three centuries, the seafaring Buginese, Makassarese, and Mandarese of South Sulawesi have played an important role in enriching and maintaining the mango germ-plasm diversity of this Province.

Modern methods of processing and preserving food (canning, drying, etc.) which make it ready for consumption without additional needs of spices or with the use of preserved instant spices tends to eliminate the needs of traditional spices such as dried mangoes. Current activities that affect the maintenance of mango populations and diversity include logging, road construction and widening,

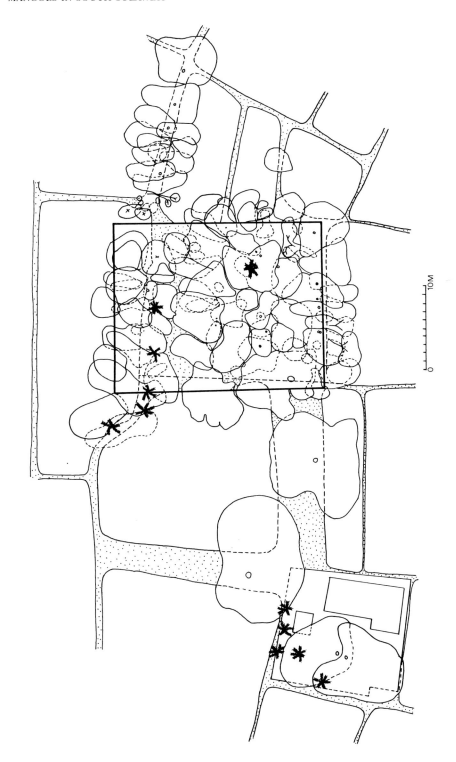

FIG. 4. An island of mango trees in the middle of a rice-field, near Ujang Pandang.

alignments of power lines and settlement activities, as well as conversion of forest into crop estates, fish ponds, and rice fields. The diversity of the local mango germ-plasm is also being threatened by the current government drives to expand the growing of a few selected mango cultivars (*manalagi, harum manis, madu,* etc.). As fruits of these cultivars currently fetch better prices than the local ones, villagers often cut down locally developed mango cultivars in order to accommodate the newly promoted cultivars in their fields. As no inventory has been made of the kinds and potential value of the South Sulawesi cultivars, such a drive may extinguish some of the potentially useful local cultivars. A thorough inventory followed by the replanting and study of the potential usage of these cultivars is therefore strongly recommended.

ACKNOWLEDGEMENTS

This research is part of a study of South Sulawesi mangoes supported by a research grant from the Biodiversity Support Program — a Consortium of the World Wildlife Fund, the Nature Conservancy, and the World Resource Institute that is funded by the US Agency for International Development, to Yayasan Indonesia Membangun, Ujung Pandang. In writing this paper I would like to express my sincere gratitude to my colleague Dr Mukhlis Paeni for his background information on the culture, traditions and history of the South Sulawesi people as well as for reading the final draft of the paper. My appreciation also goes to my students and former students who have actively participated in this study.

REFERENCES

KOMPAS. (1991). "Bapaknya si Gappa" bukan "Amana Gappa". Daily Newspaper, Jakarta, 29 December 1991.

Kostermans, A. J. G. H. & Bompard, J. M. (1993). The mangoes: their botany, nomenclature, horticulture, and utilization. 233pp. Academic Press, Harcourt Brace & Company, Publishers, London.

Lineton, I. A. (1975). An Indonesian society and its universe: a study of the Bugis of South Sulawesi (Celebes) and their role within a wider social and economic system. 230pp. Ph.D. Dissertation, Anthropology, School of Oriental and African Studies, University of London.

PEDOMAN RAKYAT. (1986). Gubernur Aminuddin: Phinisi Nusantara Bangkitkan Kebanggaan Nasional. Daily Newspaper, Ujung Pandang 16 June 1986.

Pinsuvana, M. (1976). Cooking Thai food in American kitchens. 192pp. Thai Watana Panich Press Co., Ltd.

Spillett, P. (1988). A feasibility study of the construction and sailing of a traditional Makassan prahu from Sulawesi to North Australia. 188p + App. A Bicentennial project prepared for The Historical Society of the Northern Territory, Incorp., Winnelie, N. T. 5792, Australia.

Whitmore, T. C. (1988). Tropical rain forests of the Far East. 2nd edition. 352pp. Clarendon Press, Oxford.

A Geographical Information System (GIS) for Mt Kinabalu

REED S. BEAMAN[1], ALEXIS G. THOMAS[2] & JOHN H. BEAMAN[3]

Summary. A geographical information system (GIS) is being developed for Mt Kinabalu that will include coverages for topography, rivers and streams, species distributions, satellite imagery, vegetation, geographic locations, Park boundary, geology, and land use. Georeferencing of these coverages was enhanced through linkage with a database of global positioning system (GPS) derived points. A virtually cloud-free Landsat TM image has been incorporated into the GIS. One objective of the work is to produce a detailed location map of Mt Kinabalu. Additionally, the GIS will be used to predict where particular species may be expected to occur in areas from which they are not presently known, to develop diversity indices, to examine habitat features such as slope and aspect, and especially to identify ultramafic outcrops that may have special relevance to species occurrence and vegetation types and to evolutionary and speciation processes in the flora. A long-range objective is to develop a set of GIS tools that will have general applicability for analytical and predictive purposes in biogeographic studies.

INTRODUCTION

Development of a geographical information system (GIS) for Mt Kinabalu was originally conceived to augment the botanical inventory of Mt Kinabalu, with a location map and gazetteer, and as a basis for mapping species distributions. As the technology has improved, GIS has become much more than just a mapping tool; it forms a spatial framework for analysis of biological data. As such, we are developing GIS tools to 1) predict where species occur in unexplored areas based on occurrence in known localities, 2) develop diversity indices based on surface areas and elevation, and 3) identify ultramafic habitat "islands" as part of an effort to understand plant speciation patterns on Kinabalu. The latter goal is in association with phylogenetic analyses of Mt Kinabalu taxa on which a number of collaborators are working. Identification of ultramafic outcrops is of particular interest because of the correlation between these outcrops and distribution of endemic species, particularly those hypothesized to have evolved on Kinabalu.

Coverages included in the Kinabalu GIS to date are topography, rivers and streams, the Kinabalu Park boundary, ground control points, a vegetation map published by Kitayama (1991), place names (toponyms) for specimen collections made by local collectors for the Projek Etnobotani Kinabalu, and a Landsat

---

[1]Department of Botany, University of Florida, Gainesville, Florida, 32611, U. S. A.
[2]Department of Urban and Regional Planning, University of Florida, Gainesville, Florida, 32611, U. S. A.
[3]Institute of Biodiversity and Environmental Conservation, Universiti Malaysia Sarawak, 94300 Kota Samarahan, Sarawak, Malaysia. Present address: The Herbarium, Royal Botanic Gardens, Kew, Richmond, Surrey TW9 3AB, U.K.

Thematic Mapper image from 14 June 1991. In addition to these coverages, a digital elevation model (DEM) was developed from topographic coverages.

Predictions of where a taxon occurs in unexplored areas are useful in planning expeditions into areas such as the north side of Mt Kinabalu. Likewise, Park management can use these types of data to target certain areas for protection. Predicting the occurrence of a particular taxon is based on knowing the characteristics of its habitat. Sheila Collenette (pers. comm.) probably was the first person to use remote sensing in this manner in her rediscovery of *Paphiopedilum rothschildianum* on Mt Kinabalu. Using black and white aerial photographs of Kinabalu she was able to recognize the *Casuarina* forests that are characteristic of many ultramafic areas. Site visits proved her technique worthwhile, because *P. rothschildianum* had not been known from the wild in almost 70 years (the type collection was attributed to New Guinea).

Remote sensing and GIS software (Imagine and Arc/Info, respectively) provide image enhancement and classification techniques that make these sorts of predictions possible in a systematic way. Digital elevation models (DEMs) allow us to further refine predictive models by the inclusion of elevation, slope and aspect data. Diversity indices can also be enhanced by developing models to incorporate DEM data into the diversity index. How species diversity relates to habitat diversity and the spatial extent of certain plant communities is of particular interest on Mt Kinabalu, because of the high endemicity in ultramafic areas.

SATELLITE VIEW OF KINABALU

The satellite image with which we have been working is a Landsat Thematic Mapper (TM) product. Landsat 5 is currently operational. These are in orbit c. 700 km above the earth's surface, and, between the two of them, pass over the same location every eight days. The image shown in Plate 1 is a quarter of a quarter scene, i.e. one-sixteenth of the total scene that covers 185 × 185 km. This image, recorded 14 June 1991, is the only cloud-free image available for the area, which is characterized by frequent and heavy cloud cover. The nominal ground resolution is 28.5 × 28.5 m for each pixel. The data are provided from seven bands from the electromagnetic spectrum, as designated by EOSAT: Band 1, visible blue (0.45 - 0.52 µm); band 2, visible green (0.52 - 0.60 µm); band 3, visible red (0.63 - 0.69 µm); band 4 , near infrared (0.76 - 0.90 µm); band 5, mid infrared (1.55 - 1.75 µm); band 6, thermal infrared (10.4-12.5 µm); and band 7, a second mid infrared band (2.08 - 2.35 µm). Most of these bands provide data on vegetation (Lillesand & Kiefer, 1987). For example, band 1 is useful in differentiating between soil and vegetation types. Band 2 measures green reflectance of healthy vegetation. Band 5 is sensitive to the amount of water in the leaves of plants and penetrates through smoke and haze better than bands 3, 2 or 1. The plate shown was produced by applying bands 5, 2 and 1 to a red, green and blue (RGB) color model to create a pseudo-color image. Applying the visible bands 3, 2, and 1 to RGB would create a true color image, but the former combination enhances differentiation between vegetation types, while still approximating true colors seen by the human eye. Some particularly conspicuous elements of the landscape represented include primary forest, secondary forest, and shifting agriculture. Furthermore, the bare granitic outcrop of the summit area of Kinabalu is readily apparent, as is the Mamut Copper Mine, another feature without vegetation cover. Other less conspicuous features with which we are working, such as the occurrence of ultramafic outcrops, require special techniques to enhance. Enhancement techniques we have employed so far include tasseled cap analysis (enhances brightness, wetness and greenness) and basic mineral composition and ferro-magnesium indices (e.g. occurrence of ferrous minerals can be enhanced by ratioing band 5 to band 4).

It should be noted that the four colour plates associated with this paper are presented at the end of the book, after the index.

Finally, we are also producing a location map for Mt Kinabalu, on which accurate and up-to-date placement of roads and settlements (kampungs) are important. Satellite imagery makes this easier and more accurate. Roads vs. rivers, for instance, are more conspicuous in different band combinations.

PREDICTING SPECIES RANGES USING DIGITAL ELEVATION MODELS

It is possible to convert the elevational data from contours on topographic maps into a three-dimensional digital elevation model (DEM) of a landscape. In a grid-based DEM, elevation values are stored within cells, each of which represents a discrete unit of space on earth. The cells are organized into rows and columns forming a Cartesian matrix in the same way that satellite data is stored. The Kinabalu DEM uses cell sizes of 50×50 m looking straight down at the earth. When visualized, the process is like draping an enormous fishnet over the mountain. Different elevations can be represented by different colours, thus enhancing the three-dimensional aspects of the image (Plate 2). Views from any direction can be produced. Elevation, slope, aspect, and insolation (intensity of sunlight falling on a slope) attributes can be easily extracted from a DEM. These attributes can be used to predict where a particular rare species may occur in unexplored areas, based on elevation, slope, and aspect data for known occurrences of the taxon.

A particularly exciting aspect of the DEM is that it provides a mechanism for draping other types of coverages over basic details such as elevation. For example, river systems can be shown three-dimensionally draped over the DEM. An even more dramatic example is in the ability to drape satellite imagery. Thus, a realistic three-dimensional model of Mt Kinabalu can be produced as a drape over the DEM, and additional coverages, for example the Kinabalu Park boundary, can be added as shown in Plate 3.

DIVERSITY ANALYSES THROUGH GIS

Plant diversity and distribution patterns on Mt Kinabalu were discussed by Beaman & Beaman (1990), and the hypothesis explaining the high diversity was outlined there and somewhat elaborated in Beaman *et al.* (this volume). Significant environmental variables involved in this hypothesis that are amenable to GIS analysis include extreme elevational variation from just above sea level to 4101 m, precipitous topography, diverse geological substrates, relatively open habitats (e.g.

landslides), slope, aspect and, as dependable data about species distributions on the mountain become available, relative isolation of individual populations. A major function of the GIS facility should be in analysing phylogenetic relationships of recently derived species as correlated with various environmental parameters.

In Parris *et al.* (1992) and Wood *et al.* (1993), histograms showing generic and species diversity for the pteridophytes and *Orchidaceae* indicate that the highest number of taxa occur at about 1500 m. This type of diversity index is independent of actual surface area at any particular elevation range. Among the analytical uses of the digital elevation model is that it allows us to calculate planimetric surface area for each of the elevation ranges on Kinabalu. To compare diversity for areas of unlike size it is necessary to calculate a diversity index. There is of course less overall surface area on any mountain at progressively higher elevations. The simplest type of diversity index is derived by dividing the number of taxa by the surface area. In this instance, elevation ranges in 500 m bands around Mt Kinabalu are the areas for which diversity is compared. The results are shown in Fig. 1. While the highest number of taxa occurs between 1001–1500 m, the highest number of taxa per unit area is between 2001–3000 m. This area-based diversity index is more comparable to diversity indices based on plot sampling.

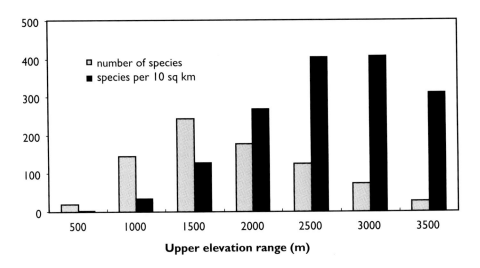

Fig. 1. Pteridophyte diversity on Mt Kinabalu. Note that although the greatest number of species occurs at 1500 m, the greatest diversity of species per unit area is between 2500 and 3000 m.

RECOGNISING ULTRAMAFIC OUTCROPS

One of the most interesting habitat types on Mt Kinabalu is that on extreme ultramafic outcrops. These occur at various elevations, from relatively low to high, and form virtually a complete collar around the mountain, as can be seen in the geological map of Jacobson (1970). Some of the rarest and most prized Kinabalu species occur predominantly or entirely on these outcrops, including *Dacrycarpus kinabaluensis, Nepenthes rajah, N. burbidgeae, N. villosa, Cleisocentron merrillianum, Paraphalaenopsis labukensis, Paphiopedilum hookerae* var. *volonteanum,* and *P. rothschildianum.* Our ability to make GIS analyses of the occurrence of such taxa should help in predicting other localities (i.e. other ultramafic outcrops) where they may occur and facilitate mechanisms for protecting them.

One technique we have used to predict habitat extents is called "seeding". Seeding is a supervised classification method that takes pixel values of the Landsat image at a known reference point (e.g. that of a specimen collection or ultramafic outcrop) and compares it to neighbouring pixel values. Before running the seeding routine, we enhanced the satellite image using a mineral composition index (see above). The seeding process starts with one pixel in the centre of a neighbourhood and continues in an expanding pattern to create a polygon of irregular shape. The boundary of a polygon is extended outward until the Euclidean distance between pixel values of the neighbors and the starting pixel exceeds a threshold value. If applied to a number of starting pixels, such as points known to be ultramafic, the result is a group of distinct polygons representing the extent of ultramafic vegetation in the area of interest. Polygons can be filled for display purposes (Plate 4). In this example, on the south side of Kinabalu, patches of ultramafic habitat were detected that are cut by river valleys. These valleys are usually filled with alluvial gravels and granitic boulders. The resulting geographic configuration of ultramafic patches can be thought of as representing an archipelago of ultramafic habitat islands, as discussed by Barkman *et al.* (this volume). Some ground truthing has been done and supports the use of this technique, but this needs to be carried out in a systematic way for determining actual conditions of known and predicted ultramafic species occurrences, which is one of our current endeavours.

CONCLUSIONS

Although our present study is restricted to Mt Kinabalu, we are attempting to develop a set of GIS tools that will be relevant to any areas in which the flora or fauna are being intensively studied from the standpoints of species occurrence, phylogeny, substrate geology, elevational distribution, and other environmental variables. We wish to emphasize that GIS methodology results in much more than just distribution maps as end products. Still, high-quality maps of selected areas can be produced by these techniques, and we hope to take advantage of this capability, as illustrated by the map in Fig. 1 in the paper by J. H. Beaman *et al.* (this volume). Ultimately, however, GIS in systematics and other biological applications should be even more useful for its analytical and predictive capabilities, as we hope to demonstrate in future publications on the biogeography of Mt Kinabalu.

ACKNOWLEDGEMENTS

We greatly appreciate the facilities and resources provided by Sabah Parks, Datuk
Lamri Ali, Director, for conducting this research. Park Warden Eric Wong supplied
the survey data for the Kinabalu Park boundary that permitted developing the Park
boundary as GIS coverage. The Sabah Department of Lands and Surveys provided
global-positioning-system data that facilitated rectifying the satellite image, which
was graciously donated by the EOSAT Corporation. Dr. Jean Claude Thomas has
provided much useful advice and made available facilities of the U. S. Geological
Survey for some of the research. Dr. Scot Smith, Department of Mapping and
Surveying, University of Florida, has been particularly helpful in facilitating our use
of the Arc/Info software. This research has been supported by U. S. National
Science Foundation Grant DEB-9400888 to Michigan State University.

LITERATURE CITED

Barkman, T. J., Repin, R., Beaman, R. S., & Beaman, J. H. (this volume). Ultramafic
    habitat patches as islands on Mt Kinabalu.
Beaman, J. H. and Beaman, R. S. (1990). Diversity and distribution patterns in the
    flora of Mt Kinabalu. Pp. 147–160 in P. Baas *et al.* The Plant Diversity of Malesia.
    Kluwer Academic Publishers, Dordrecht, Netherlands.
Beaman, J. H., Beaman, R. S., & Anderson, C. E. (this volume). Botanical inventory
    of Mt Kinabalu: a progress report.
Jacobson, G. (1970). Gunong Kinabalu area, Sabah, Malaysia: Geological Survey
    Malaysia, report 8. 111 pp. Government Printing Office, Kuching, Sarawak,
    Malaysia.
Kitayama, K. (1991). Vegetation of Mt Kinabalu Park, Sabah, Malaysia: Map of
    Physiognomically Classified Vegetation, Scale 1:100,000. East-West Center,
    Honolulu, Hawaii.
Lillesand, T. M., & Kiefer R. W. (1987). Remote Sensing and Image Interpretation.
    J. Wiley, New York.
Parris, B. S., Beaman, R. S. & Beaman, J. H. (1992). The Plants of Mt Kinabalu: 1.
    Ferns and Fern Allies. Royal Botanic Gardens, Kew. viii + 165 pp. + 5 pl.
Wood, J. J., Beaman, R. S. & Beaman, J. H. (1993). The Plants of Mt Kinabalu: 2.
    Orchids. Royal Botanic Gardens, Kew. xii + 411 pp. + 84 pl.

Subject Index

Taxonomic Index

437

PLATE 1. Landsat TM 5 image of Mt Kinabalu recorded 14 June 1991. This figure was produced using image bands 5, 2, 1. The Kinabalu Park boundary is shown in red.

PLATE 2. Digital elevation model (DEM) of Mt Kinabalu as seen from the southeast. The DEM was created from topographic coverage at 500-ft contour intervals. An elevation-determined colour ramp is used in the model, with purple at the low elevations grading to red upward.

PLATE 3. Drape of the Landsat TM image of Mt Kinabalu over the DEM, as seen from the northwest. This figure was produced using image bands 7, 2, 1. The Kinabalu Park boundary shown in red is an additional coverage. Black holes in the draped image are areas where surface visibility was miscalculated owing to three-dimensional surface-resolution limitations. The surface resolution is 50 × 50 m while the image resolution is 28.5 × 28.5 m.

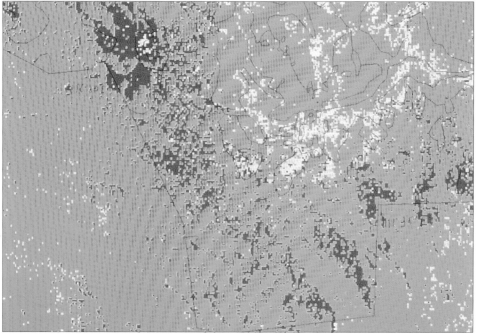

PLATE 4. Ultramafic island polygons (shown in red) on the south side of Mt Kinabalu, using image "seeding".